KINEMATICS AND DYNAMICS OF MACHINERY

KINEMATICS AND DYNAMICS OF MACHINERY

Elena R. Fiori
Ingo Barth

Kruger Brentt
Publishers
2025

Kruger Brentt Publishers UK. LTD.
Company Number 9728962
Regd. Office: 68 St Margarets Road, Edgware, Middlesex HA8 9UU

ISBN: 978-1-78715-382-0

Disclaimer:

Every possible effort has been made to ensure that the information contained in this book is accurate at the time of going to press, and the publisher and author cannot accept responsibility for any errors or omissions, however caused. No responsibility for loss or damage occasioned to any person acting, or refraining from action, as a result of the material in this publication can be accepted by the editor, the publisher or the author. The Publisher is not associated with any product or vendor mentioned in the book. The contents of this work are intended to further general scientific research, understanding and discussion only. Readers should consult with a specialist where appropriate.

Every effort has been made to trace the owners of copyright material used in this book, if any. The author and the publisher will be grateful for any omission brought to their notice for acknowledgement in the future editions of the book.

For information on all our publications visit our website at http://krugerbrentt.com/

PREFACE

The study of kinematics and dynamics lies at the heart of mechanical engineering, providing the fundamental principles and analytical tools necessary for the design, analysis, and optimization of machinery and mechanical systems. Understanding the motion, forces, and energy transfer within machines is essential for engineers and designers seeking to create efficient, reliable, and innovative mechanical devices for a wide range of applications. Kinematics and dynamics are not only foundational to mechanical engineering education but also critical for advancing technological innovation and solving real-world engineering challenges.

This handbook, titled "**Kinematics and Dynamics of Machinery**" is a comprehensive exploration of the principles, theories, and applications of kinematics and dynamics in mechanical systems. Drawing upon the expertise of mechanical engineers, researchers, and educators, this volume offers theoretical insights, practical techniques, and case studies for understanding and applying kinematic and dynamic principles in machinery design and analysis.

From the fundamentals of rigid body kinematics and particle dynamics to the advanced concepts of mechanism synthesis and vibration analysis, the "Kinematics and Dynamics of Machinery" covers a wide range of topics essential for mastering the field. Through clear explanations, mathematical derivations, and engineering examples, readers will gain a deep understanding of the principles and methods used to analyze and optimize the performance of mechanical systems.

Moreover, this handbook explores the applications of kinematics and dynamics across different domains of mechanical engineering, including robotics, automotive engineering, aerospace systems, and manufacturing. By examining case studies, design challenges, and real-world applications, we highlight the ways in which kinematic and dynamic analyses are used to solve practical engineering problems and drive technological innovation.

As we navigate the complexities of modern engineering design and manufacturing, the importance of kinematics and dynamics in machinery cannot be overstated. It is our hope that the "Kinematics and Dynamics of Machinery" handbook will serve as a valuable resource for students, researchers, and practitioners in the field of mechanical engineering. Whether you are studying the fundamentals of mechanical motion, designing complex machinery, or optimizing system performance, this handbook invites you to explore the principles and practices of kinematics and dynamics and contribute to the advancement of mechanical engineering knowledge and practice.

Elena R. Fiori
Ingo Barth

CONTENTS

Chapter 1 ANALYSIS AND OPTIMIZATION OF PERFORMANCE MEASURES FOR PARALLEL KINEMATICS MANIPULATORS

1. INTRODUCTION: AN OVERVIEW

Serial kinematics mechanisms (SKMs) are extensively utilized for many purposes. Despite the numerous advantages of SKMs, these serial mechanisms also suffer from several downsides including reduced stiffness, the accumulation of pose error, limited agility, a low payload-to-weight ratio, and complex inverse kinematics. Therefore, in order to address these limitations, parallel kinematics mechanisms (PKMs) are employed, specifically for activities that require greater performance, such as high-speed and high-precision applications. Although PKMs offer numerous benefits, they also come with certain limitations, including a limited workspace, intricate forward kinematics, and the potential for singularity issues. In order to mitigate these limitations, optimization using a range of techniques is frequently employed to enhance their disadvantages while preserving their benefits. The optimization process can be classified as either single-objective or multi-objective, depending on the number of objectives being optimized. Typically, many objectives need to be optimized in most circumstances. Moreover, certain aims often contradict with one another. For instance, the majority of PKMs often necessitate a wider workspace as well as a more rigid structure with reduced mass. Expanding the workspace typically necessitates longer linkages, leading to decreased stiffness and increased mass. In the field of multi-objective optimization, the selection of objectives is determined by their priority, which is contingent upon the specific application.

In this chapter, a thorough examination of several performance metrics is provided. These indices are defined and presented in a complete manner, and they are applicable to numerous scenarios. Next, a comprehensive examination of the optimization approaches employed in the design of various systems to meet specific objectives or multiple purposes is conducted. Given the nonlinearity of parallel link manipulator systems and the competing nature of numerous performance indicators, it is crucial to find a solution that optimizes these indices. Trial and error methods may not be effective, thus mathematical approaches are necessary.

2. PERFORMANCE MEASURES

There are numerous measurements or indices available to assess the performance of a robot. Patel and Sobh [1] categorized them as either local or global, kinematic or dynamic, and intrinsic or extrinsic. Local metrics are contingent upon the specific configuration of the robot, while global measures assess the performance of the robot throughout the entire workspace. The global performance index (GPI) can be obtained by integrating the local performance index P over the workspace W as given by Eq. (1). If the local performance index cannot be expressed analytically, discrete integration as given by Eq. (2) can be used. In the latter case, the accuracy of the integration depends on the number of points *n* being used for evaluation. If the inclusion of large number of points is very computationally expensive, less representative sampling points can be used.

$$GPI = \frac{\int_W P \, . dW}{\int_W dW} \quad (1)$$

$$GPI = \frac{1}{n}\sum_{i=1}^{n} P_i \quad (2)$$

Kinematic measures pertain to the motion characteristics of the robot, while dynamic measures are associated with the robot's dynamic features. The intrinsic metrics pertain to the inherent properties of the robot, independent of its specific task, whereas the extrinsic measures are specifically tied to the robot's task. Commonly employed performance metrics encompass workspace, proximity to or avoidance of singularity, dexterity, manipulability, stiffness, precision, repeatability, and reliability. Several of them are examined and explained below. It is important to take into account the performance measures throughout the design process of the robot. Optimal design typically involves the optimization of one or more performance measurements, which serve as the objective function(s) to be optimized.

2.1. Workspace

The workspace refers to the collection of points or locations that the end effector is capable of reaching. The mobility of the robot is determined by its degrees of freedom (DOF), which refer to the amount and types of movements it can do. However, the robot's mobility is limited by factors such as the length of its linkages, the range of its joints, and the potential interference between its components. The assessment of the workplace typically encompasses its dimensions (area/volume) and configuration. The shape can be quantified using many metrics, such as the aspect ratio of the typical workspace. Typically, a larger and more well-designed workspace is necessary. An alternative method of describing the workspace is through the utilization of the workspace volume index [2] or footprint ratio [3], which is defined as the ratio between the volume of the workspace and the volume of the machine.

The first thing to determine in order to achieve better workspace before optimizing the geometrical parameters is selecting better topology. For example, many mechanism designs include sliders (gliders) in order to get larger workspace, such as in Hexaglide, Linapod, Pentaglide, sliding H4, and Triaglide.

The robot workspace is commonly classified into several types as follows [2]:

- Constant orientation workspace (translation workspace) which defines reachable points with constant orientation of the moving platform.
- Orientation workspace which defines reachable orientations while the center tool point is fixed.
- Reachable workspace (maximal workspace) which defines reachable points with at least one orientation of the moving platform.
- Inclusive workspace which is reachable workspace in a given orientation range.
- Dexterous workspace which defines reachable points with any orientation of the moving platform.
- Total orientation workspace which is dexterous workspace in a given orientation range.
- Useful workspace (sometimes also called used workspace) which defines part of the workspace to be used for a specified application. It is usually regular workspace such as rectangle, circle, cuboid, sphere, or cylinder.
- The term "useful workspace" refers to the portion of a workspace that may be effectively employed for a specific application or task. Baek et al. [4] introduced a technique for identifying the largest possible rectangle that can fit within the working area of both serial kinematics manipulators (SKM) and parallel kinematics manipulators (PKM).

In order to fully represent the workspace, a six-dimensional space is necessary. Nevertheless, graphical representation is limited to three-dimensional space. Therefore, it is customary to depict the position workspace and the orientation workspace as distinct entities. The workspace of a planar mechanism can be represented in a two-dimensional plot, while the workspace of a spherical or spatial mechanism can be shown in a three-dimensional plot. The workspace plot can be displayed using either the Cartesian or polar coordinate system for 2D plots, and using either the Cartesian, cylindrical, or spherical coordinate system for 3D plots. Generating the visual depiction of the operational area is more straightforward in serial kinematic machines (SKMs) but might be challenging in parallel kinematic machines (PKMs). In the second scenario, the visual representation of the workspace is limited to parallel kinematic machines (PKMs) that have a maximum of 3 degrees of freedom (DOF). In order to visually represent the workspace, it is necessary to fix n - 3 degrees of freedom (DOF) for PKMs with more than 3 DOF. The workspace varies depending on which degree of freedom (DOF) is fixed [2].

Typically, there are three primary methods for determining and graphing the workspace: the geometric approach, the discretization numerical technique, and the non-discretization numerical approach. The estimation of the workspace of PKMs was initially carried out using a geometrical technique. Bajpai and Roth [5] examined the workspace of parallel kinematic machines (PKMs) and the impact of leg length on the workspace. After a span of three years, Gosselin [6] introduced his research on determining the constant-orientation workspace of a 6-DOF PKM. This was followed by Merlet [7], who provided the determination of the orientation workspace, and Kim et al. [8], who proposed an algorithm for determining the reachable and dexterous workspace. Gosselin and Jean [9], followed by Merlet et al. [10], provided the determination of the workspace for planar parallel kinematic machines (PKMs), which require specific treatments. All of the previously described works employ a geometric methodology.

The geometrical approach to determining the borders of a PKM workspace involves finding the intersection of the boundaries of each open-loop chain that makes up the PKM. This strategy is both efficient and precise. In order to streamline and expedite the process, CAD software can also be employed, as suggested by Arrouk et al. [11]. A significant limitation of this approach is its limited scope as various robot configurations may require distinct methodologies to implement this approach. Put simply, this strategy typically needs to be customized for the specific robot being examined. Another limitation of this strategy is the challenge of incorporating all the limitations.

The discretization numerical approach involves creating a discretized bounding space that encompasses all potential points in the workspace. This space is then tested using inverse kinematics and constraints to determine if it falls within the workspace. This method is commonly referred to as binary representation as it allocates binary numbers throughout the evaluation process. Specifically, a value of "1" is assigned to indicate reachability and hence inclusion in the plot, while a value of "0" is assigned to indicate unreachability and exclusion from the plot.

The primary benefit of this approach lies in its versatility to be applied to all forms of PKMs, as well as its inherent intuitiveness. Furthermore, this technique has the capability to incorporate all of the limitations. Nevertheless, the precision of this method relies on the magnitude of the discretization intervals. Additionally, it is not possible to detect small empty spaces inside the workspace unless the discretization steps are sufficiently small to accurately capture the presence of these empty spaces. The Monte Carlo method is a technique comparable to the discretization approach. It involves inputting a large number of discrete active joint points within the joint range into the forward kinematics. This process allows for the plotting of the end effector position points. To establish the bounds of the workspace or calculate its volume, additional processing of the Monte Carlo findings can be carried out by placing the workspace points in a discretized bounding space, as done in the discretization approach.

A recent study by Bonev [14] utilized the discretization numerical technique to establish the three-dimensional orientation workspace of 6-DOF PKMs. Castelli et al. [15] introduced an algorithm that utilizes the discretization approach to calculate the workspace, workspace borders, workspace volume, and workspace shape index of SKMs and PKMs. Dash et al. [16] introduced a discretization technique for determining the attainable workspace and identifying empty spaces in parallel kinematic mechanisms (PKM).

Beyond the aforementioned two approaches, several works proposed non-discretization numerical methods to determine the workspace of PKMs. Some of the methods are as follows [2, 17]:

- Jacobian rank deficiency method [18] but only practical for the determination of constant orientation workspace.
- Numerical continuation method [19, 20] can avoid singularity points but only practical for the determination of constant orientation workspace.
- Constrained optimization method [21] which is modified from the numerical continuation method.
- Boundary search method [17] which is based on constrained non-linear programming.
- Principle that the velocity vector of the moving platform cannot have a component along the normal of the boundary [22], but this method does not work for mechanisms with prismatic joints as well as it is difficult to include the mechanical limits and interference between links.
- Interval analysis method [23] which can deal with almost any constraints and any number of DOF.

In a recent study, Bohigas et al. [24] introduced a branch-and-prune technique that can accurately identify all the boundary points of the workspace for universal lower-DOF (3-DOF or lower) SKMs and PKMs. This methodology surpasses the constraint of the numerical continuation method. In addition, Gao and Zhang [25] introduced a simplified boundary searching (SBS) technique that combines a geometric approach, discretization method, and inverse kinematics model of a parallel mechanism. Saputra et al. [26] introduced a swarm optimization method to ascertain the workspace of parallel kinematic machines (PKM).

2.2. Jacobian matrix

The Jacobian matrix represents the connection between the velocities of the moving platform in the task space and the velocities of the active joints. Moreover, it also establishes the correlation between the exerted force on the joint and the applied force on the task. The reason for discussing it here is its relevance to numerous kinematic performance measurements.

Given that the velocity kinematics of the robot is expressed by:

$$A q + B x = 0 \tag{3}$$

where $\dot{x}$ is the end effector twist and $\dot{q}$ is the actuator twist, then A is called forward Jacobian matrix, B is inverse Jacobian matrix, and the total Jacobian matrix J is given by:

$$J = -B^{-1} A \tag{4}$$

If the Jacobian matrix is unit-consistent (homogeneous), it can be directly utilized in the calculation of Jacobian-based performance metrics, such as the Jacobian condition number and manipulability. The issue arises when the Jacobian matrix lacks unit consistency (non-homogeneity), resulting in a lack of relevant physical interpretations. In order to tackle this issue, the literature has identified various methods for standardizing inconsistent (non-homogeneous) Jacobian matrices. These methods include:

- Using natural length or characteristic length [27–29]
- Using scaling matrix [30, 31]
- Using weighting factor [32]
- By using power transition concept [33]
- Point-based method [34–37]
- General and systematic method [38]
- Homogeneous extended Jacobian matrix [39]

2.3. Singularity

Gosselin and Angeles [40] provide the most basic categorization of singularity in PKMs. The singularity in parallel kinematic mechanisms (PKMs) can be classified into two categories based on the behavior of the active joints and the end-effector. These types are mathematically characterized by the singularity of the two Jacobian matrices in the robot's kinematics, as specified in Equation (3). There are three distinct types of singularity:

- Type 1 singularity (also called: direct kinematic singularity, forward kinematic singularity, serial singularity, or sub-mobility) occurs when the forward Jacobian matrix A is singular. When this kind of singularity occurs, it is not possible to generate some velocities of the end-effector. In other words, very small changes in the joint space do not affect the end-effector pose. In these configurations, the mechanism loses one or more degrees of freedom.
- Type 2 singularity (also called: inverse kinematic singularity, parallel singularity, or over-mobility) occurs when the inverse Jacobian matrix B is singular. It corresponds to the appearance of uncontrollable mobility of the end-effector because it is possible to move it while the joints are locked. At the corresponding configurations, the mechanism gains one or more uncontrollable DOF. In other words, the end-effector can move without the joints moving. Equivalently, the stiffness of the PKM is locally lost.
- Type 3 singularity (also called: combined singularity) occurs when both the forward Jacobian matrix A and the inverse Jacobian matrix B are singular. When this kind of singularity occurs, the end-effector can move when the joints are locked, and at the same time the end-effector pose does not change due to very small changes in the joints.

In addition, singular configurations can be determined by analyzing the Jacobian matrices or by using a geometric technique.

Zlatanov et al. [41] provided a broader analysis of the singularity, taking into account the passive joints in the evaluation. The singularity is categorized into three types: redundant input (RI) singularity, which is equivalent to serial singularity; redundant output (RO) singularity, which encompasses parallel singularity and constraint singularity; and actuator singularity, which refers to non-zero passive joint velocities when the actuators are locked and the end-effector has zero velocities [42] [43]. In addition, certain publications have also addressed the concept of higher order singularity, albeit it does not provide significant practical advantages [2].

2.4. Jacobian condition number

The Jacobian condition number, derived from its mathematical formulation, varies from 1 to infinity. A value of infinity denotes singularity, while a value of 1 indicates isotropy of the Jacobian matrix. Alternatively, it can also be represented by its reciprocal value, known as the inverse Jacobian condition number. This value ranges from 0 to 1, with 0 indicating singularity and 1 indicating the isotropy of the Jacobian matrix. An ill-conditioned Jacobian matrix refers to a situation when the matrix is near to singularity. Conversely, the Jacobian matrix is referred to as well-conditioned when it is significantly distant from singularity. Moreover, in the case of an isotropic Jacobian matrix, the amplification of velocity and force is equal in all directions.

The commonly used norms to define the Jacobian condition number are as follows:

- 2-norm, which is given by the ratio of the maximum and minimum singular values of the Jacobian matrix.
- Frobenius norm, which is very advantageous because it is an analytical function of the robot parameters and hence will not give serious concern if its gradient is evaluated [44], as well as it avoids the computation of the singular values.
- Weighted Frobenius norm, which can be rendered to specific context by adjusting its weights [45] in addition to all of the mentioned advantages of the Frobenius norm.

The Jacobian condition number quantifies the level of kinematic dexterity, also known as dexterity. It signifies proximity to a singularity, homogeneity in motion (isotropy), agility, and precision. Kinematic dexterity refers to a robot's capacity to effortlessly manipulate the end-effector in any direction. The kinematic isotropy of a robot directly reflects its dexterity, as a higher level of isotropy implies that the robot can move as easily in all directions. However, this information is still incomplete regarding dexterity as it just indicates the degree of equality in different directions, but not the level of ease. It is possible that the motion in all directions either requires minimal effort or requires significant effort. The upcoming review of manipulability will provide a comprehensive understanding of kinematic dexterity.

The Jacobian condition number can also be understood as a measure of how much the error in the task space will increase as a result of a small inaccuracy in the joint space. As the state of the Jacobian matrix worsens, even a tiny error in the joint space will result in a bigger error in the task space. The Jacobian condition number serves as a measure of the precision of the manipulator, based on this fact.

The Jacobian condition number is a characteristic that describes the sensitivity of a function's output to changes in its input variables in a specific region. The answer varies based on the specific arrangement or positioning of the robot. The global condition number (or global condition index, GCI) is employed to assess the overall quality of a system. The Global Condition Index (GCI) is calculated by integrating the Local Condition Index (LCI) across the entire workspace. The Jacobian condition number is often used as a measure of dexterity, and as a result, the Global Condition Index (GCI) is often referred to as the Global Dexterity Index (GDI). A map displaying the values of LCI over the workplace is frequently known as a dexterity map.

Due to the reliance on the Jacobian matrix, the manipulability encounters a problem of unit inconsistency when combining translation and rotation. For this situation, it is necessary to homogenize the Jacobian matrix.

2.5. Manipulability

The concept of manipulability measure was initially proposed by Yoshikawa [46] as a local measure, known as the local manipulability index (LMI). This means that the measure is influenced by the robot's configuration, namely its posture, as it is derived from the Jacobian matrix. The global manipulability of a system can be assessed using the global manipulability measure (GMI), which is obtained by integrating the local manipulability measure throughout the entire workspace. Another metric is the uniformity of manipulability, which quantifies the consistency of manipulability throughout the workspace [47]. Like the Jacobian matrix, the manipulability also encounters a problem of unit inconsistency when there is a combination of translation and rotation. Similarly, when considering the Jacobian condition number, it is necessary to homogenize the Jacobian matrix in such a scenario.

Manipulability is a metric that quantifies the efficiency of converting input effort into output performance. Put simply, it signifies the degree to which velocity and force are conveyed and magnified. The measure of manipulability offers more detailed insights into the amplification of velocity and force compared to the condition number of the Jacobian matrix. The latter merely indicates the isotropy of the velocity and force amplification, without specifying the magnitude. In contrast, the earlier provides information about both the magnitude and the isotropy of the velocity and force amplification.

There are two well-known types of manipulability: twist manipulability, which involves velocity, and wrench manipulability, which involves force. The former is typically depicted by a velocity manipulability ellipse/ellipsoid, whereas the latter is represented by a force manipulability ellipse/ellipsoid. The velocity manipulability ellipsoid represents the velocity minimum, velocity maximum, and velocity isotropy through the length of its axes, while the manipulability is represented by the volume of the ellipsoid. The manipulability ellipsoid's major axis represents the direction of least resistance for the mechanism's movement, while the minor axis represents the direction of maximum stiffness, meaning the mechanism's actuators can exert the least effort to resist forces in this direction. The concept of force manipulability relies on the duality relationship between velocity and force, specifically between differential kinematics and statics. In addition to being represented by an ellipse or ellipsoid, manipulability can also be represented by a manipulability polytope.

Tanev and Stoyanov [48] have introduced the use of normalized manipulability index which is bounded between zero and unity. Doty et al. [49] proposed weighted twist manipulability and weighted wrench manipulability. Furthermore, a new manipulability measure for parallel robot was introduced by Hong and Kim [50].

2.6. Stiffness

In addition to the workspace, the stiffness or rigidity of a robot's structure is crucial as it directly impacts the precision and consistency of the robot's movements. Stiffness refers to the capacity of the robot's structure to withstand deformation caused by applied forces. A stiffness matrix establishes the relationship between a vector representing deformation and a vector representing wrench. Compliance, often known as flexibility, is another phrase that is synonymous with stiffness. A structure with high stiffness implies that the structure possesses minimal compliance. A compliance matrix is the reciprocal of the stiffness matrix, and vice versa. The stiffness comprises both static stiffness and dynamic stiffness. High stiffness in a machine tool allows for high-speed and high-feed machining, while also ensuring good precision, accuracy, and surface polish.

The rigidity of a mechanism is contingent upon the robot's topology, geometry, and the material from which the mechanism is constructed. The overall stiffness is determined by the combined stiffness of the stationary base, the moving platform, the joints, and the linkages. The stiffness of the joints include both the active joints (actuators) and the passive joints. Several studies have examined the impact of passive joints on the stiffness of robots. The stiffness of the links is often characterized by their axial stiffness, bending stiffness, or a combination of both in the transverse direction. In order to simplify the stiffness model, it is common to apply the rigid body assumption to one or more components of the robot. For instance, joints can be seen as elastic, whilst links are presumed to be stiff, or vice versa. A more accurate model typically include the elasticity of both the joints and the links. Hybrid machine tools often incorporate a parallel mechanism for the spindle platform and a serial mechanism for the worktable. This design choice is based on the fact that the spindle platform is typically the most adaptable component of the machine. In addition, certain studies have proposed the utilization of passive legs to enhance stiffness, as seen in the Tricept and Georg V designs.

Stiffness is a characteristic that is specific to a certain area or region. The answer varies based on the specific arrangement or positioning of the robot. Global stiffness measurements are employed for global evaluation. Moreover, the stiffness varies depending on the direction in which it is assessed, as well as the direction of the wrench. Hence, stiffness can be distinguished in various orientations, namely translational directions (translational stiffness) or rotational directions (rotational stiffness).

In literature, compliance, which is the reciprocal of stiffness, is occasionally employed as an alternative to stiffness. Various forms of stiffness have been employed in academic literature, such as engineering stiffness, generalized stiffness matrix, and Cartesian stiffness matrix. The engineering stiffness is a measure of the resistance to deformation in a single direction, determined by calculating the displacement in that direction caused by an applied force [51]. The generalized stiffness matrix, as described by Quennouelle and Gosselin [52], comprises three components: the stiffness of the unconstrained joints, the stiffness resulting from dependent coordination and internal wrench, and the stiffness resulting from external wrench. Put simply, the generalized stiffness matrix is the sum of the three stiffness components.

The Cartesian stiffness matrix is the predominant representation of stiffness that is commonly employed. Ruggiu [53] demonstrates that the Cartesian stiffness matrix of a mechanism can be derived from the Hessian of its potential energy expression. The Cartesian stiffness matrix exhibits both symmetry and positive definiteness or positive semi-definiteness. Nevertheless, several studies have determined that the Cartesian stiffness matrix of the elastic structure, which connects two rigid bodies, is typically asymmetric. However, it does become symmetric if the connection is not exposed to any preloading. Klimchik [54] and Quennouelle and Gosselin [52] have provided various formulations for the Cartesian stiffness matrix. The authors suggested a Cartesian stiffness matrix that considers non-zero external loads, non-constant Jacobian matrices, stiff passive joints, and extra compliances. Moreover, the Cartesian matrix can be directly correlated with the generalized stiffness matrix by employing the Jacobian matrix.

The Cartesian stiffness matrix in robots with both translational and rotational degrees of freedom will lack unit consistency. Therefore, assessing other measures of stiffness, such as the stiffness condition number, becomes meaningless. In order to address this issue, various strategies have been suggested, which include the following:

- Homogenizing the Jacobian matrix (such as using characteristic length) and subsequently using the homogenized Jacobian matrix to calculate the stiffness matrix [55, 56].
- Eigenscrew decomposition of the stiffness or compliance matrix [57–60].
- Principal axis decomposition through congruence transformation was proposed by making use of the eigenvectors of the translational entry in the stiffness matrix [61].
- Decomposition of the dynamic inertia matrix by transforming variables into dimensionless parameters, which can be applied to the stiffness matrix [62, 63].
- Decoupling of the stiffness matrix into translational parts and rotational parts [64–66].
- Furthermore, to model the robot stiffness beyond using continuous model which works only for simple system, the following three different models are widely used in the literature:
- Jacobian matrix-based model; also called lumped parameter model or virtual joint method (VJM) model. A one-dimensional VJM was introduced by Gosselin [67], followed by Anatol et al. [68] who introduced multi-dimensional VJM. This model is widely used and preferred in robotics since it is analytical, and therefore the same expression works for all configurations of the robot and it requires lower computational cost. However, it gives lower accuracy but still acceptable. For that reason, this method is good for initial estimates of the robot stiffness as well as for design optimization purpose.
- Finite element model (FEM). As opposite of the lumped parameter model, this model discretizes the mechanism into many elements and therefore can also be called distributed model which implies more closeness to the realistic, continuous model. It is widely used in structural mechanics due to its high accuracy. However, it requires high computational cost. Furthermore, it needs new mesh at every different configuration of the robot which makes it not practical. Due to its high accuracy, this model is usually used to verify another less accurate model such as VJM model.
- Matrix structural analysis (MSA) model. This model is actually a special case of FEM model because it uses one-dimensional finite elements such as beam elements instead of two or three dimensional elements such as brick elements. As a result, the computational cost decreases. This model gives trade-off between accuracy and computational cost [69].

After that, modifications and improvements on the aforementioned methods have been conducted, such as follows:

- Online FEM by utilizing MSA using generalized springs [70].
- VJM combined with FEM-based identification technique: high accuracy with low computational cost [71].
- Virtual spring approach: spring compliance is evaluated based on FEM concept; high accuracy with low computational cost [71, 72].

Evaluation of stiffness can also be conducted by experimental method, i.e., from measurement. In this case, the stiffness is obtained from the relation between measured wrench and measured displacement. Another way to evaluate the stiffness is by estimation or identification of the stiffness model. Least squares estimation algorithm or other estimation algorithms can be utilized in the estimation based on measurement data.

As a performance measure, the robot stiffness is represented in the following different ways in the literature:

- Graphical representations including stiffness maps, by which the stiffness distribution can be plotted [67, 73], and other graphical representations such as iso-stiffness curves or surfaces (global) [2].
- Trace of the stiffness matrix.
- Weighted trace of the stiffness matrix [74].
- (Minimum, average, or maximum) eigenvalues (and eigenvectors) of the stiffness matrix [51]. For example, the evaluation of minimum and maximum eigenvalues across the workspace by Li and Xu [56].
- Mean value of the eigenvalues [70].
- Determinant of stiffness matrix, which is the product of the stiffness matrix eigenvalues), and indicates the area/volume of a stiffness ellipse/ellipsoid. It also indicates how far from singularity.
- Norm of the stiffness matrix, which can be its Euclidian norm, Frobenius norm, or Chebyshev norm [75].
- Center of stiffness or equivalently center of compliance [76].
- Global compliance index which is given by mean value and deviation of generalized compliance matrix [77].
- Virtual work stiffness index which is able to avoid the problem caused by different units of translation and orientation
- Collinear stiffness value (CSV) [78].

2.7. Stiffness condition number

The stiffness condition number is a metric that quantifies the local stiffness of a system. The answer varies based on the specific arrangement of the robot. Similar to the Jacobian condition number, the stiffness condition number can range from 1 to infinity. Alternatively, the reciprocal of the stiffness condition number, which ranges from 0 to 1, can also be utilized. Stiffness ellipses/ellipsoids are frequently employed as a visual representation because they effectively depict the isotropy or uniformity of stiffness at any given place in the workspace, as indicated by the stiffness condition number.

The stiffness condition number can be evaluated using alternative definitions of norms, similar to the Jacobian condition number. The often employed norms include the 2-norm, Frobenius norm, and weighted Frobenius norm. The criteria for choosing any of them has been previously explained during the discussion on the Jacobian condition number.

The global stiffness condition number is often quantified using the global stiffness index (GSI), which is defined as the reciprocal of the condition number of the stiffness matrix integrated over the reachable workspace, divided by the volume of the workspace. It illustrates the consistency of rigidity throughout the entire workspace.

3. DESIGN OPTIMIZATION

Optimization can be categorized based on the number of objectives being optimized. It can be either single-objective, also known as single-criteria, or multi-objective, also known as multi-criteria. A straightforward approach to design optimization involves employing trial and error. This method is selecting several values for design parameters based on intuition, knowledge, or experience, and subsequently comparing the resulting objective values. Nevertheless, this approach lacks a systematic approach and fails to encompass all potential values of the design parameters, therefore potentially resulting in suboptimal solutions. Performance atlas and optimization algorithms are frequently employed in the design optimization of mechanisms in literature. The performance atlas visually displays the correlation between the design parameters (link length) and the performance metrics. Various performance atlases, including the atlas of workspace, atlas of GSI, and atlas of LSI, have been utilized in academic literature. The utilization of a performance atlas is simple and uncomplicated for single-objective optimization. Nevertheless, conducting a multi-objective optimization necessitates examining multiple atlases, which can be inconvenient, especially when some objectives contradict each other.

In addition to utilizing performance atlas, a variety of techniques have been employed to optimize PKMs and HKMs. The search concepts classify those techniques into two primary categories: gradient-based optimization techniques and population-based optimization strategies. The initial category pertains to a local search algorithm. The system is deterministic and its linearity or nonlinearity depends on the specific challenge at hand. The second category is characterized by stochastic behavior and does not require gradient information. The genetic algorithm is a widely used population-based technique that employs evolutionary optimization principles, namely the concept of natural selection or survival of the fittest. The genetic algorithm can be utilized for optimizing both single-objective and multi-objective problems. Various techniques have been developed for the latter application, including VEGA, NPGA, NPGA-II, NSGA, NSGA-II, PAES, PESA, PESA-II, SPEA, SPEA-II, SPEA-II+, and numerous others [79–84]. In addition to the genetic algorithm, various global optimization algorithms have been suggested in the literature, including controlled random search (CRS) [85], differential evolution (DE) [86, 87], particle swarm optimization (PSO) [88–90], quantum particle swarm optimization (QPSO) [91], and artificial neural network (ANN) [74]. The following will provide a more comprehensive analysis of both methods of optimization.

3.1. Single-objective optimization

While single-objective optimization is very simple, several strategies can be employed to seek for the optimal solution. When optimizing a parallel mechanism, the primary focus is typically on the workspace, as the main disadvantage of parallel mechanisms is their restricted workspace.

In addition to commonly used gradient-based optimization techniques, several algorithms have been suggested for the single-objective optimization of parallel kinematic mechanisms (PKMs). Hosseini et al. [32] employed a genetic algorithm to enhance the efficiency of the dexterous workspace of a Tricept PKM. Kordjazi et al. [92] employed a combination of genetic algorithm and fuzzy logic algorithm to enhance the isotropy of the Jacobian matrix in parallel kinematic mechanisms (PKM). Their findings demonstrated that this approach yielded superior results compared to utilizing genetic algorithm alone. Arana [93] presented an approach for expanding the operational range of parallel manipulators through the use of non-singular transitions. Ghao and Zhang [25] suggested employing particle swarm optimization (PSO) to optimize the workspace of a 3-DOF spatial parallel mechanism.

3.2. Multi-objective optimization

Typically, many objectives need to be optimized in most circumstances. Moreover, certain aims often contradict with one another. For instance, the majority of PKMs often necessitate a wider workspace as well as a more rigid structure with reduced mass. Expanding the workspace typically necessitates longer linkages, leading to decreased stiffness and increased mass. Multi-objective optimization encompasses both bi-objective optimization, which involves optimizing two objectives, and many-objective optimization, which involves optimizing more than two objectives. The former is less complex than the latter. Introducing more than two objectives typically results in increased computing expenses. Furthermore, it has greater complexity in terms of visual depiction. When dealing with more than three objectives, it is only possible to create graphical plots for three design factors that vary, while the remaining parameters must be kept constant. One additional method to decrease the number of objectives in the optimization process is to include a performance index threshold value as a constraint in the optimization. Nevertheless, this method is only appropriate if the requirement is solely to meet the minimum standard.

In multi-objective optimization, the selection of different objectives (criteria) is dependent on the priority assigned to each objective, which is determined by the specific application. There are two primary techniques that are frequently employed for multi-objective optimization:

- Scalarization method which is commonly conducted by putting the weighted multiple objective functions into one composite objective function. While transforming the problem into one objective function gives simplicity, the determination of appropriate weights is difficult, even for those who are familiar with the problem. This approach can be conducted through gradient-based optimization methods as well as single-objective evolutionary methods (such as single-objective genetic algorithm).
- Pareto method which will give non-dominated solutions. This method can be conducted through multi-objective evolutionary methods (such as multi-objective genetic algorithm).

PKMs (Parallel Kinematic Machines) can be optimized for several purposes based on specific application requirements. To do this, several techniques have been suggested. Hodgins [94] enhanced the efficiency, rigidity, and maneuverability of a modified Delta robot by employing a weighted sum optimization technique. Kelaiaia et al. [95] enhanced both the kinematic dexterity and dynamic dexterity through the utilization of genetic algorithms. Wu [96] enhanced the Geometric Condition Index (GCI) and Geometric Dilution Index (GDI) of a 3RRR spherical parallel mechanism, which serves as an orienting device, through the application of a genetic algorithm. Bounab [97] enhanced the efficiency of a delta mechanism by utilizing a genetic algorithm to maximize both the dexterous regular workspace and stiffness. Shijun [3] enhanced the performance of GDI, GSI, and the proportion of the workspace to the work volume by employing a genetic algorithm. Gao et al. [74] enhanced the rigidity and agility by the utilization of a genetic algorithm and an artificial neural network. In their study, Abbasnejad et al. [91] utilized particle swarm optimization (PSO) and quantum particle swarm optimization (QPSO) to improve the workspace and dexterity. They found that QPSO exhibits faster convergence compared to PSO when considering a weighted sum objective. In addition, Gao and Zhang [98] proposed a comprehensive index to combine four distinct aims.

4. SUGGESTIONS

The presence of mixed degrees of freedom (DOFs) leads to inconsistencies in the indices, while many parallel kinematic mechanisms (PKMs) require mixed DOFs to perform the necessary duties. Hence, the authors propose that any future introduction of a new index should effectively address this problem in a more organic manner, ensuring that the physical understanding of the index remains robust and intuitive. In addition, the authors have not come across any published literature that discusses the optimization of a significant number of performance measures. Although several attempts have been made to address up to three or four objective functions, it remains both practically useful and problematic to manage a larger number of objective functions. Recognizing the challenge of assigning suitable weights in the scalarization approach, even for a small number of two to four objective functions, it will undoubtedly be more arduous to allocate appropriate weights to a larger number of objectives. Moreover, it may no longer be feasible to employ the Pareto approach for a larger number of objectives. Thus, the authors propose the implementation of a novel approach or technique to effectively optimize a greater quantity of objective functions.

5. CONCISE SUMMARY

This chapter presented a thorough summary of the literature about several kinematic performance indices that designers may find useful when designing parallel kinematics devices. Designers prioritize kinematic performance metrics, such as workspace, Jacobian condition number, manipulability, stiffness magnitude, and stiffness condition number, in order to optimize the designs of parallel kinematic mechanisms. Nevertheless, some indices present inconsistent outcomes, such as the expansion of workspace dimensions potentially resulting in a decline in rigidity and/or a fall in speed. Hence, the utilization of optimization techniques is the remedy for accommodating competing requirements by assigning suitable weights to diverse objectives that accurately reflect the interests and priorities of the designer. However, formulating the appropriate goal function necessitates a significant amount of expertise and deep understanding of the topic. The selection of objectives and their corresponding weights will significantly influence the outcomes of the design parameters, and therefore should be made with great care.

This paper aimed to offer a detailed overview of the requirements for parallel kinematic mechanism designers to optimize their design and enhance mechanism performance.

REFERENCES

1. Patel S, Sobh T. Manipulator performance measures—A comprehensive literature survey. Journal of Intelligent and Robotic Systems. 2015;77 (3-4):547-570
2. Merlet J-P. Parallel Robots. 2nd ed. Dordrecht: Springer; 2006
3. Shijun Y. A Serial-Parallel Hybrid Robot for Machining of Complex Surfaces. National University of Singapore, Singapore; 2015
4. Baek J, Iurascu C-C, Park FC. Finding the maximally inscribed rectangle in a Robot's workspace. KSME International Journal. 2001;15(8):1119-1131
5. Bajpai A, Roth B. Workspace and mobility of a closed-loop manipulator. The International Journal of Robotics Research. 1986;5:131-142
6. Gosselin CM. Determination of the workspace of 6-DOF parallel manipulators. ASME Journal of Mechanical Design. 1989;112:331-336
7. Merlet J-P. Determination of the orientation workspace of parallel manipulators. Journal of Intelligent and Robotic Systems. 1995;13:143-160
8. Kim DI, Chung WK, Youm Y. Geometrical approach for the workspace of 6-DOF parallel manipulators. In: IEEE International Conference on Robotics and Automation (ICRA); Albuquerque, IEEE, New Mexico; 1997
9. Gosselin CM, Jean M. Determination of the workspace of planar parallel manipulators with joint limits. Robotics and Autonomous Systems. 1996;17:129-138
10. Merlet J-P, Gosselin CM, Mouly N. Workspaces of planar parallel manipulators. Mechanism and Machine Theory. 1998;33(1-2):7-20
11. Arrouk KA, Bouzgarrou BC, Gogu G. Workspace determination and representation of planar parallel manipulators in a CAD environment. In: New Trends in Mechanism Science. Netherlands: Springer; 2010. p. 605-612
12. Alciatore DG, Ng C-CD. Determining manipulator workspace boundaries using the Monte Carlo method and least squares segmentation. ASME Robotics: Kinematics, Dynamics and Controls. 1994;72:141-146
13. Rastegar J, Perel D. Generation of manipulator workspace boundary geometry using the Monte Carlo method and interactive computer graphics. ASME Trends and Developments in Mechanisms, Machines, and Robotics. 1990;3:299-305

14. Bonev IA, Ryu J. A new approach to orientation workspace analysis of 6-DOF parallel manipulators. Mechanism and Machine Theory. 2001;36(1):15-28

15. Castelli G, Ottaviano E, Ceccarelli M. A fairly general algorithm to evaluate workspace characteristics of serial and parallel manipulators#. Mechanics Based Design of Structures and Machines. 2008;36(1):14-33

16. Dash AK, et al. Workspace generation and planning singularity-free path for parallel manipulators. Mechanism and Machine Theory. 2005;40(7):776-805

17. Wang Z, et al. A study on workspace, boundary workspace analysis, and workpiece positioning for parallel machine tools. Mechanism and Machine Theory. 2001;36(5):605-622

18. D-Y Jo, Haug EJ. Workspace analysis of closed loop mechanisms with unilateral constraints. In: ASME Design Automation Conference; Montreal; 1989. pp. 53-60.

19. Haug EJ, et al. Numerical algorithms for mapping boundaries of manipulator workspaces. ASME Journal of Mechanical Design. 1995;118:228-234

20. Jo D-Y, Haug EJ. Workspace analysis of multibody mechanical systems using continuation methods. Journal of Mechanisms, Transmissions, and Automation in Design. 1989;111:581-589

21. Snyman JA, Plessis LJd, Duffy J. An optimization approach to the determination of the boundaries of manipulator workspaces. ASME Journal of Mechanical Design. 2000;122(4):447-456

22. Kumar V. Characterization of workspaces of parallel manipulators. ASME Journal of Mechanical Design. 1992;114(3):368-375

23. Kaloorazi MHF, Masouleh MT, Caro S. Determining the maximal singularity-free circle or sphere of parallel mechanisms using interval analysis. Robotica. 2014;34(01):135-149

24. Bohigas O, Ros L, Manubens M. A complete method for workspace boundary determination on general structure manipulators. IEEE Transactions on Robotics. 2012;28(5):993-1006

25. Gao Z, Zhang D. Workspace representation and optimization of a novel parallel mechanism with three-degrees-of-freedom. Sustainability. 2011;3(12):2217-2228

26. Saputra VB, Ong SK, Nee AYC. A swarm optimization approach for solving workspace determination of parallel manipulators. Robotica. 2014;33(03):649-668

27. Angeles J. Fundamentals of Robotic Mechanical Systems, Mechanical Engineering Series. New York: Springer; 2002

28. Ma, Angeles J. Optimum architecture design of platform manipulators. In: IEEE International Conference on Advanced Robotics; Pisa; 1991. pp. 1130-1135

29. Ranjbaran F, et al. The mechanical design of a seven-axes manipulator with kinematic isotropy. Journal of Intelligent and Robotic Systems. 1995;14:21-41

30. Stocco LJ, Salcudean SE, Sassani F. Matrix normalization for optimal robot design, in IEEE International Conference on Robotics and Automation. Leuven, Belgium; 1998

31. Stocco LJ, Saculdean SE, Sassani F. On the use of scaling matrices for task-specific robot design. IEEE Transactions on Robotics and Automation, 1999;15:958-965

32. Hosseini MA, Daniali H-RM, Taghirad HD. Dexterous workspace optimization of a Tricept parallel manipulator. Advanced Robotics. 2011;25(13-14):1697-1712

33. Mansouri I, Ouali M. A new homogeneous manipulability measure of robot manipulators, based on power concept. Mechatronics. 2009;19(6):927-944

34. Altuzarra O, et al. Point-based Jacobian formulation for computational kinematics of manipulators. Mechanism and Machine Theory. 2006;41(12):1407-1423

35. Gosselin CM. The optimum design of robotic manipulators using dexterity indices. Robotics and Autonomous Systems. 1992;9(4):213-226

36. Kim S-G, Ryu J. New dimensionally homogeneous Jacobian matrix formulation by three end-effector points for optimal design of parallel manipulators. IEEE Transactions on Robotics and Automation. 2003;19(4):731-736

37. Kong M, et al. A novel approach to deriving the unit-homogeneous Jacobian matrices of mechanisms. In: IEEE International Conference of Mechatronics and Automation; 2007

38. Liu H, Huang T, Chetwynd DG. A method to formulate a dimensionally homogeneous Jacobian of parallel manipulators. IEEE Transactions on Robotics. 2011;27(1):150-156

39. Nurahmi L, Caro S. Dimensionally Homogeneous Jacobian and Condition Number. Applied Mechanics and Materials. 2016;836:42-47

40. Gosselin CM, Angeles J. Singularity analysis of closed-loop kinematic chains. IEEE Transactions on Robotics and Automation. 1990;6(3):281-290

41. Zlatanov D, Fenton RG, Benhabib B. A unifying framework for classification and interpretation of mechanism singularities. Journal of Mechanical Design. 1995;117(4):566

42. Zlatanov D, Bonev IA, Gosselin CM. Constraint singularities of parallel mechanisms. In: IEEE International Conference of Robotics and Automation (ICRA); 2002

43. Han C, Park. Kinematic sensitivity analysis of the 3-UPU parallel manipulator. Mechanism and Machine Theory. 2002;37(8):787-798

44. Altuzarra O, et al. Optimal dimensioning for parallel manipulators: Workspace, dexterity, and energy. Journal of Mechanical Design. 2011;133(4):041007

45. Khan WA, Angeles J. The Kinetostatic optimization of robotic manipulators. Journal of Mechanical Design. 2006;128:168-178

46. Yoshikawa T. Manipulability of robot mechanisms. The International Journal of Robotics Research. 1985;4(2):3-9

47. Pham HH, Chen I-M. Optimal synthesis for workspace and manipulability of parallel flexure mechanism. In: 11th World Congress on Theory of Machines and Mechanisms; Tianjin; 2004

48. Tanev T, Stoyanov B. On the performance indexes for robot manipulators. Problems of Engineering Cybernetics and Robotics. 2000;49:64-71

49. Doty KL, et al. Robot Manipulability. IEEE Transactions on Robotics and Automation. 1995;11(3):462-468

50. Hong K-S, Kim J-G. Manipulability analysis of a parallel machine tool: Application to optimal link length design. Journal of Robotic Systems. 2000;17(8):403-415

51. El-Khasawneh BS, Ferreira PM. Computation of stiffness and stiffness bounds for parallel link manipulators. International Journal of Machine Tools and Manufacture. 1999;39(2):321-342

52. Quennouelle C, Gosselin CM. A General Formulation for the Stiffness Matrix of Parallel Mechanisms. 2012. arXiv preprint arXiv:1212.0950

53. Ruggiu M. On the Lagrangian and Cartesian stiffness matrices of parallel mechanisms with elastic joints. Advanced Robotics. 2012;26(1-2):137-153

54. Klimchik A. Enhanced Stiffness Modeling of Serial and Parallel Manipulators for Robotic-Based Processing of High Performance Materials. Ecole Centrale de Nantes, France; 2011

55. Li Y, Xu Q. GA-based multi-objective optimal design of a planar 3-DOF cable-driven parallel manipulator. In: IEEE International Conference on Robotics and Biomimetics; Kunming, IEEE, China; 2006

56. Li Y, Xu Q. Stiffness analysis for a 3-PUU parallel kinematic machine. Mechanism and Machine Theory. 2008;43(2):186-200

57. Ciblak N, Lipkin H. Synthesis of Cartesian stiffness for robotic applications. In: IEEE International Conference of Robotics and Automation. IEEE, Detroit, MI.1999

58. Dai JS, Ding X. Compliance analysis of a three-legged rigidly-connected platform device.Journal of Mechanical Design. 2006;128(4):755

59. Ding X, Selig JM. On the compliance of coiled springs. International Journal of Mechanical Sciences. 2004;46(5):703-727

60. Huang S, Schimmels JM. The Eigenscrew decomposition of spatial stiffness matrices. IEEE Transactions on Robotics and Automation. 2000;16(2):146-156

61. Chen G, et al. The principal axes decomposition of spatial stiffness matrices. IEEE Transactions on Robotics. 2015;31(1):191-207

62. Taghvaeipour A, Angeles J, Lessard L. On the elastostatic analysis of mechanical systems. Mechanism and Machine Theory. 2012;58:202-216

63. Wu G. Stiffness analysis and optimization of a co-axial spherical parallel manipulator. Modeling, Identification and Control: A Norwegian Research Bulletin. 2014;35(1):21-30

64. Angeles J. On the nature of the Cartesian stiffness matrix. Ingenierıa Mecanica. 2010;3(5):163-170

65. Wu G, Bai S, Hjornet P. On the stiffness of three or four DOF parallel pick-and-place robots with four identical limbs. In: IEEE International Conference on Robotics and Automation (ICRA); IEEE, Stockholm, Sweden; 2016

66. Wu G, Zou P. Stiffness analysis and comparison of a Biglide parallel grinder with alternative spatial modular parallelograms. Robotica. 2016;35(06):1310-1326

67. Gosselin CM. Stiffness mapping for parallel manipulators. IEEE Transactions on Robotics and Automation. 1990;6(3):377-382

68. Pashkevich A, Chablat D, Wenger P. Stiffness analysis of multi-chain parallel robotic systems. Journal of Automation, Mobile Robotics and Intelligent Systems. 2009;3(3):75-82

69. Ceccarelli M. Stiffness analysis of parallel manipulator using matrix structural analysis. In: Proceedings of EUCOMES 08. 2008. Springer, Dordrecht

70. Taghvaeipour A, Angeles J, Lessard L. Online Computation of the Stiffness Matrix in Robotic Structures Using Finite Element Analysis. Department of Mechanical Engineering and Centre for Intelligent Machines, McGill University, Montreal; 2010

71. Pashkevich A, Chablat D, Wenger P. Stiffness analysis of overconstrained parallel manipulators. Mechanism and Machine Theory. 2009;44(5):966-982

72. Pashkevich A, Klimchik A, Chablat D. Enhanced stiffness modeling of manipulators with passive joints. Mechanism and Machine Theory. 2011;46(5):662-679

73. Mekaouche A, Chapelle F, Balandraud X. FEM-based generation of stiffness maps. IEEE Transactions on Robotics. 2015;31(1):217-222

74. Gao Z, Zhang D, Ge Y. Design optimization of a spatial six degree-of-freedom parallel manipulator based on artificial intelligence approaches. Robotics and Computer-Integrated Manufacturing. 2010;26(2):180-189

75. Carbone G, Ceccarelli M. A comparison of indices for stiffness performance evaluation. In: 12th International federation for the promotion of mechanism and machine science (IFToMM) World Congress; Besançon, France; 2007

76. Patterson T, Lipkin H. A classification of robot compliance. ASME Journal of Mechanical Design. 1993;115(3):581-584

77. Xi F, et al. Global kinetostatic modelling of tripod-based parallel kinematic machine. Mechanism and Machine Theory. 2004;39(4):357-377

78. Shneor Y, Portman VT. Stiffness of 5-axis machines with serial, parallel, and hybrid kinematics: Evaluation and comparison. CIRP Annals - Manufacturing Technology. 2010;59(1):409-412

79. Amouzgar K. Multi-Objective Optimization Using Genetic Algorithms. Stockholm, Sweden: Tekniska Hogskolan; 2012

80. Coello CA. An updated survey of GA-based multiobjective optimization techniques. ACM Computing Surveys. 2000;32(2):109-143

81. Coello CA, Lamont GB, Veldhuizen DAV. Evolutionary Algorithms for Solving Multi-Objective Problems. 2nd ed. New York, NY, USA: Springer; 2007

82. Deb K. Multi-Objective Optimization Using Evolutionary Algorithms. New York, NY, USA: John Wiley & Sons, Inc.; 2001

83. Konak A, Coit DW, Smith AE. Multi-objective optimization using genetic algorithms: A tutorial. Reliability Engineering and System Safety. 2006;91:992-1007

84. Tan KC, Khor EF, Lee TH. Multiobjective Evolutionary Algorithms and Applications. London: Springer-Verlag; 2005

85. Lou YJ, Liu GF, Li ZX. Randomized optimal Design of Parallel Manipulators. IEEE Transactions on Automation Science and Engineering. 2008;4(2):625-649

86. Bulatović RR, Dordeević SR. On the optimum synthesis of a four-bar linkage using differential evolution and method of variable controlled deviations. Mechanism and Machine Theory. 2009;44(1):235-246

87. Shiakolas PS, Koladiya D, Kebrle J. On the optimum synthesis of six-bar linkages using differential evolution and the goemetric centroid of precision positions technique. Mechanism and Machine Theory. 2005;40(3):319-335

88. Li Y, Xu Q. Design and analysis of a totally decoupled flexurebased XY parallel micromanipulator. IEEE Transactions on Robotics. 2009;25(3):645-657

89. Li Y, Xu Q. A totally decoupled piezo-driven XYZ flexure parallel micropositioning stage for micro/nanomanipulation. IEEE Transactions on Automation Science and Engineering. 2011;8(2):265-279

90. Xu Q, Li Y. Error analysis and optimal design of a class of translational parallel kinematic machine using particle swarm optimization. Robotica. 2008;27(01):67

91. Abbasnejad G, Daniali HM, Fathi A. Architecture optimization of 4PUS+1PS parallel manipulator. Robotica. 2010;29(05):683-690

92. Kordjazi H, Akbarzadeh A, Rostami AS. Isotropy design and optimization of a planar parallel manipulator with combination of fuzzy logic and genetic algorithm. In: 8th World Congress on Structural and Multidisciplinary Optimization. International Society for Structural and Multidisciplinary Optimization; Lisbon, Portugal; 2009

93. Arana MU. Methodology to Enlarge the Workspace of Parallel Manipulators by Means of Non-singular Transitions. Bilbao, Spain: University of the Basque Country (UPV/EHU); 2012

94. Hodgins J. Design, Optimization, and Prototyping of a 3 Translational DOF Parallel Robot. University of Ontario Institute of Technology: Canada; 2012

95. Kelaiaia R, Company O, Zaatri A. Multiobjective optimization of parallel kinematic mechanisms by the genetic algorithms. Robotica. 2011;30(05):783-797

96. Wu G. Multiobjective optimum design of a 3-RRR spherical parallel manipulator with kinematic and dynamic dexterities. Modeling, Identification and Control: A Norwegian Research Bulletin. 2012;33(3):111-122

97. Bounab B. Multi-objective optimal design based kineto-elastostatic performance for the delta parallel mechanism. Robotica. 2014;34(02):258-273

98. Gao Z, Zhang D. Performance analysis, mapping, and multiobjective optimization of a hybrid robotic machine tool. IEEE Transactions on Industrial Electronics. 2015;62(1):423-433

2
Chapter
THE TRANSFORMATION OF INERTIA VALUES IN MARITIME APPLICATIONS

1. INTRODUCTION: AN OVERVIEW

Contemporary simulation tools provide a more in-depth investigation of diverse engineering problems during the initial phase of design. Maritime kinematics primarily examines the response of offshore structures to natural environmental factors such as wind, waves, and currents. Thorough understanding of weights and kinematics is necessary for accurately sizing efficient and secure systems. Moreover, simulations are progressively employed for the development of controller-based automation and help systems. Aside from accurately calculating structural movements, the speed at which a simulation can perform calculations is a crucial quality characteristic for such applications, particularly when real-time restrictions are involved. Various methodologies are typically employed to establish the kinematics model of a unanchored structure in a maritime environment. The equations of motion can be defined in either the inertial reference frame or the body-fixed reference frame. The distinction between dry mechanics and maritime mechanics lies in the presence of hydrodynamic added mass, which introduces additional hydrodynamic effects [1-3]. The hydrodynamic mass force is a force resulting from the inertia of a fluid. The difference in acceleration between an incompressible fluid and a structure causes a pressure field, which in turn generates a hydrodynamic force. This force is calculated by multiplying the difference in acceleration by a 'virtual' mass. The hydrodynamic additional mass is mostly influenced by the direction of movement and the structure's geometry [4].

In order to accurately describe translational and rotational motions, it is required to consider non-scalar and directed inertia values, which account for the phenomenon of additional or virtual masses. Typically, this leads to a common method of describing floating structures from a perspective that is fixed to the body, such as using the Kirchhoff motion equations for floating bodies [1, 3].

The authors are primarily focused on investigating mechanically coupled and numerous rigid body systems, which poses a unique challenge in this field [2, 5]. In literature prior to the turn of the millennium, only Paschen's algorithm was discovered to describe the behavior of genuine 3D systems in fishery and mine-hunting systems. These systems consist of flexible, non-elastic numerical elements and stiff bodies, with the added masses partially ignored [6, 7]. Subsequently, several marine multibody systems were characterized using techniques based on structural mechanics [8] or solely described in terms of planar motion, for example [9–13]. In order to explain the reciprocal relationships, it is necessary to establish the equations of motion for all relevant entities inside the same coordinate system, ideally the inertial or earth-fixed system. This encompasses the conversion of all vector-based elements, such as the inertia matrices (mass matrix, moments of inertia). Hence, the utilization of a transformation technique is necessary to depict the system in the inertial reference frame. The presented transformation of inertia values, also referred to as Kane's approach [14], was initially introduced by Korte and Takagi [2, 5] for the analysis of forces and motions in fishing systems including purse seines. The net structure incorporates nodes that are concentrated and connected using damped mass-spring elements, following the notion of concentrated masses. In order to establish a connection between the inertia values of the nodes, it is necessary to change the hydrodynamic added masses [15].

In 2015, Korte et al. demonstrated the use of the inertia value transformation for multibody systems. Their application focused on a pure spinning body that did not have any elasticity [5]. The scenario under analysis involves the transfer of offshore service personnel from a crew transfer vessel (CTV) to a wind turbine, which is a tangible structure impacted by the environment. The CTV is restricted at the front of the vessel to facilitate the docking of the wind turbine in the ocean. The inertial reference system is situated precisely at the point of contact of the structure. The ship is capable of performing a perfect revolution around each of the three axes of the inertial system. The equations of motion for the system are formulated within the inertial reference frame in order to determine the forces that restrict its motion. The ship's erratic motions were found to result from the temporal integration of the equations within the inertial reference frame. The simulation exhibited instability. While working on the multibody situation, it was found that introducing an extra transformation of the rotational accelerations in the opposite direction results in a stable solution for the simulation.

In order to assess the overall suitability of the inertia value transformation for maritime purposes, Korte et al. established the equations of motion for a ship that is freely floating and has six degrees of freedom (6DOF) in the inertial reference frame [17]. The motion equations were developed using the principles of momentum and angular momentum. The outcomes of the unmoored vessel were contrasted with conventional simulations conducted in the body-fixed reference frame and kinematics (see to [1]).

This article discusses the usefulness of transforming inertia data in maritime applications when moving them between a body-fixed frame and an inertial reference frame. The discussion is centered around the motion behavior of the free-floating structure. This section presents the modeling of irregular seas as an environmental factor affecting structural motions. Subsequently, the six degrees of freedom of a floating body, along with the conventional motion equations in the body-fixed reference system, are presented. The process of obtaining the equations of motion in the inertial reference frame through the transformation of inertia values is included in the main section. The process of converting the equations from an inertial reference frame to a body-fixed reference frame is demonstrated. The discussion focuses on comparing the simulation results of a free-floating CTV in waves using both methodologies. The previous section presents a proof that demonstrates the conservation of energy in the transformation of inertia values for a purely spinning rigid body.

2. WAVE MODELLING

Natural irregular seas are classified into two types: wind seas and swell, based on their creation. In naval practice, real sea state phenomena are referred to as the combination of a random wind sea and two detectable swells.

Wind seas are generated by the direct action of wind. The waves exhibit a somewhat compact and abrupt nature. Friction causes the wind to transfer energy to the surface of the water. The presence of capillary waves with a small wavelength causes the free surface area to expand, resulting in an amplification of the energy transfer effect. This results in the generation of larger and more persistent gravity waves, which have an impact on the water column up to a depth equal to half the wavelength. The waves propagate in the direction of the wind. The properties of wind waves are influenced by the speed of the wind, the duration of the wind, and the fetch. The wave components of wind seas are situated within the higher-frequency range of a wave spectrum. As a result of the presence of high-frequency waves with low amplitudes, the water column oscillates uniformly. The wind seas diminish due to the cessation of wind stimulation. The correlation between wind categories (Beaufort wind scale) and sea-state categories (Douglas sea scale) is seen in Figure 1.

Contrary to wind seas, which is described as evolving waves, is swell the full homogeneously oscillating water column. Swell is a decaying, nearly sinusoidal wave that has moved away from the formation area, e.g. a storm region. The frequencies are in the lower range of a wave spectrum. The energy density of waves is inversely proportional to the wave frequency. Long waves with a small frequency are more energetic than short waves.

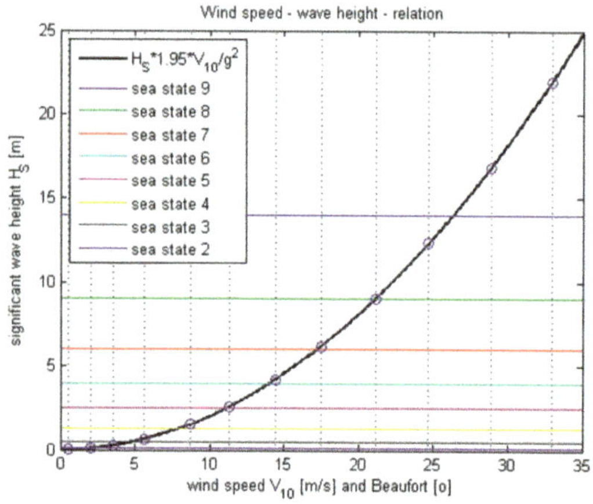

Figure 1. Relation of wind speed (Beaufort wind scale) and wave height (Douglas sea scale) for fully developed seas (modified to practice).

Irregular waves are mathematically described as superposition of a finite number of wave components with different wave frequencies, different amplitudes and different phases. The representation is carried out with energy density spectra, which have a unique distribution of the wave components for each sea area. The spectrum contains the energy $S_{n(\omega)}$ of all included wave frequencies. Figure 2 shows the structure of a wave spectrum.

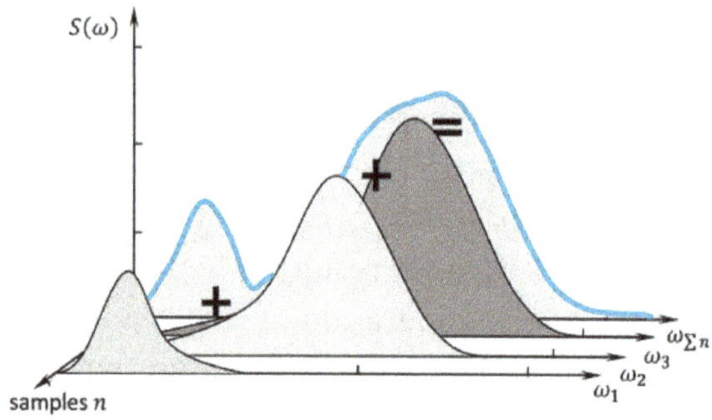

Figure 2. Structure of a wave spectrum as superposition of different wave components.

Relevant wave parameters can be derived from the wave spectrum. The mean wave frequency ω_0 is located at the maximum $S(\omega)$ of the spectrum. The peak period T_P is

$$T_P = \frac{2\pi}{\omega_0} \tag{1}$$

Further, wave properties can be determined with the so-called spectral moments m_n, compared with [1]. Their definition is

$$m_n = \int_0^\infty \omega^n S(\omega) d\omega \tag{2}$$

The moments up to the second order, n=2, are of interest to characterise the sea state. The characteristic wave height of the spectrum is

$$H_{m0} = 4\sqrt{m_0} \tag{3}$$

and the significant wave height is

$$H_S = 3.81 \cdot \sqrt{m_0} \tag{4}$$

The period of the characteristic wave height H_{m0} is defined as

$$T_1 = \frac{m_0}{m_1} \tag{5}$$

The mean zero upcrossing period is

$$T_Z = \sqrt{\frac{m_0}{m_2}} \tag{6}$$

Variance σ^2 and standard deviation σ are

$$\sigma^2 = m_0 \tag{7}$$

$$\sigma = \sqrt{m_0} = 0.263 \cdot H_S \tag{8}$$

Multiple mathematical wave spectra are accessible for computer simulations of irregular waves, including the Bretschneider spectrum, the Phillips spectrum, the Pierson-Moskowitz spectrum, and the Joint North Sea Wave Project (JONSWAP) spectrum. This section introduces the most significant spectra.

2.1. Pierson-Moskowitz spectrum

The Pierson-Moskowitz (PM) spectrum was introduced for fully developed seas and is commonly used since 1964 [18]. It is based on long-term observation data of a weather ship in the period from 1955 to 1960. It assumes the fully developed sea, including wind sea and swell, on the North Atlantic with unlimited water depth and fetch as well as steady wind for a long time. The spectrum is

$$S_{PM}(\omega) = A \cdot \omega^{-5} \cdot e^{-B\omega^{-4}} \tag{9}$$

with parameters A and B

$$A = 0.0081 \cdot g^2 \tag{10}$$

$$B = 0.74\left(\frac{g}{V_{wind}}\right)^4 = \frac{2.814}{H_S^2} \tag{11}$$

The relation between significant wave height and wind speed is:

$$H_S = \frac{1.95}{g^2} V_{wind}^2 \tag{12}$$

2.2. Modified Pierson-Moskowitz spectrum

After experiences with the Pierson-Moskowitz spectrum, the maritime community recommended a magnification of the spectrum [19, 20]. According to this recommendation, the parameters *A* and *B* were modified and can be determined from the observable weather parameters H_S and T_Z:

$$A = \frac{4\pi^3 H_S^2}{T_Z^4} \tag{13}$$

$$B = \frac{16\pi^3}{T_Z^4} \tag{14}$$

Figure 3 presents a juxtaposition of the Pierson-Moskowitz spectrum and the Modified Pierson-Moskowitz (MPM) spectrum across various wind speeds. One can observe an augmentation in the energy density at the highest frequency of the MPM, along with a displacement of the peak frequency towards lower frequencies.

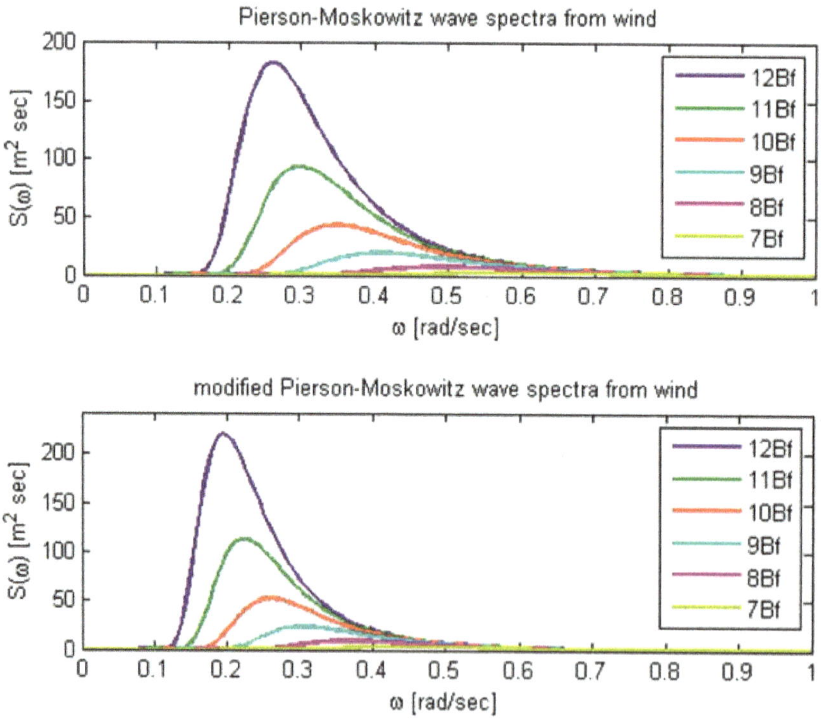

Figure 3. Comparison of Pierson-Moskowitz spectrum (top) and modified Pierson-Moskowitz spectrum (bottom) for several wind speeds.

2.3. JONSWAP spectrum

The Pierson-Moskowitz spectra were created for water depths that have no limitations. Nevertheless, the marine areas that are subjected to the highest levels of human activity are typically located in locations characterized by shallow water depths and limited fetch.

The wave structures in sea areas with restricted fetch and water depth were studied in the North Sea as part of the international collaborative project called the Joint North Sea Wave Project (JONSWAP) in 1968/1969. Over a span of around 10 weeks, measurements were conducted and assessed at locations extending up to 160 km offshore in the vicinity of the island of Sylt. The JONSWAP spectrum is derived from the PM spectrum and includes a magnification factor γ to adjust the distribution of peak values.

Hasselmann's mathematical definition of the JONSWAP spectrum is [20]

$$S_{JON}(\omega) = \alpha g^2 \omega^{-5} \cdot e^{\left[-\frac{5}{4}\left(\frac{\omega_0}{\omega}\right)^4\right]} \cdot \gamma^r \qquad (15)$$

with

$$r = e^{\left[-\frac{(\omega-\omega_0)^2}{2\sigma^2\omega_0^2}\right]} \qquad (16)$$

$$\gamma = 3.3 \qquad (17)$$

$$\alpha = 0.076\left(\frac{V_{10}^2}{F_W g}\right)^{0.22} \qquad (18)$$

$$\omega_0 = 22\left(\frac{g^2}{F_W V_{10}}\right)^{\frac{1}{3}} \qquad (19)$$

$$\sigma = \begin{cases} 0.07, & \omega \leq \omega_0 \\ 0.09, & \omega > \omega_0 \end{cases} \qquad (20)$$

Fetch is defined in metres and increases the wave energy linearly. A direct comparison of the JONSWAP spectrum with the MPM spectrum for a large-scale wind sea on the North Sea illustrates the lower total energy in Figure 4. The characteristic frequencies ω_0 of the individual wind classes are higher than those of the PM spectrum. Due to limited water depth, the wave heights are reduced. Both, the maximum of the spectrum and the area under the curve are considerable smaller than those of the reference spectrum. A comparison of seaways with different fetch lengths confirms this trend (cf. Figure 4). A JONSWAP spectra are used for the here presented simulations.

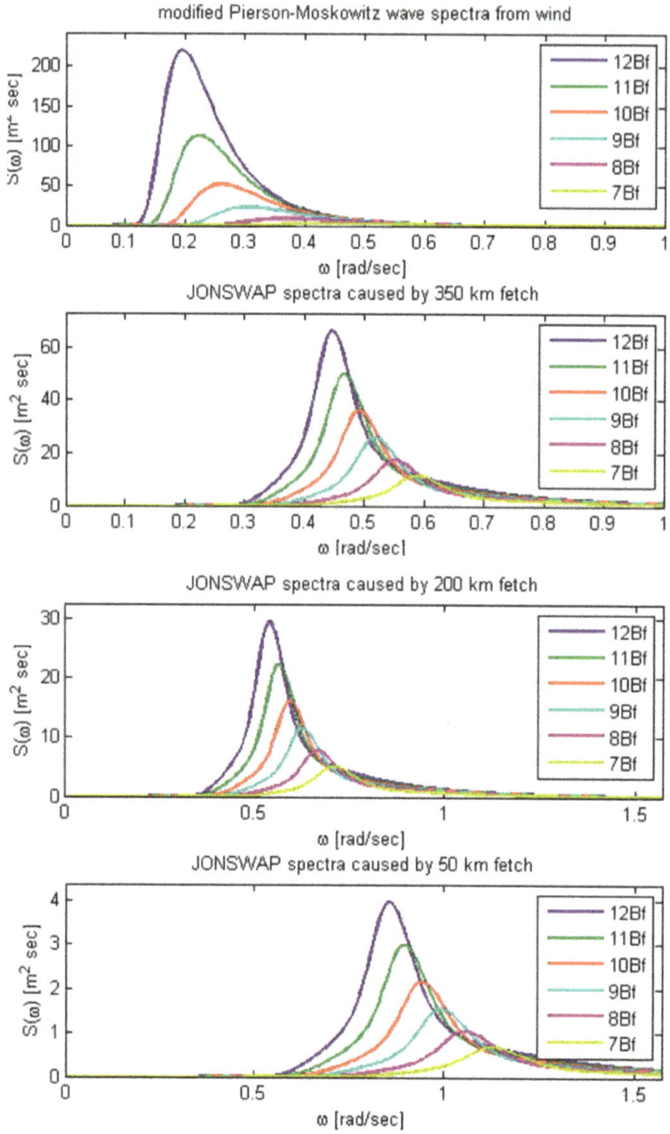

Figure 4. Comparison of modified Pierson-Moskowitz spectra and JONSWAP spectra for 350 km, 200 km and 50 km fetch and several wind speeds.

3. 6DOF MOTIONS OF A FREE-FLOATING OFFSHORE STRUCTURE

In the following section, the six degrees of freedom of a ship are introduced. Figure 5 shows a sketch of a ship hull with a body-fixed coordinate system. Body-fixed coordinate system means that the coordinate system moves with the ship. Due to advantages in determining the moments of inertia, the origin of the system is preferably located in the ships centre of gravity. The motions of a ship are described in the body-fixed coordinate system. However, the ship's position and the orientation are referenced in an earth-fixed coordinate system.

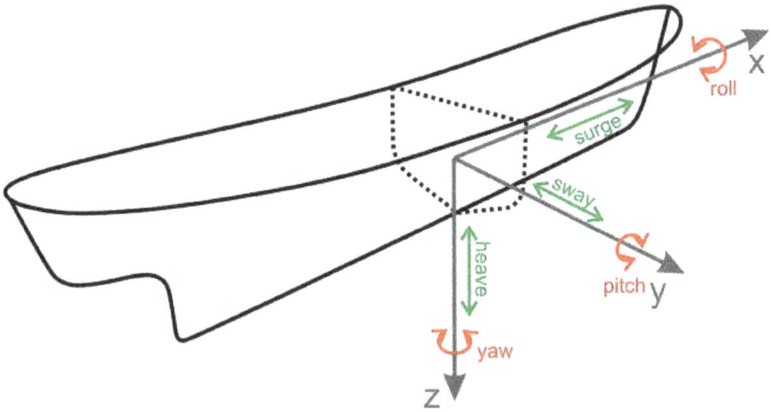

Figure 5. DOF motions of a free-floating ship.

The translational motions are as follows:

- Surge: Translation along the longitudinal x-axis
- Sway: Translation along the transversal y-axis
- Heave: Translation along the vertical z-axis

The rotational motions are as follows:

- Roll: Rotation around the longitudinal x-axis
- Pitch: Rotation around the transversal y-axis
- Yaw: Rotation around the vertical z-axis

Another distinction is the classification into horizontal (surge, sway, yaw) and vertical (heave, roll, pitch) motions. The reason for this differentiation is the restoring forces caused by gravitation. These exist only for the vertical degrees of freedom. After a perturbation of the equilibrium, a ship always tends to return to it. This does not apply for horizontal motions. The result is the drift of a free-floating structure away from the original position as well as a change of the heading in case of wind, waves or current. For a driven ship or a ship in dynamic positioning mode, the ship's actuators (propeller, rudder, thruster, etc.) control the horizontal motions.

4. MOTION EQUATIONS IN THE BODY-FIXED REFERENCE SYSTEM: KIRCHHOFF EQUATIONS

As already mentioned, hydrodynamic inertial effects in the form of a hydrodynamic added mass have to be taken into account in marine applications. The mass matrix m_b (b for body) in the traditionally used body-fixed reference system is defined in Eq. (22). In the matrix, m is the physical mass and $m_{h,ij}$ is the direction-dependent content of the hydrodynamic inertia. The mass matrix is constant in the body-fixed frame. This applies analogously for the matrix of moments of inertia:

$$m_{b,ij} = m + m_{h,ij} \tag{21}$$

$$\overline{\text{m}}_{\text{b}} = \begin{bmatrix} m + m_{h,11} & 0 & 0 \\ 0 & m + m_{h,22} & 0 \\ 0 & 0 & m + m_{h,33} \end{bmatrix} \tag{22}$$

In 1869, Kirchhoff published his work about the 'Movement of a rotating Body in a Fluid' [3]. In this work, he defined the motion equations of a floating body in a body-fixed reference system in analogy to Euler's gyro equation. The Kirchhoff equations are a system of each three equations for translation (Eq. (23)) and rotation (Eq. (24)):

$$\frac{d\,\vec{\text{P}}}{dt} + \vec{\omega} \times \vec{\text{P}} = (X, Y, Z)^{\text{T}} \tag{23}$$

$$\frac{d\,\vec{\text{L}}}{dt} + \vec{\omega} \times \vec{\text{L}} + \vec{v} \times \vec{\text{P}} = (K, M, N)^{\text{T}} \tag{24}$$

The so-called living forces and moments depict all external forces and moments including all hydrodynamic effects, weight, buoyancy and their effect and can be found at the right-hand side of the equations inclusively. The determination of these external forces and moments is especially difficult for the horizontal degrees of freedom with a lack of restoring forces. For the simulations in this work, a simplified model is used, which considers weight, buoyancy as well as potential damping. At present, viscous effects are neglected.

The accelerations have to be integrated twice to analyse the position and orientation of the free-floating ship. The first integration is executed in the body-fixed system.

To calculate the position $(x,y,z)^{\text{T}}$ and orientation $(\Phi,\Theta,\Psi)^{\text{T}}$ of a floating body, the velocities have to be transformed into the inertia system. The peculiarity is the differing transformation of the rotational vector from the body-fixed system into the inertia system and vice versa, e.g. using the transformation with Euler's angles (Eq. (26), cf. [1]):

$$\vec{\omega}_e = \overline{\text{C}}_{rot,\,be}\vec{\omega}_b \tag{25}$$

with

$$\overline{\text{C}}_{rot,\,be} = \begin{bmatrix} 1 & s\Phi t\Theta & c\Phi t\Theta \\ 0 & c\Phi & -s\Phi \\ 0 & {}^{s\Phi}/_{c\Theta} & {}^{c\Phi}/_{c\Theta} \end{bmatrix} = \left(\overline{\text{C}}_{rot,\,eb}\right)^{-1} \tag{26}$$

(e for earth or inertial, b for body, c for cosine, s for sine and t for tangent).

5. MOTION EQUATIONS IN THE INERTIAL REFERENCE SYSTEM

In the following section, the motion equations of a free-floating body in the earth-fixed reference system are shown. The equations are formally derived from momentum and angular momentum theorem. The transformation of the inertia values is discussed, and it is shown that a transformation of the motion equations leads to the Kirchhoff equations. It is known that the twice integration of the rotational equations results in an unstable solution. By using an additional and opposite directed transformation of the rotational accelerations, the simulations can be stabilised for longer time series.

5.1. Inertia value transformation

Kane introduced the transformation of inertia values in 1985 [14]. The used transformation matrix from the body-fixed into the earth-fixed reference system is

$$\overline{\mathbf{C}}_{be} = \begin{bmatrix} c\Theta c\Psi & s\Phi s\Theta c\Psi - c\Phi s\Psi & c\Phi s\Theta c\Psi + s\Phi s\Psi \\ c\Theta s\Psi & s\Phi s\Theta s\Psi + c\Phi c\Psi & c\Phi s\Theta s\Psi - s\Phi c\Psi \\ -s\Theta & s\Phi c\Theta & c\Phi c\Theta \end{bmatrix} \tag{27}$$

with

$$\overline{\mathbf{C}}_{be}\overline{\mathbf{C}}_{be}^{\mathsf{T}} = \begin{bmatrix} 1 & 0 & 0 \\ 0 & 1 & 0 \\ 0 & 0 & 1 \end{bmatrix} = \overline{\mathbf{C}}_{be}^{\mathsf{T}}\overline{\mathbf{C}}_{be} \tag{28}$$

and

$$\overline{\mathbf{C}}_{eb} = \overline{\mathbf{C}}_{be}^{\mathsf{T}} \tag{29}$$

The transformation of the inertia matrix is [14, 21]

$$\overline{\mathbf{J}}_{e} = \overline{\mathbf{C}}_{be}\overline{\mathbf{J}}_{b}\overline{\mathbf{C}}_{be}^{\mathsf{T}} \tag{30}$$

Analogously, it follows the transformation of the mass matrix including added mass [2]:

$$\overline{\mathbf{m}}_{e} = \overline{\mathbf{C}}_{be}\overline{\mathbf{m}}_{b}\overline{\mathbf{C}}_{be}^{\mathsf{T}} \tag{31}$$

The cross product-operator relation for matrices can be found in the literature [22, 23]:

$$\dot{\overline{\mathbf{C}}}_{be} \cdot \overline{\mathbf{C}}_{be}^{\mathsf{T}} = \tilde{\omega}_{e} = \begin{bmatrix} 0 & -\dot{\Psi} & \dot{\Theta} \\ \dot{\Psi} & 0 & -\dot{\Phi} \\ -\dot{\Theta} & \dot{\Phi} & 0 \end{bmatrix} = \left(\vec{\omega}_{e}\times\right) \tag{32}$$

5.2. Rotation

The rotational equations follow from angular momentum theorem, which is defined in the inertial reference frame [21, 23]:

$$\vec{L}_e = \bar{J}_e \vec{\omega}_e \tag{33}$$

Taking the temporal change of the inertia matrix $\bar{J}_e$ into account follows for the moment:

$$\frac{d\vec{L}_e}{dt} = \vec{M}_e = \bar{J}_e \dot{\vec{\omega}}_e + \dot{\bar{J}}_e \vec{\omega}_e \tag{34}$$

The temporal derivative of the inertia matrix is

$$\dot{\bar{J}}_e = \dot{\bar{C}}_{be} \bar{J}_b \bar{C}_{be}{}^{T} + \bar{C}_{be} \bar{J}_b \dot{\bar{C}}_{be}{}^{T} \tag{35}$$

Substituting Eq. (35) into Eq. (34) follows:

$$\vec{M}_e = \bar{J}_e \dot{\vec{\omega}}_e + \left(\dot{\bar{C}}_{be} \bar{J}_b \bar{C}_{be}{}^{T} + \bar{C}_{be} \bar{J}_b \dot{\bar{C}}_{be}{}^{T} \right) \vec{\omega}_e \tag{36}$$

Using the cross product-operator relation follows the momentum equation in the inertial reference system:

$$\vec{M}_e = \bar{J}_e \dot{\vec{\omega}}_e + \tilde{\omega}_e \bar{J}_e \vec{\omega}_e = \bar{J}_e \dot{\vec{\omega}}_e + \vec{\omega}_e \times \left(\bar{J}_e \vec{\omega}_e \right) \tag{37}$$

The transformation of Eq. (37) into the body-fixed reference frame is

$$\vec{M}_b = \bar{C}_{eb} \vec{M}_e = \bar{C}_{eb} \bar{J}_e \bar{C}_{eb}{}^{T} \cdot \bar{C}_{eb} \dot{\vec{\omega}}_e + \bar{C}_{eb} \tilde{\omega}_e \bar{C}_{eb}{}^{T} \cdot \bar{C}_{eb} \bar{J}_e \bar{C}_{eb}{}^{T} \cdot \bar{C}_{eb} \vec{\omega}_e \tag{38}$$

It follows with

$$\vec{M}_b = \bar{J}_b \dot{\vec{\omega}}_b + \tilde{\omega}_b \bar{J}_b \vec{\omega}_b = \bar{J}_b \dot{\vec{\omega}}_b + \vec{\omega}_b \times \left(\bar{J}_b \vec{\omega}_b \right) \tag{39}$$

Eq. (39) is identical to Euler's gyroscope equation, which is defined in the body-fixed system and corresponds to the Kirchhoff rotation equation (Eq. (24)) if the term $\vec{v} \times \vec{p}$ is neglected.

As already mentioned, the integration of the equations in the inertial system results in unstable behaviour. The floating body makes unpredictable, chaotic movements. It is assumed that the instability is a consequence of the numerical inaccuracy as well as the twice integration of the equations. It was found within the project Mine Hunting 2000 that an additional and opposing transformation stabilises the simulation for longer time [6]. Due to the missing inertia value algorithm at that time, it could not be explained why. The authors assumed a numerical reason, which cannot be proved. That phenomenon of stabilisation was rebuilt for simulations of docked CTV at wind turbine tower, compare [24] with [16].

During the presented motion of the free-floating vessel, the changed rotational accelerations stabilise the system for longer periods (cf. Eq. (40)):

$$\vec{\omega}_e = \int \overline{C}_{eb}\dot{\vec{\omega}}_b dt \equiv \int \overline{C}_{eb}\dot{\vec{\omega}}_e dt \tag{40}$$

By using the additional transformation in the multibody application described in Ref. [16], the motions of the vessel on the wind turbine tower were stable at all times.

5.3. Translation

Consequently, the derivation of the translational equations follows the derivation of the rotational equations from the last section. The hydrodynamic added mass force is generally an external force, which defined the right-hand side of the motion equations. For the derivation of the motion equations, the hydrodynamic mass is considered as intrinsic property of the free-floating body, which has to be taken into account for every accelerated marine system. Analogously, as for the rotational inertia matrix, the mass matrix of a rotating body changes. This temporal change is equal to zero in case of pure translation.

The momentum theorem is defined in the inertial reference system:

$$\vec{P}_e = \overline{m}_e \vec{v}_e \tag{41}$$

The force is the temporal derivative of the momentum:

$$\frac{dP_e}{dt} = \vec{F}_e = \overline{m}_e \dot{\vec{v}}_e + \dot{\overline{m}}_e \vec{v}_e \tag{42}$$

With the transformation matrix $\overline{C}_{be}$, the transformation of the mass matrix into the inertial system follows:

$$\overline{m}_e = \overline{C}_{be}\overline{m}_b\overline{C}_{be}^{T} \tag{43}$$

The other direction is

$$\overline{m}_b = \overline{C}_{be}^{T}\overline{m}_e\overline{C}_{be} \tag{44}$$

The temporal derivative of the mass matrix is

$$\dot{\overline{m}}_e = \dot{\overline{C}}_{be}\overline{m}_b\overline{C}_{be}^{T} + \overline{C}_{be}\overline{m}_b\dot{\overline{C}}_{be}^{T} \tag{45}$$

By substituting Eq. (43) into Eq. (45) follows

$$\dot{\overline{m}}_e = \dot{\overline{C}}_{be}\overline{C}_{be}^{T}\overline{m}_e + \overline{m}_e\overline{C}_{be}\dot{\overline{C}}_{be}^{T} \tag{46}$$

and by using the cross product-operator relation

$$\dot{\overline{m}}_e = \tilde{\omega}_e\overline{m}_e + \overline{m}_e\left(-\tilde{\omega}_e\right) \tag{47}$$

Substituting Eq. (47) into Eq. (42) results in the translational equation for a floating body in the inertial reference frame:

$$\vec{F}_e = \overline{\mathrm{m}}_e \dot{\vec{v}}_e + \tilde{\omega}_e \left(\overline{\mathrm{m}}_e \vec{v}_e \right) - \left(\overline{\mathrm{m}}_e \tilde{\omega}_e \right) \vec{v}_e = \overline{\mathrm{m}}_e \dot{\vec{v}}_e + \vec{\omega}_e \times \left(\overline{\mathrm{m}}_e \vec{v}_e \right) - \overline{\mathrm{m}}_e \vec{\omega}_e \times \vec{v}_e \tag{48}$$

Eq. (48) contains only values of the inertial reference system. Due to change of the orientation and the non-scalar characteristic of the mass matrix, Eq. (48) contains two additional terms. These terms disappear for each application where the hydrodynamic added mass is not considered and the mass is scalar, e.g. in aerospace industries or robotics. In case of pure translation, these terms are equal to zero too, and the equation results in Newton's second law.

In the following equation, the transformation of Eq. (48) into the body-fixed reference system is shown. The relation of the velocity in both systems is

$$\vec{v}_e = \overline{\mathrm{C}}_{be} \vec{v}_b \tag{49}$$

In case of simultaneous rotation and translation, the rotation matrix changes. The acceleration in inertial frame is

$$\dot{\vec{v}}_e = \dot{\overline{\mathrm{C}}}_{be} \vec{v}_b + \overline{\mathrm{C}}_{be} \dot{\vec{v}}_b \tag{50}$$

(Remark: In case of pure translation, $\dot{\vec{v}}_e = \mathrm{C}_{be} \dot{\vec{v}}_b$).

By inserting identity matrix, Eq. (28) follows:

$$\dot{\vec{v}}_e = \dot{\overline{\mathrm{C}}}_{be} \overline{\mathrm{C}}_{be}^{\mathrm{T}} \overline{\mathrm{C}}_{be} \vec{v}_b + \overline{\mathrm{C}}_{be} \dot{\vec{v}}_b \tag{51}$$

This is identical to

$$\dot{\vec{v}}_e = \tilde{\omega}_e \vec{v}_e + \overline{\mathrm{C}}_{be} \dot{\vec{v}}_b = \vec{\omega}_e \times \vec{v}_e + \overline{\mathrm{C}}_{be} \dot{\vec{v}}_b \tag{52}$$

Substituting Eq. (52) into Eq. (48), the term $\left(\overline{\mathrm{m}}_e \tilde{\omega}_e \right) \vec{v}_e$ disappears, and Eq. (48) becomes

$$\vec{F}_e = \overline{\mathrm{m}}_e \overline{\mathrm{C}}_{be} \dot{\vec{v}}_b + \tilde{\omega}_e \left(\overline{\mathrm{m}}_e \vec{v}_e \right) \tag{53}$$

The transformation of Eq. (53) into the body-fixed reference frame is

$$\vec{F}_b = \overline{\mathrm{C}}_{eb} \vec{F}_e = \overline{\mathrm{C}}_{eb} \overline{\mathrm{m}}_e \overline{\mathrm{C}}_{eb}^{\mathrm{T}} \cdot \overline{\mathrm{C}}_{eb} \overline{\mathrm{C}}_{be} \dot{\vec{v}}_b + \overline{\mathrm{C}}_{eb} \tilde{\omega}_e \overline{\mathrm{C}}_{eb}^{\mathrm{T}} \cdot \overline{\mathrm{C}}_{eb} \overline{\mathrm{m}}_e \overline{\mathrm{C}}_{eb}^{\mathrm{T}} \overline{\mathrm{C}}_{eb} \vec{v}_e \tag{54}$$

Replacing the values from inertial system with values from body-fixed system follows:

$$\vec{F}_b = \overline{\mathrm{m}}_b \dot{\vec{v}}_b + \tilde{\omega}_b \overline{\mathrm{m}}_b \vec{v}_b = \overline{\mathrm{m}}_b \dot{\vec{v}}_b + \vec{\omega}_b \times \left(\overline{\mathrm{m}}_b \vec{v}_b \right) \tag{55}$$

Eq. (55) corresponds to the translational Kirchhoff equation (Eq. (23)) in the body-fixed system.

6. SIMULATIONS IN BODY-FIXED AND INERTIAL REFERENCE FRAME

The following section shows simulation comparisons of a free-floating vessel in both reference systems. 6DOF motion simulations are performed for various wave conditions. The hydrodynamic model of the simulation is implemented in MATLAB/Simulink.

6.1. Parameterisation of the ship

The simulated ship is a crew transfer vessel, which is used for the transfer of offshore service staff in the German Bight. Figure 6 shows a snapshot of the ship's CAD model. It is a catamaran hull with a length of $L_{oA=22.0\ m}$, a breadth of B=8.3m and a mass of m=60t. For the simulation, the hydrodynamic parameters like hydrodynamic added masses and moments of inertia, as well as the potential damping coefficients of the CTV, are required. They were determined within the project 'Safe Offshore Operations (SOOP)' using the potential radiation and diffraction programme WAMIT (cf. [24]). The ship is discretised station-wise to calculate wave-induced forces and moments. The calculated forces are buoyancy and weight as well as potential damping force.

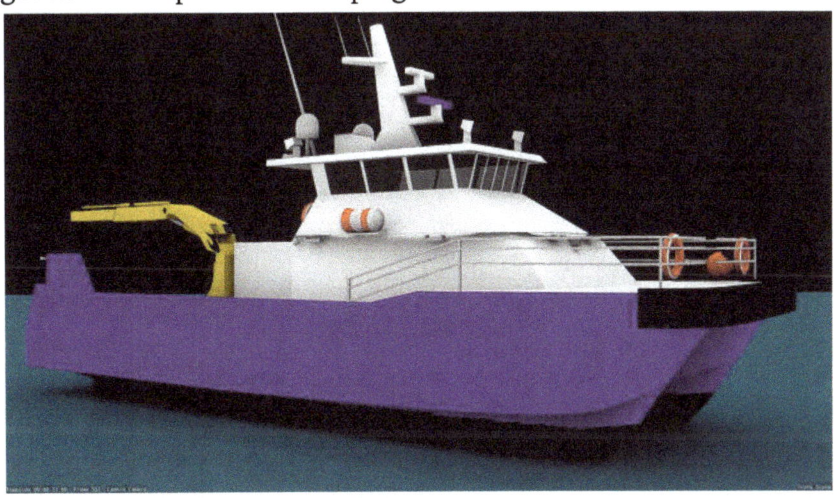

Figure 6. CAD snapshot of the simulated crew transfer vessel.

6.2. DOF simulations

The following figures show the comparison of the CTV 6DOF body motions in irregular seas. The ship is free floating with no initial velocity. Figure 7 shows the simulation results for the case of head seas and Figure 8 for beam seas. The wind wave parameters are a significant wave height of $H_{S=2.1m}$ and a peak period of $\omega_{0=0.7\ rad/s}$. Swell is neglected.

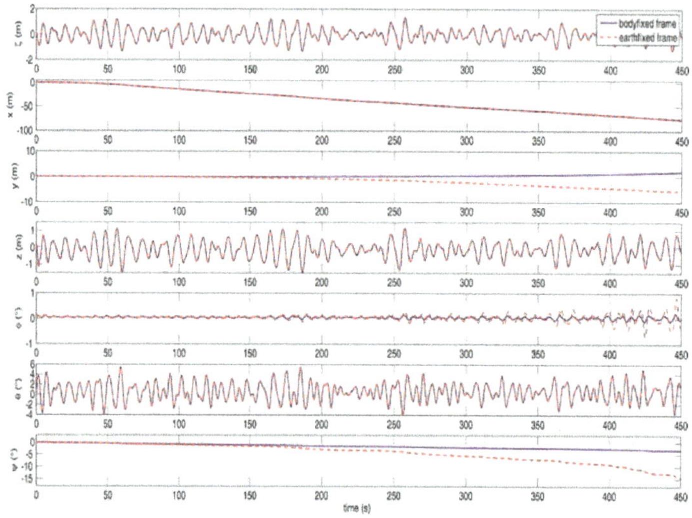

Figure 7. Comparison of the ship position and orientation for head seas. Solid line for body-fixed reference frame. Dashed line for inertial reference frame.

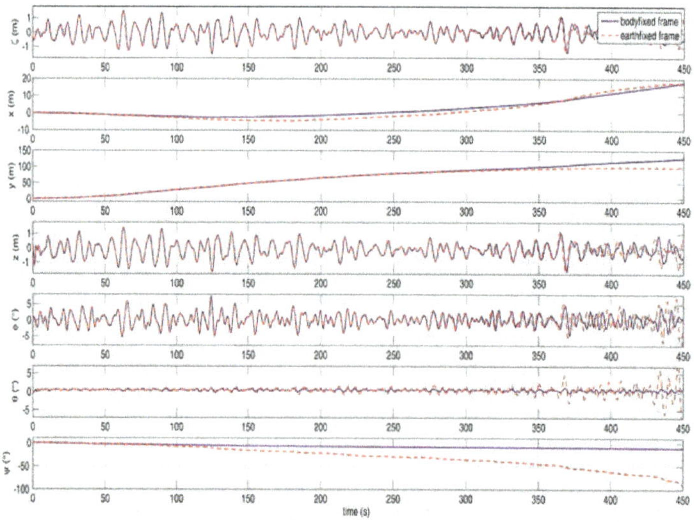

Figure 8. Comparison of the ship position and orientation for beam seas. Solid line for body-fixed reference frame. Dashed line for inertial reference frame.

The figures show from top to bottom the registered wave elevation at the ships centre of gravity, the x-, y- and z-position as well as the orientation angles Φ, Θ and Ψ. Blue curves are for the body-fixed reference frame and red curves for simulations in the inertial reference frame.

At the beginning of the presented simulations, the motion behaviour in both reference systems shows an identical behaviour. In further process, however, an increasing deviation of the ship's heading in the inertial reference is registered. The result is a changing encounter angle of ship and waves. For the case of head seas, the ship begins to move in transverse direction and to roll. In the beam seas, simulation leads the changed encounter angle in a pitch motion. In total, it can be seen that the comparison of vertical motions shows an identical behaviour. As already mentioned, the yaw motion is controlled in case of a driven ship. It is assumed that the motions in both reference systems are identical in this case. Analysis of yaw motion is the subject of the present work. It has to be pointed out that the simulations in the inertial reference system were unstable after longer simulation time. Instability occurs abruptly. Intensive analysis of the instability reason is also the subject of the present work.

7. PROOF OF ENERGY CONSERVATION FOR ROTATORY INERTIA VALUE TRANSFORMATION

Problem: Common practice in calculating rotational motions of rigid bodies applies the so-called *angular velocity transformation* between a first integration over time in body-fixed frame and a second within inertial frame (see [1]). This is contrary to the so-called *linear velocity transformation*, which uses the ordinary vector transformation algorithm as applied in our solution [2], e.g. Euler's angle transformation. The derivation of the rotation matrix $\overline{\mathbf{C}}_{rot,be}$ (Eq. (26)) is carried out by observations and describes the rotation around the three body-fixed coordinate axes sequentially in time.

In 1995, Blass and Gurevich postulates 'Matrix Transformation Is Complete for the Average Case' [25]:

$$\overline{A}_e = \overline{\mathbf{C}}_b^e \cdot \overline{A}_b \cdot \left(\overline{\mathbf{C}}_b^e\right)^{-1} \tag{56}$$

whereby $\overline{\mathbf{C}}^e_b$ describes the ordinary transformation matrix for vector values from coordinate system b, here body fixed, to the inertial or earth-fixed ones e. Which transformation is correct, way I in comparison to Eq. (25) or way II regarding Eqs. (30) and (49)?

I: if $\vec{\omega}_e = \overline{\mathbf{C}}_{rot,be} \cdot \vec{\omega}_b$ is true; also,

$\vec{\mathbf{M}}_e = \overline{\mathbf{C}}_{rot,be} \cdot \overline{\mathbf{J}}_b \cdot \left(\overline{\mathbf{C}}_{rot,be}\right)^{-1} \cdot \overline{\mathbf{C}}_{rot,be} \cdot \vec{\omega}_b$ might be possible being true, but wrong in our opinion, because $\left|\overline{\mathbf{C}}_{rot,be}\right| \neq 1$ for each $\Theta \neq 0$

or II: if $\vec{\omega}_e = \overline{\mathbf{C}}_b^e \cdot \vec{\omega}_b$ is true, contrary to the above

$\vec{\mathbf{M}}_e = \overline{\mathbf{C}}_b^e \cdot \overline{\mathbf{J}}_b \cdot \left(\overline{\mathbf{C}}_b^e\right)^{\mathrm{T}} \cdot \overline{\mathbf{C}}_b^e \cdot \vec{\omega}_b$ have to be true!

Matrix mathematics cannot give answer, because both assumptions are valid. Authors expect calculus II as valid due to the correct transformation of the moment vector. Its correctness may prove only by the general use of one physical principal of conservation. For that purpose, the energy conservation shall be applied.

Therefore, the pure rotating energy from one mass point of the rigid body (see *step 1*) has to be compared with the kinetic energy of same mass point and motion applying continuum physics within FEM methods (see *step 2*). Furthermore and in accordance with authors who claim of wrong description of rotatory kinetic energy within the body-fixed frame, all used rotation speed components are time derivatives of inertial-fixed Euler's angles. Within the calculation of *step 3*, it has to be shown that all coefficients of both energy equations are equal, otherwise the proof fails.

7.1. Calculation of rotatory kinetic energy of a mass point from a rotating body (*step 1*)

Vector value transformation based on Eulerian angles $\vec{\Pi}_e = (\Phi \;\; \Theta \;\; \Psi)^T$ using the well known transformation matrix, compare with [1] (Eq. (27)). The components C_{ij} of rotation matrix are

$$\overline{C}_b^e = \begin{bmatrix} C_{11} & C_{12} & C_{13} \\ C_{21} & C_{22} & C_{23} \\ C_{31} & C_{32} & C_{33} \end{bmatrix} = \begin{bmatrix} c\Theta c\Psi & s\Phi s\Theta c\Psi - c\Phi s\Psi & c\Phi s\Theta c\Psi + s\Phi s\Psi \\ c\Theta s\Psi & s\Phi s\Theta s\Psi + c\Phi c\Psi & c\Phi s\Theta s\Psi - s\Phi c\Psi \\ -s\Theta & s\Phi c\Theta & c\Phi c\Theta \end{bmatrix} \tag{57}$$

The rotatory inertia value $\overline{J}_{dm}$ of the mass point calculates

$$\overline{J}_{dm} \cdot dm = dm \cdot \begin{bmatrix} R_y^2 + R_z^2 & -R_x R_y & -R_x R_z \\ -R_x R_y & R_x^2 + R_z^2 & -R_y R_z \\ -R_x R_z & -R_y R_z & R_x^2 + R_y^2 \end{bmatrix} \tag{58}$$

by angular speed

$$\vec{\dot{\Pi}}_e = \vec{\omega}_e = \begin{pmatrix} \dot{\Phi} & \dot{\Theta} & \dot{\Psi} \end{pmatrix}^T \tag{59}$$

and lever arm $\vec{R}_m = \begin{pmatrix} R_x & R_y & R_z \end{pmatrix}^T$ from the bodies' centre of gravity to the mass point in body-fixed frame b. By vectorial depiction, the rotatory kinetic energy can be formulated as follows, e.g. Ginsberg [21]:

$$E_{rot} = \frac{1}{2} dm \cdot \vec{\omega}_e^{\,T} \cdot \left(\overline{C}_b^e \cdot \overline{J}_{dm} \cdot \overline{C}_e^b \right) \cdot \vec{\omega}_e \tag{60}$$

The first calculation step can be $\overline{C}_e^b \cdot \vec{\omega}_e$ (cf. Eq. (61)):

$$\overline{C}_e^b \cdot \vec{\omega}_e = \begin{bmatrix} C_{11}\dot{\Phi} + C_{21}\dot{\Theta} + C_{31}\dot{\Psi} \\ C_{12}\dot{\Phi} + C_{22}\dot{\Theta} + C_{32}\dot{\Psi} \\ C_{13}\dot{\Phi} + C_{23}\dot{\Theta} + C_{33}\dot{\Psi} \end{bmatrix} \tag{61}$$

Now, $\overline{J}_{dm} \cdot \overline{C}_e^b \cdot \vec{\omega}_e$ can be estimated:

$$\begin{bmatrix} R_y^2+R_z^2 & -R_xR_y & -R_xR_z \\ -R_xR_y & R_x^2+R_z^2 & -R_yR_z \\ -R_xR_z & -R_yR_z & R_x^2+R_y^2 \end{bmatrix} \cdot \begin{bmatrix} C_{11}\dot\Phi + C_{21}\dot\Theta + C_{31}\dot\Psi \\ C_{12}\dot\Phi + C_{22}\dot\Theta + C_{32}\dot\Psi \\ C_{13}\dot\Phi + C_{23}\dot\Theta + C_{33}\dot\Psi \end{bmatrix} =$$

$$\begin{bmatrix} C_{11}R_y^2\dot\Phi + C_{11}R_z^2\dot\Phi + C_{21}R_y^2\dot\Theta + C_{21}R_z^2\dot\Theta + C_{31}R_y^2\dot\Psi + C_{31}R_z^2\dot\Psi \ldots \\ -C_{12}R_xR_y\dot\Phi - C_{22}R_xR_y\dot\Theta - C_{32}R_xR_y\dot\Psi - C_{13}R_xR_z\dot\Phi - C_{23}R_xR_z\dot\Theta - C_{33}R_xR_z\dot\Psi \\ -C_{11}R_xR_y\dot\Phi - C_{21}R_xR_y\dot\Theta - C_{31}R_xR_y\dot\Psi + C_{12}R_x^2\dot\Phi + C_{12}R_z^2\dot\Phi + C_{22}R_x^2\dot\Theta \ldots \\ +C_{22}R_z^2\dot\Theta + C_{32}R_x^2\dot\Psi + C_{32}R_z^2\dot\Psi - C_{13}R_yR_z\dot\Phi - C_{23}R_yR_z\dot\Theta - C_{33}R_yR_z\dot\Psi \\ -C_{11}R_xR_z\dot\Phi - C_{21}R_xR_z\dot\Theta - C_{31}R_xR_z\dot\Psi - C_{12}R_yR_z\dot\Phi - C_{22}R_yR_z\dot\Theta - C_{32}R_yR_z\dot\Psi \ldots \\ C_{13}R_x^2\dot\Phi + C_{13}R_y^2\dot\Phi + C_{23}R_x^2\dot\Theta + C_{23}R_y^2\dot\Theta + C_{33}R_x^2\dot\Psi + C_{33}R_y^2\dot\Psi \end{bmatrix} \tag{62}$$

In the following equation, the transposed vector from Eq. (61) is multiplied with the vector of Eq. (62):

$$\left(\overline{\mathbf{C}}_e^b \cdot \vec{\omega}_e\right)^{\mathrm{T}} \cdot \overline{\mathbf{J}}_{dm} \cdot \overline{\mathbf{C}}_e^b \cdot \vec{\omega}_e = E_1 + E_2 + E_3 \tag{63}$$

getting three components of a sum E_1 to E_3:

$$\begin{aligned} E_1 = {} & C_{11}^2R_y^2\dot\Phi^2 + C_{11}^2R_z^2\dot\Phi^2 + C_{11}C_{21}R_y^2\dot\Phi\dot\Theta + C_{11}C_{21}R_z^2\dot\Phi\dot\Theta + C_{11}C_{31}R_y^2\dot\Phi\dot\Psi \ldots \\ & + C_{11}C_{31}R_z^2\dot\Phi\dot\Psi - C_{11}C_{12}R_xR_y\dot\Phi^2 - C_{11}C_{22}R_xR_y\dot\Phi\dot\Theta - C_{11}C_{32}R_xR_y\dot\Phi\dot\Psi \ldots \\ & - C_{11}C_{13}R_xR_z\dot\Phi^2 - C_{11}C_{23}R_xR_z\dot\Phi\dot\Theta - C_{11}C_{33}R_xR_z\dot\Phi\dot\Psi + C_{11}C_{21}R_y^2\dot\Phi\dot\Theta + C_{21}^2R_z^2\dot\Theta^2 \ldots \\ & + C_{11}C_{21}R_z^2\dot\Phi\dot\Theta + C_{21}^2R_y^2\dot\Theta^2 + C_{21}C_{31}R_y^2\dot\Theta\dot\Psi + C_{21}C_{31}R_z^2\dot\Theta\dot\Psi - C_{12}C_{21}R_xR_y\dot\Phi\dot\Theta \ldots \\ & - C_{21}C_{22}R_xR_y\dot\Theta^2 - C_{21}C_{32}R_xR_y\dot\Theta\dot\Psi - C_{13}C_{21}R_xR_z\dot\Phi\dot\Theta - C_{21}C_{23}R_xR_z\dot\Theta^2 \ldots \\ & - C_{21}C_{33}R_xR_z\dot\Theta\dot\Psi + C_{11}C_{31}R_y^2\dot\Phi\dot\Psi + C_{11}C_{31}R_z^2\dot\Phi\dot\Psi + C_{21}C_{31}R_y^2\dot\Theta\dot\Psi + C_{21}C_{31}R_z^2\dot\Theta\dot\Psi \ldots \\ & + C_{31}^2R_y^2\dot\Psi^2 + C_{31}^2R_z^2\dot\Psi^2 - C_{12}C_{31}R_xR_y\dot\Phi\dot\Psi - C_{22}C_{31}R_xR_y\dot\Theta\dot\Psi - C_{32}C_{31}R_xR_y\dot\Psi^2 \ldots \\ & - C_{13}C_{31}R_xR_z\dot\Phi\dot\Psi - C_{23}C_{31}R_xR_z\dot\Theta\dot\Psi - C_{31}C_{33}R_xR_z\dot\Psi^2 \end{aligned}$$

$$\begin{aligned} E_2 = {} & -C_{11}C_{12}R_xR_y\dot\Phi^2 - C_{12}C_{21}R_xR_y\dot\Phi\dot\Theta - C_{12}C_{31}R_xR_y\dot\Phi\dot\Psi + C_{12}^2R_x^2\dot\Phi^2 + C_{12}^2R_z^2\dot\Phi^2 \ldots \\ & + C_{12}C_{22}R_x^2\dot\Phi\dot\Theta + C_{12}C_{22}R_z^2\dot\Phi\dot\Theta + C_{12}C_{32}R_x^2\dot\Phi\dot\Psi + C_{12}C_{32}R_z^2\dot\Phi\dot\Psi - C_{12}C_{13}R_yR_z\dot\Phi^2 \ldots \\ & - C_{12}C_{23}R_yR_z\dot\Phi\dot\Theta - C_{12}C_{33}R_yR_z\dot\Phi\dot\Psi - C_{11}C_{22}R_xR_y\dot\Phi\dot\Theta - C_{21}C_{22}R_xR_y\dot\Theta^2 \ldots \\ & - C_{22}C_{31}R_xR_y\dot\Theta\dot\Psi + C_{12}C_{22}R_x^2\dot\Phi\dot\Theta + C_{12}C_{22}R_z^2\dot\Phi\dot\Theta + C_{22}^2R_x^2\dot\Theta^2 + C_{22}^2R_z^2\dot\Theta^2 \ldots \\ & + C_{22}C_{32}R_x^2\dot\Theta\dot\Psi + C_{22}C_{32}R_z^2\dot\Theta\dot\Psi - C_{13}C_{22}R_yR_z\dot\Phi\dot\Theta - C_{22}C_{23}R_yR_z\dot\Theta^2 \ldots \\ & - C_{22}C_{33}R_yR_z\dot\Theta\dot\Psi - C_{11}C_{32}R_xR_y\dot\Phi\dot\Psi - C_{21}C_{32}R_xR_y\dot\Theta\dot\Psi - C_{31}C_{32}R_xR_y\dot\Psi^2 \ldots \\ & + C_{12}C_{32}R_x^2\dot\Phi\dot\Psi + C_{12}C_{32}R_z^2\dot\Phi\dot\Psi + C_{22}C_{32}R_x^2\dot\Theta\dot\Psi + C_{22}C_{32}R_z^2\dot\Theta\dot\Psi + C_{32}^2R_x^2\dot\Psi^2 \ldots \\ & + C_{32}^2R_z^2\dot\Psi^2 - C_{13}C_{32}R_yR_z\dot\Phi\dot\Psi - C_{23}C_{32}R_yR_z\dot\Theta\dot\Psi - C_{32}C_{33}R_yR_z\dot\Psi^2 \end{aligned}$$

$$\begin{aligned}E_3 = & -C_{11}C_{13}R_xR_z\dot{\Phi}^2 - C_{13}C_{21}R_xR_z\dot{\Phi}\dot{\Theta} - C_{13}C_{31}R_xR_z\dot{\Phi}\dot{\Psi} - C_{12}C_{13}R_yR_z\dot{\Phi}^2\ldots\\ & - C_{13}C_{22}R_yR_z\dot{\Phi}\dot{\Theta} - C_{13}C_{32}R_yR_z\dot{\Phi}\dot{\Psi} + C_{13}^2R_x^2\dot{\Phi}^2 + C_{13}^2R_y^2\dot{\Phi}^2 + C_{13}C_{23}R_x^2\dot{\Phi}\dot{\Theta}\ldots\\ & + C_{13}C_{23}R_y^2\dot{\Phi}\dot{\Theta} + C_{13}C_{33}R_x^2\dot{\Phi}\dot{\Psi} + C_{13}C_{33}R_y^2\dot{\Phi}\dot{\Psi} - C_{11}C_{23}R_xR_z\dot{\Phi}\dot{\Theta}\ldots\\ & - C_{21}C_{23}R_xR_z\dot{\Theta}^2 - C_{23}C_{31}R_xR_z\dot{\Theta}\dot{\Psi} - C_{12}C_{23}R_yR_z\dot{\Phi}\dot{\Theta} - C_{22}C_{23}R_yR_z\dot{\Theta}^2\ldots\\ & - C_{23}C_{32}R_yR_z\dot{\Theta}\dot{\Psi} + C_{13}C_{23}R_x^2\dot{\Phi}\dot{\Theta} + C_{13}C_{23}R_y^2\dot{\Phi}\dot{\Theta} + C_{23}^2R_x^2\dot{\Theta}^2 + C_{23}^2R_y^2\dot{\Theta}^2\ldots\\ & + C_{23}C_{33}R_x^2\dot{\Theta}\dot{\Psi} + C_{23}C_{33}R_y^2\dot{\Theta}\dot{\Psi} - C_{11}C_{33}R_xR_z\dot{\Phi}\dot{\Psi} - C_{21}C_{33}R_xR_z\dot{\Theta}\dot{\Psi}\ldots\\ & - C_{31}C_{33}R_xR_z\dot{\Psi}^2 - C_{12}C_{33}R_yR_z\dot{\Phi}\dot{\Psi} - C_{22}C_{33}R_yR_z\dot{\Theta}\dot{\Psi} - C_{32}C_{33}R_yR_z\dot{\Psi}^2\ldots\\ & + C_{13}C_{33}R_x^2\dot{\Phi}\dot{\Psi} + C_{13}C_{33}R_y^2\dot{\Phi}\dot{\Psi} + C_{23}C_{33}R_x^2\dot{\Theta}\dot{\Psi} + C_{23}C_{33}R_y^2\dot{\Theta}\dot{\Psi} + C_{33}^2R_x^2\dot{\Psi}^2\ldots\\ & + C_{33}^2R_y^2\dot{\Psi}^2\end{aligned}$$

Finally, the sum of these components E_1, E_2 and E_3 has to be multiplied by the factor ½dm½dmto get the rotatory kinetic energy. Because that factor is implemented also within the second energy equation (*step 2*), the energy components can be summarised and compared directly with Eq. (68). By sorting all components to the lever arm and rotating speed combinations, we get Eq. (64):

$$\begin{aligned}\frac{2E_{rot}}{dm} = & (C_{32}^2 + C_{33}^2)R_x^2\dot{\Psi}^2 - 2C_{31}C_{32}R_xR_y\dot{\Psi}^2 - 2C_{31}C_{33}R_xR_z\dot{\Psi}^2 - 2C_{32}C_{33}R_yR_z\dot{\Psi}^2\ldots\\ & + (C_{31}^2 + C_{33}^2)R_y^2\dot{\Psi}^2 + (C_{31}^2 + C_{32}^2)R_z^2\dot{\Psi}^2 + 2(C_{22}C_{32} + C_{23}C_{33})R_x^2\dot{\Theta}\dot{\Psi}\ldots\\ & - 2(C_{22}C_{31} + C_{21}C_{32})R_xR_y\dot{\Theta}\dot{\Psi} - 2(C_{21}C_{33} + C_{23}C_{31})R_xR_z\dot{\Theta}\dot{\Psi}\ldots\\ & - 2(C_{22}C_{33} + C_{23}C_{32})R_yR_z\dot{\Theta}\dot{\Psi} + 2(C_{23}C_{33} + C_{21}C_{31})R_y^2\dot{\Theta}\dot{\Psi}\ldots\\ & + 2(C_{22}C_{32} + C_{21}C_{31})R_z^2\dot{\Theta}\dot{\Psi} + 2(C_{13}C_{33} + C_{12}C_{32})R_x^2\dot{\Phi}\dot{\Psi}\ldots\\ & - 2(C_{12}C_{31} + C_{11}C_{32})R_xR_y\dot{\Phi}\dot{\Psi} - 2(C_{11}C_{33} + C_{13}C_{31})R_xR_z\dot{\Phi}\dot{\Psi}\ldots\\ & - 2(C_{12}C_{33} + C_{13}C_{32})R_yR_z\dot{\Phi}\dot{\Psi} + 2(C_{13}C_{33} + C_{11}C_{31})R_y^2\dot{\Phi}\dot{\Psi}\ldots\\ & + 2(C_{12}C_{32} + C_{11}C_{31})R_z^2\dot{\Phi}\dot{\Psi} + (C_{22}^2 + C_{23}^2)R_x^2\dot{\Theta}^2 - 2C_{21}C_{22}R_xR_y\dot{\Theta}^2\ldots\\ & - 2C_{21}C_{23}R_xR_z\dot{\Theta}^2 - 2C_{22}C_{23}R_yR_z\dot{\Theta}^2 + (C_{21}^2 + C_{23}^2)R_y^2\dot{\Theta}^2 + (C_{21}^2 + C_{22}^2)R_z^2\dot{\Theta}^2\ldots\\ & + 2(C_{13}C_{23} + C_{12}C_{22})R_x^2\dot{\Phi}\dot{\Theta} - 2(C_{11}C_{22} + C_{12}C_{21})R_xR_y\dot{\Phi}\dot{\Theta}\ldots\\ & - 2(C_{11}C_{23} + C_{13}C_{21})R_xR_z\dot{\Phi}\dot{\Theta} - 2(C_{12}C_{23} + C_{13}C_{22})R_yR_z\dot{\Phi}\dot{\Theta}\ldots\\ & + 2(C_{11}C_{21} + C_{13}C_{23})R_y^2\dot{\Phi}\dot{\Theta} + 2(C_{11}C_{21} + C_{12}C_{22})R_z^2\dot{\Phi}\dot{\Theta} + (C_{12}^2 + C_{13}^2)R_x^2\dot{\Phi}^2\ldots\\ & - 2C_{11}C_{12}R_xR_y\dot{\Phi}^2. - 2C_{11}C_{13}R_xR_z\dot{\Phi}^2 - 2C_{12}C_{13}R_yR_z\dot{\Phi}^2 + (C_{11}^2 + C_{13}^2)R_y^2\dot{\Phi}^2\ldots\\ & + (C_{11}^2 + C_{12}^2)R_z^2\dot{\Phi}^2\end{aligned} \quad (64)$$

7.2. Calculation of kinetic energy from a mass point of a pure rotating body *(step 2)*

The energy can also be calculated by the tangential speed vector, preferably within the inertial frame (Eq. (65)):

$$E_{kin} = \frac{1}{2} dm \left[\vec{\omega}_e \times \left(\overline{\mathbf{C}}_b^e \cdot \vec{\mathbf{R}}_m \right) \right]^{\mathrm{T}} \cdot \left[\vec{\omega}_e \times \left(\overline{\mathbf{C}}_b^e \cdot \vec{\mathbf{R}}_m \right) \right] \tag{65}$$

The cross product can be calculated by a so-called skew operator $\tilde{\omega}_e$ (Eq. (32)). So, it is possible to calculate the tangential speed vector of the mass point within the inertial frame:

$$\vec{\mathbf{v}}_{dm} = \tilde{\omega}_e \cdot \overline{\mathbf{C}}_b^e \cdot \vec{\mathbf{R}}_m = \begin{bmatrix} -C_{21}R_x\dot{\Psi} - C_{22}R_y\dot{\Psi} - C_{23}R_z\dot{\Psi} + C_{31}R_x\dot{\Theta} + C_{32}R_y\dot{\Theta} + C_{33}R_z\dot{\Theta} \\ C_{11}R_x\dot{\Psi} + C_{12}R_y\dot{\Psi} + C_{13}R_z\dot{\Psi} - C_{31}R_x\dot{\Phi} - C_{32}R_y\dot{\Phi} - C_{33}R_z\dot{\Phi} \\ -C_{11}R_x\dot{\Theta} - C_{12}R_y\dot{\Theta} - C_{13}R_z\dot{\Theta} + C_{21}R_x\dot{\Phi} + C_{22}R_y\dot{\Phi} + C_{23}R_z\dot{\Phi} \end{bmatrix} \tag{66}$$

From Eq. (66) the square of speed can be calculated (see Eq. (67)):

$$\frac{2E_{kin}}{dm} = v_{dm}^2 = \vec{\mathbf{v}}_{dm}^{\mathrm{T}} \cdot \vec{\mathbf{v}}_{dm} = E_4 + E_5 + E_6 \tag{67}$$

with.

$$\begin{aligned} E_4 = {} & C_{21}^2 R_x^2 \dot{\Psi}^2 + 2C_{21}C_{22}R_xR_y\dot{\Psi}^2 + 2C_{21}C_{23}R_xR_z\dot{\Psi}^2 - 2C_{21}C_{31}R_x^2\dot{\Theta}\dot{\Psi} \ldots \\ & - 2C_{21}C_{32}R_xR_y\dot{\Theta}\dot{\Psi} - 2C_{21}C_{33}R_xR_z\dot{\Theta}\dot{\Psi} + C_{22}^2R_y^2\dot{\Psi}^2 + 2C_{22}C_{23}R_yR_z\dot{\Psi}^2 \ldots \\ & - 2C_{22}C_{31}R_xR_y\dot{\Theta}\dot{\Psi} - 2C_{22}C_{32}R_y^2\dot{\Theta}\dot{\Psi} - 2C_{22}C_{33}R_yR_z\dot{\Theta}\dot{\Psi} + C_{23}^2R_z^2\dot{\Psi}^2 \ldots \\ & - 2C_{23}C_{31}R_xR_z\dot{\Theta}\dot{\Psi} - 2C_{23}C_{32}R_yR_z\dot{\Theta}\dot{\Psi} - 2C_{23}C_{33}R_z^2\dot{\Theta}\dot{\Psi} + C_{31}^2R_x^2\dot{\Theta}^2 \ldots \\ & + 2C_{31}C_{32}R_xR_y\dot{\Theta}^2 + 2C_{31}C_{33}R_xR_z\dot{\Theta}^2 + C_{32}^2R_y^2\dot{\Theta}^2 + 2C_{32}C_{33}R_yR_z\dot{\Theta}^2 + C_{33}^2R_z^2\dot{\Theta}^2 \end{aligned}$$

$$\begin{aligned} E_5 = {} & C_{11}^2 R_x^2 \dot{\Psi}^2 + 2C_{11}C_{12}R_xR_y\dot{\Psi}^2 + 2C_{11}C_{13}R_xR_z\dot{\Psi}^2 - 2C_{11}C_{31}R_x^2\dot{\Phi}\dot{\Psi} \ldots \\ & - 2C_{11}C_{32}R_xR_y\dot{\Phi}\dot{\Psi} - 2C_{11}C_{33}R_xR_z\dot{\Phi}\dot{\Psi} + C_{12}^2R_y^2\dot{\Psi}^2 + 2C_{12}C_{13}R_yR_z\dot{\Psi}^2 \ldots \\ & - 2C_{12}C_{31}R_xR_y\dot{\Phi}\dot{\Psi} - 2C_{12}C_{32}R_y^2\dot{\Phi}\dot{\Psi} - 2C_{12}C_{33}R_yR_z\dot{\Phi}\dot{\Psi} + C_{13}^2R_z^2\dot{\Psi}^2 \ldots \\ & - 2C_{13}C_{31}R_xR_z\dot{\Phi}\dot{\Psi} - 2C_{13}C_{32}R_yR_z\dot{\Phi}\dot{\Psi} - 2C_{13}C_{33}R_z^2\dot{\Phi}\dot{\Psi} + C_{31}^2R_x^2\dot{\Phi}^2 \ldots \\ & + 2C_{31}C_{32}R_xR_y\dot{\Phi}^2 + 2C_{31}C_{33}R_xR_z\dot{\Phi}^2 + C_{32}^2R_y^2\dot{\Phi}^2 + 2C_{32}C_{33}R_yR_z\dot{\Phi}^2 + C_{33}^2R_z^2\dot{\Phi}^2 \end{aligned}$$

$$\begin{aligned}E_6 = {} & C_{11}^2R_x^2\dot{\Theta}^2 + 2C_{11}C_{12}R_xR_y\dot{\Theta}^2 + 2C_{11}C_{13}R_xR_z\dot{\Theta}^2 - 2C_{11}C_{21}R_x^2\dot{\Phi}\dot{\Theta}\ldots\\ & - 2C_{11}C_{22}R_xR_y\dot{\Phi}\dot{\Theta} - 2C_{11}C_{23}R_xR_z\dot{\Phi}\dot{\Theta} + C_{12}^2R_y^2\dot{\Theta}^2 + 2C_{12}C_{13}R_yR_z\dot{\Theta}^2\ldots\\ & - 2C_{12}C_{21}R_xR_y\dot{\Phi}\dot{\Theta} - 2C_{12}C_{22}R_y^2\dot{\Phi}\dot{\Theta} - 2C_{12}C_{23}R_yR_z\dot{\Phi}\dot{\Theta} + C_{13}^2R_z^2\dot{\Theta}^2\ldots\\ & - 2C_{13}C_{21}R_xR_z\dot{\Phi}\dot{\Theta} - 2C_{13}C_{22}R_yR_z\dot{\Phi}\dot{\Theta} - 2C_{13}C_{23}R_z^2\dot{\Phi}\dot{\Theta} + C_{21}^2R_x^2\dot{\Phi}^2\ldots\\ & + 2C_{21}C_{22}R_xR_y\dot{\Phi}^2 + 2C_{21}C_{23}R_xR_z\dot{\Phi}^2 + C_{22}^2R_y^2\dot{\Phi}^2 + 2C_{22}C_{23}R_yR_z\dot{\Phi}^2 + C_{23}^2R_z^2\dot{\Phi}^2\end{aligned}$$

Summarising E_4, E_5 and E_6, follows Eq. (68).

$$\begin{aligned}\frac{2E_{kin}}{dm} = {} & (C_{11}^2 + C_{21}^2)R_x^2\dot{\Psi}^2 + 2(C_{11}C_{12} + C_{21}C_{22})R_xR_y\dot{\Psi}^2 + 2(C_{11}C_{13} + C_{21}C_{23})R_xR_z\dot{\Psi}^2\ldots\\ & + 2(C_{12}C_{13} + C_{22}C_{23})R_yR_z\dot{\Psi}^2 + (C_{12}^2 + C_{22}^2)R_y^2\dot{\Psi}^2 + (C_{13}^2 + C_{23}^2)R_z^2\dot{\Psi}^2\ldots\\ & - 2C_{21}C_{31}R_x^2\dot{\Theta}\dot{\Psi} - 2(C_{21}C_{32} + C_{22}C_{31})R_xR_y\dot{\Theta}\dot{\Psi} - 2(C_{21}C_{33} + C_{23}C_{31})R_xR_z\dot{\Theta}\dot{\Psi}\ldots\\ & - 2(C_{22}C_{33} + C_{23}C_{32})R_yR_z\dot{\Theta}\dot{\Psi} - 2C_{22}C_{32}R_y^2\dot{\Theta}\dot{\Psi} - 2C_{23}C_{33}R_z^2\dot{\Theta}\dot{\Psi} - 2C_{11}C_{31}R_x^2\dot{\Phi}\dot{\Psi}\ldots\\ & - 2(C_{11}C_{32} + C_{12}C_{31})R_xR_y\dot{\Phi}\dot{\Psi} - 2(C_{11}C_{33} + C_{13}C_{31})R_xR_z\dot{\Phi}\dot{\Psi}\ldots\\ & - 2(C_{12}C_{33} + C_{13}C_{32})R_yR_z\dot{\Phi}\dot{\Psi} - 2C_{12}C_{32}R_y^2\dot{\Phi}\dot{\Psi} - 2C_{13}C_{33}R_z^2\dot{\Phi}\dot{\Psi} + (C_{11}^2 + C_{31}^2)R_x^2\dot{\Theta}^2\ldots\\ & + 2(C_{11}C_{12} + C_{31}C_{32})R_xR_y\dot{\Theta}^2 + 2(C_{11}C_{13} + C_{31}C_{33})R_xR_z\dot{\Theta}^2\ldots\\ & + 2(C_{12}C_{13} + C_{32}C_{33})R_yR_z\dot{\Theta}^2 + (C_{12}^2 + C_{32}^2)R_y^2\dot{\Theta}^2 + (C_{13}^2 + C_{33}^2)R_z^2\dot{\Theta}^2\ldots\\ & - 2C_{11}C_{21}R_x^2\dot{\Phi}\dot{\Theta} - 2(C_{11}C_{22} + C_{12}C_{21})R_xR_y\dot{\Phi}\dot{\Theta} - 2(C_{11}C_{23} + C_{13}C_{21})R_xR_z\dot{\Phi}\dot{\Theta}\ldots\\ & - 2(C_{12}C_{23} + C_{13}C_{22})R_yR_z\dot{\Phi}\dot{\Theta} - 2C_{12}C_{22}R_y^2\dot{\Phi}\dot{\Theta} - 2C_{13}C_{23}R_z^2\dot{\Phi}\dot{\Theta} + (C_{21}^2 + C_{31}^2)R_x^2\dot{\Phi}^2\ldots\\ & + 2(C_{21}C_{22} + C_{31}C_{32})R_xR_y\dot{\Phi}^2 + 2(C_{21}C_{23} + C_{31}C_{33})R_xR_z\dot{\Phi}^2\ldots\\ & + 2(C_{22}C_{23} + C_{32}C_{33})R_yR_z\dot{\Phi}^2 + (C_{22}^2 + C_{32}^2)R_y^2\dot{\Phi}^2 + (C_{23}^2 + C_{33}^2)R_z^2\dot{\Phi}^2\end{aligned} \tag{68}$$

7.3. Comparison of the coefficients from both energy calculations for the mass point of pure rotating body (step *3*)

To show that energy is identical in both approaches (*step 1* and *step 2*), the coefficients of all 36 combinations of lever arms and angular velocities are compared individually with each other in *step 3*. The coefficients of the rotation energy (cf. Eq. (64)) of *step 1* can be found on the left-hand side and the coefficients of kinetic energy (cf. Eq. (68)) of *step2* on the right-hand side. As described previously, the condition of identical energy is only proved, if all coefficients ar equal.

$$\begin{aligned}& C_{32}^2 + C_{33}^2 = C_{11}^2 + C_{21}^2\\ R_x^2\dot{\Psi}^2 :\ & s^2\Phi c^2\Theta + c^2\Phi c^2\Theta = c^2\Theta c^2\Psi + c^2\Theta s^2\Psi\\ & c^2\Theta = c^2\Theta\end{aligned} \tag{69}$$

$$-2C_{31}C_{32} = 2(C_{11}C_{12} + C_{21}C_{22})$$

$$R_xR_y\dot{\Psi}^2: \begin{aligned} s\Theta c\Theta s\Phi &= s\Phi s\Theta c\Theta c^2\Psi - c\Phi c\Theta s\Psi c\Psi + s\Phi s\Theta c\Theta s^2\Psi \ldots \\ &\quad + c\Phi c\Theta s\Psi c\Psi \\ s\Theta c\Theta s\Phi &= s\Phi s\Theta c\Theta \end{aligned} \tag{70}$$

$$-2C_{31}C_{33} = 2(C_{11}C_{13} + C_{21}C_{23})$$

$$R_xR_z\dot{\Psi}^2: \begin{aligned} s\Theta c\Theta c\Phi &= c\Psi s\Psi c\Theta s\Phi + c^2\Psi c\Theta s\Theta c\Phi - s\Psi c\Psi c\Theta s\Phi \ldots \\ &\quad + s^2\Psi c\Theta s\Theta c\Phi \\ s\Theta c\Theta c\Phi &= c\Theta s\Theta c\Phi \end{aligned} \tag{71}$$

$$-2C_{32}C_{33} = 2(C_{12}C_{13} + C_{22}C_{23})$$

$$R_yR_z\dot{\Psi}^2: \begin{aligned} -c^2\Theta s\Phi c\Phi &= -s^2\Psi c\Phi s\Phi - s\Psi c\Psi c^2\Phi s\Theta + s\Psi c\Psi s^2\Phi s\Theta \ldots \\ &\quad + c^2\Psi s^2\Theta s\Phi c\Phi - c^2\Psi c\Phi s\Phi + s\Psi c\Psi c^2\Phi s\Theta \ldots \\ &\quad - s\Psi c\Psi s^2\Phi s\Theta + s^2\Psi s^2\Theta s\Phi c\Phi \\ &= -1 \cdot c\Phi s\Phi + s^2\Theta s\Phi c\Phi \\ -c^2\Theta s\Phi c\Phi &= -c^2\Theta s\Phi c\Phi \end{aligned} \tag{72}$$

$$C_{31}^2 + C_{33}^2 = C_{12}^2 + C_{22}^2$$

$$R_y^2\dot{\Psi}^2: \begin{aligned} s^2\Theta + c^2\Theta c^2\Phi &= s^2\Psi c^2\Phi - 2s\Psi c\Psi s\Theta c\Phi s\Phi + c^2\Psi s^2\Theta s^2\Phi \ldots \\ &\quad + c^2\Psi c^2\Phi + 2s\Psi c\Psi s\Theta s\Phi c\Phi + s^2\Phi s^2\Theta s^2\Psi \\ &= c^2\Phi + s^2\Theta s^2\Phi = c^2\Phi \cdot 1 + s^2\Theta s^2\Phi \\ &= c^2\Phi c^2\Theta + c^2\Phi s^2\Theta + s^2\Theta \cdot 1 - s^2\Theta c^2\Phi \\ s^2\Theta + c^2\Theta c^2\Phi &= c^2\Phi c^2\Theta + s^2\Theta \end{aligned} \tag{73}$$

$$C_{31}^2 + C_{32}^2 = C_{13}^2 + C_{23}^2$$

$$R_z^2\dot{\Psi}^2: \begin{aligned} s^2\Theta + c^2\Theta s^2\Phi &= s^2\Psi s^2\Phi + 2s\Psi c\Psi s\Theta s\Phi c\Phi + c^2\Psi s^2\Theta c^2\Phi \ldots \\ &\quad + c^2\Psi s^2\Phi - 2s\Psi c\Psi s\Theta s\Phi c\Phi + s^2\Theta s^2\Psi c^2\Phi \\ &= s^2\Phi + s^2\Theta c^2\Phi = s^2\Phi \cdot 1 + s^2\Theta c^2\Phi \\ &= s^2\Phi c^2\Theta + s^2\Phi s^2\Theta + s^2\Theta \cdot 1 - s^2\Theta s^2\Phi \\ s^2\Theta + c^2\Theta s^2\Phi &= s^2\Phi c^2\Theta + s^2\Theta \end{aligned} \tag{74}$$

$$2(C_{22}C_{32} + C_{23}C_{33}) = -2C_{21}C_{31}$$

$$R_x^2\dot{\Theta}\dot{\Psi}: c\Psi c\Theta s\Phi c\Phi + s\Psi s\Theta c\Theta s^2\Phi - c\Psi c\Theta s\Phi c\Phi. + s\Psi s\Theta c\Theta c^2\Phi = s\Psi c\Theta s\Theta \tag{75}$$

$$s\Psi s\Theta c\Theta = s\Psi c\Theta s\Theta$$

$$R_xR_y\dot{\Theta}\dot{\Psi}:\ -2(C_{22}C_{31}+C_{21}C_{32})=-2(C_{21}C_{32}+C_{22}C_{31}) \quad (76)$$

$$R_xR_z\dot{\Theta}\dot{\Psi}:\ -2(C_{21}C_{33}+C_{23}C_{31})=-2(C_{21}C_{33}+C_{23}C_{31}) \quad (77)$$

$$R_yR_z\dot{\Theta}\dot{\Psi}:\ -2(C_{22}C_{33}+C_{23}C_{32})=-2(C_{22}C_{33}+C_{23}C_{32}) \quad (78)$$

$$R_y^2\dot{\Theta}\dot{\Psi}:\ \begin{aligned} 2(C_{23}C_{33}+C_{21}C_{31}) &= -2C_{22}C_{32} \\ -c\Psi c\Theta s\Phi c\Phi + s\Psi s\Theta c\Theta c^2\Phi - s\Psi s\Theta c\Theta &= -c\Psi c\Theta c\Phi s\Phi - s\Psi c\Theta s\Theta s^2\Phi \\ s\Psi s\Theta c\Theta(1-s^2\Phi-1) &= -s\Psi c\Theta s\Theta s^2\Phi \\ -s\Psi c\Theta s\Theta s^2\Phi &= -s\Psi c\Theta s\Theta s^2\Phi \end{aligned} \quad (79)$$

$$R_z^2\dot{\Theta}\dot{\Psi}:\ \begin{aligned} 2(C_{22}C_{32}+C_{21}C_{31}) &= -2C_{23}C_{33} \\ c\Psi c\Theta s\Phi c\Phi + s\Psi s\Theta c\Theta s^2\Phi - s\Psi s\Theta c\Theta &= c\Psi c\Theta s\Phi c\Phi - s\Psi s\Theta c\Theta c^2\Phi \\ s\Psi s\Theta c\Theta(1-c^2\Phi-1) &= -s\Psi c\Theta s\Theta c^2\Phi \\ -s\Psi c\Theta s\Theta c^2\Phi &= -s\Psi c\Theta s\Theta c^2\Phi \end{aligned} \quad (80)$$

$$R_x^2\dot{\Phi}\dot{\Psi}:\ \begin{aligned} 2(C_{13}C_{33}+C_{12}C_{32}) &= -2C_{11}C_{31} \\ \begin{pmatrix} s\Psi c\Theta s\Phi c\Phi + c\Psi s\Theta c\Theta c^2\Phi - s\Psi c\Theta s\Phi c\Phi \ldots \\ +c\Psi s\Theta c\Theta s^2\Phi \end{pmatrix} &= c\Psi c\Theta s\Theta \\ c\Psi s\Theta c\Theta &= c\Psi c\Theta s\Theta \end{aligned} \quad (81)$$

$$R_xR_y\dot{\Phi}\dot{\Psi}:\ -2(C_{12}C_{31}+C_{11}C_{32})=-2(C_{11}C_{32}+C_{12}C_{31}) \quad (82)$$

$$R_xR_z\dot{\Phi}\dot{\Psi}:\ -2(C_{11}C_{33}+C_{13}C_{31})=-2(C_{11}C_{33}+C_{13}C_{31}) \quad (83)$$

$$R_yR_z\dot{\Phi}\dot{\Psi}:\ -2(C_{12}C_{33}+C_{13}C_{32})=-2(C_{12}C_{33}+C_{13}C_{32}) \quad (84)$$

$$R_y^2\dot{\Phi}\dot{\Psi}:\ \begin{aligned} 2(C_{13}C_{33}+C_{11}C_{31}) &= -2C_{12}C_{32} \\ s\Psi c\Theta s\Phi c\Phi + c\Psi s\Theta c\Theta c^2\Phi - c\Psi c\Theta s\Theta &= s\Psi c\Theta s\Phi c\Phi - c\Psi s\Theta c\Theta s^2\Phi \\ c\Psi s\Theta c\Theta(1-s^2\Phi-1) &= -c\Psi s\Theta c\Theta s^2\Phi \\ -c\Psi s\Theta c\Theta s^2\Phi &= -c\Psi s\Theta c\Theta s^2\Phi \end{aligned} \quad (85)$$

$$R_z^2\dot{\Phi}\dot{\Psi}:\ \begin{aligned} 2(C_{12}C_{32}+C_{11}C_{31}) &= -2C_{13}C_{33} \\ -s\Psi c\Theta s\Phi c\Phi + c\Psi s\Theta c\Theta s^2\Phi - c\Psi c\Theta s\Theta &= -s\Psi c\Theta s\Phi c\Phi - c\Psi s\Theta c\Theta c^2\Phi \\ c\Psi s\Theta c\Theta(1-c^2\Phi-1) &= -c\Psi s\Theta c\Theta c^2\Phi \\ -c\Psi s\Theta c\Theta c^2\Phi &= -c\Psi s\Theta c\Theta c^2\Phi \end{aligned} \quad (86)$$

$$R_x^2\dot{\Theta}^2: \quad \begin{aligned} \begin{pmatrix} c^2\Psi c^2\Phi + 2s\Psi c\Psi s\Theta s\Phi c\Phi + s^2\Phi s^2\Theta s^2\Psi \ldots \\ +c^2\Psi s^2\Phi - 2s\Psi c\Psi s\Theta s\Phi c\Phi + s^2\Theta s^2\Psi c^2\Phi \end{pmatrix} &= c^2\Psi c^2\Theta + s^2\Theta \\ c^2\Psi + s^2\Theta s^2\Psi &= \\ c^2\Psi \cdot 1 + s^2\Theta s^2\Psi &= \\ c^2\Psi s^2\Theta + c^2\Psi c^2\Theta + s^2\Theta \cdot 1 - s^2\Theta c^2\Psi &= \\ c^2\Psi c^2\Theta + s^2\Theta &= c^2\Psi c^2\Theta + s^2\Theta \end{aligned} \tag{87}$$

$$-2C_{21}C_{22} = 2(C_{11}C_{12} + C_{31}C_{32})$$

$$R_xR_y\dot{\Theta}^2: \quad \begin{aligned} -s\Psi c\Psi c\Theta c\Phi - s^2\Psi c\Theta s\Theta s\Phi &= -c\Psi s\Psi c\Theta c\Phi + c^2\Psi s\Theta c\Theta s\Phi - s\Theta c\Theta s\Phi \\ -s^2\Psi c\Theta s\Theta s\Phi &= (1 - s^2\Psi - 1)s\Theta c\Theta s\Phi \\ -s^2\Psi c\Theta s\Theta s\Phi &= -s^2\Psi s\Theta c\Theta s\Phi \end{aligned} \tag{88}$$

$$-2C_{21}C_{23} = 2(C_{11}C_{13} + C_{31}C_{33})$$

$$R_xR_z\dot{\Theta}^2: \quad \begin{aligned} s\Psi c\Psi c\Theta s\Phi - s^2\Psi s\Theta c\Theta c\Phi &= s\Psi c\Psi c\Theta s\Phi + c^2\Psi c\Theta s\Theta c\Phi - s\Theta c\Theta c\Phi \\ -s^2\Psi s\Theta c\Theta c\Phi &= (1 - s^2\Psi - 1)s\Theta c\Theta c\Phi \\ -s^2\Psi s\Theta c\Theta c\Phi &= -s^2\Psi s\Theta c\Theta c\Phi \end{aligned} \tag{89}$$

$$-2C_{22}C_{23} = 2(C_{12}C_{13} + C_{32}C_{33})$$

$$R_yR_z\dot{\Theta}^2: \quad \begin{aligned} \begin{pmatrix} c^2\Psi s\Phi c\Phi - s\Psi c\Psi s\Theta c^2\Phi \ldots \\ +s\Psi c\Psi s\Theta s^2\Phi - s^2\Psi s^2\Theta s\Phi c\Phi \end{pmatrix} &= \begin{pmatrix} -s^2\Psi s\Phi c\Phi - s\Psi c\Psi s\Theta c^2\Phi \ldots \\ +s\Psi c\Psi s\Theta s^2\Phi + c^2\Psi s^2\Theta s\Phi c\Phi \ldots \\ +c^2\Theta s\Phi c\Phi \end{pmatrix} \\ c^2\Psi s\Phi c\Phi - s^2\Psi s^2\Theta s\Phi c\Phi &= \begin{pmatrix} -s^2\Psi s\Phi c\Phi + c^2\Psi s^2\Theta s\Phi c\Phi \ldots \\ +c^2\Theta s\Phi c\Phi \end{pmatrix} \\ s\Phi c\Phi(c^2\Psi - s^2\Psi s^2\Theta) &= s\Phi c\Phi(-s^2\Psi + c^2\Psi s^2\Theta + c^2\Theta) \\ c^2\Psi - s^2\Psi s^2\Theta &= -s^2\Psi + c^2\Psi s^2\Theta + c^2\Theta \\ 1 - s^2\Psi - s^2\Theta + c^2\Psi s^2\Theta &= \\ -s^2\Psi + c^2\Theta + c^2\Psi s^2\Theta &= -s^2\Psi + c^2\Psi s^2\Theta + c^2\Theta \end{aligned} \tag{90}$$

$$C_{21}^2 + C_{23}^2 = C_{12}^2 + C_{32}^2$$

$$
R_y^2\dot{\Theta}^2 : \quad
\begin{aligned}
&\begin{pmatrix} s^2\Psi c^2\Theta + c^2\Psi s^2\Phi \ldots \\ -2s\Psi c\Psi s\Theta s\Phi c\Phi \ldots \\ + s^2\Psi s^2\Theta c^2\Phi \end{pmatrix} = \begin{pmatrix} s^2\Psi c^2\Phi - 2s\Psi c\Psi s\Theta s\Phi c\Phi \ldots \\ + c^2\Psi s^2\Theta s^2\Phi + c^2\Theta s^2\Phi \end{pmatrix} \\
&\begin{pmatrix} s^2\Psi c^2\Theta + c^2\Psi s^2\Phi \ldots \\ + s^2\Psi s^2\Theta c^2\Phi \end{pmatrix} = \begin{pmatrix} s^2\Psi c^2\Phi + c^2\Psi s^2\Theta s^2\Phi \ldots \\ + c^2\Theta s^2\Phi \end{pmatrix} \\
&\begin{pmatrix} s^2\Psi c^2\Theta \cdot 1 + c^2\Psi \cdot 1 \cdot s^2\Phi \ldots \\ + s^2\Psi s^2\Theta c^2\Phi \end{pmatrix} = \begin{pmatrix} s^2\Psi \cdot 1 \cdot c^2\Phi + c^2\Psi s^2\Theta s^2\Phi \ldots \\ + 1 \cdot c^2\Theta s^2\Phi \end{pmatrix} \\
&\begin{pmatrix} s^2\Psi c^2\Theta s^2\Phi_1 + s^2\Psi c^2\Theta c^2\Phi_2 \ldots \\ + c^2\Psi s^2\Theta s^2\Phi_3 + c^2\Psi c^2\Theta s^2\Phi_4 \ldots \\ + s^2\Psi s^2\Theta c^2\Phi_5 \end{pmatrix} = \begin{pmatrix} s^2\Psi s^2\Theta c^2\Phi_5 + s^2\Psi c^2\Theta c^2\Phi_2 \ldots \\ + c^2\Psi s^2\Theta s^2\Phi_3 + s^2\Psi c^2\Theta s^2\Phi_1 \ldots \\ + c^2\Psi c^2\Theta s^2\Phi_4 \end{pmatrix}
\end{aligned}
\tag{91}
$$

$$
R_z^2\dot{\Theta}^2 : \quad
\begin{aligned}
&C_{21}^2 + C_{22}^2 = C_{13}^2 + C_{33}^2 \\
&\begin{pmatrix} s^2\Psi c^2\Theta + c^2\Psi c^2\Phi \ldots \\ + 2s\Psi c\Psi s\Theta s\Phi c\Phi + s^2\Psi s^2\Theta s^2\Phi \end{pmatrix} = \begin{pmatrix} s^2\Psi s^2\Phi + 2s\Psi c\Psi s\Theta s\Phi c\Phi \ldots \\ + c^2\Psi s^2\Theta c^2\Phi + c^2\Theta c^2\Phi \end{pmatrix} \\
&s^2\Psi c^2\Theta + c^2\Psi c^2\Phi + s^2\Psi s^2\Theta s^2\Phi = s^2\Psi s^2\Phi + c^2\Psi s^2\Theta c^2\Phi + c^2\Theta c^2\Phi \\
&\begin{pmatrix} s^2\Psi c^2\Theta \cdot 1 + c^2\Psi \cdot 1 \cdot c^2\Phi \ldots \\ + s^2\Psi s^2\Theta s^2\Phi \end{pmatrix} = \begin{pmatrix} s^2\Psi \cdot 1 \cdot s^2\Phi + c^2\Psi s^2\Theta c^2\Phi \ldots \\ + 1 \cdot c^2\Theta c^2\Phi \end{pmatrix} \\
&\begin{pmatrix} s^2\Psi c^2\Theta s^2\Phi_1 + s^2\Psi c^2\Theta c^2\Phi_2 \ldots \\ + c^2\Psi s^2\Theta c^2\Phi_3 + c^2\Psi c^2\Theta c^2\Phi_4 \ldots \\ + s^2\Psi s^2\Theta s^2\Phi_5 \end{pmatrix} = \begin{pmatrix} s^2\Psi s^2\Theta s^2\Phi_5 + s^2\Psi c^2\Theta s^2\Phi_1 \ldots \\ + c^2\Psi s^2\Theta c^2\Phi_3 + s^2\Psi c^2\Theta c^2\Phi_2 \ldots \\ + c^2\Psi c^2\Theta c^2\Phi_4 \end{pmatrix}
\end{aligned}
\tag{92}
$$

$$
R_x^2\dot{\Phi}\dot{\Theta} : \quad
\begin{aligned}
&2(C_{13}C_{23} + C_{12}C_{22}) = -2C_{11}C_{21} \\
&\begin{pmatrix} -s\Psi c\Psi s^2\Phi + s^2\Psi s\Theta s\Phi c\Phi - c^2\Psi s\Theta s\Phi c\Phi + s\Psi c\Psi s^2\Theta c^2\Phi \ldots \\ -s\Psi c\Psi c^2\Phi - s^2\Psi s\Theta s\Phi c\Phi + c^2\Psi s\Theta s\Phi c\Phi + s\Psi c\Psi s^2\Theta s^2\Phi \end{pmatrix} = -c\Psi s\Psi c^2\Theta \\
&-s\Psi c\Psi s^2\Phi + s\Psi c\Psi s^2\Theta c^2\Phi - s\Psi c\Psi c^2\Phi + s\Psi c\Psi s^2\Theta s^2\Phi = \\
&-s\Psi c\Psi + s\Psi c\Psi s^2\Theta = \\
&s\Psi c\Psi\left(-1 + 1 - c^2\Theta\right) = \\
&-s\Psi c\Psi c^2\Theta = -c\Psi s\Psi c^2\Theta
\end{aligned}
\tag{93}
$$

$$
R_x R_y \dot{\Phi}\dot{\Theta} : \ -2(C_{11}C_{22} + C_{12}C_{21}) = -2(C_{11}C_{22} + C_{12}C_{21}) \tag{94}
$$

$$
R_x R_z \dot{\Phi}\dot{\Theta} : \ -2(C_{11}C_{23} + C_{13}C_{21}) = -2(C_{11}C_{23} + C_{13}C_{21}) \tag{95}
$$

$$
R_y R_z \dot{\Phi}\dot{\Theta} : \ -2(C_{12}C_{23} + C_{13}C_{22}) = -2(C_{12}C_{23} + C_{13}C_{22}) \tag{96}
$$

$$
2(C_{11}C_{21} + C_{13}C_{23}) = -2C_{12}C_{22}
$$

$$
R_y^2\dot{\Phi}\dot{\Theta}:
\begin{aligned}
&\begin{pmatrix} c\Psi s\Psi c^2\Theta - s\Psi c\Psi s^2\Phi + s^2\Psi s\Theta s\Phi c\Phi\ldots \\ - c^2\Psi s\Theta s\Phi c\Phi + s\Psi c\Psi s^2\Theta c^2\Phi \end{pmatrix} = \begin{pmatrix} s\Psi c\Psi c^2\Phi + s^2\Psi s\Theta s\Phi c\Phi\ldots \\ -c^2\Psi s\Theta s\Phi c\Phi - s\Psi c\Psi s^2\Theta s^2\Phi \end{pmatrix} \\
&c\Psi s\Psi c^2\Theta - s\Psi c\Psi s^2\Phi + s\Psi c\Psi s^2\Theta c^2\Phi = s\Psi c\Psi c^2\Phi - s\Psi c\Psi s^2\Theta s^2\Phi \\
&s\Psi c\Psi\left(c^2\Theta - s^2\Phi + s^2\Theta c^2\Phi\right) = s\Psi c\Psi\left(c^2\Phi - s^2\Theta s^2\Phi\right) \\
&c^2\Theta - s^2\Phi + s^2\Theta c^2\Phi = c^2\Phi - s^2\Theta s^2\Phi \\
&1 - s^2\Theta - 1 + c^2\Phi + s^2\Theta c^2\Phi = \\
&c^2\Phi + s^2\Theta\left(-1 + c^2\Phi\right) = \\
&c^2\Phi + s^2\Theta\left(-1 + 1 - s^2\Phi\right) = \\
&c^2\Phi - s^2\Theta s^2\Phi = c^2\Phi - s^2\Theta s^2\Phi
\end{aligned}
\tag{97}
$$

$$2(C_{11}C_{21} + C_{12}C_{22}) = -2C_{13}C_{23}$$

$$
\begin{aligned}
&R_z^2\dot{\Phi}\dot{\Theta}:\begin{pmatrix} s\Psi c\Psi c^2\Theta - s\Psi c\Psi c^2\Phi\ldots \\ - s^2\Psi s\Theta s\Phi c\Phi + c^2\Psi s\Theta s\Phi c\Phi\ldots \\ + s\Psi c\Psi s^2\Theta s^2\Phi \end{pmatrix} = \begin{pmatrix} s\Psi c\Psi s^2\Phi - s^2\Psi s\Theta s\Phi c\Phi\ldots \\ +c^2\Psi s\Theta s\Phi c\Phi - s\Psi c\Psi s^2\Theta c^2\Phi \end{pmatrix} \\
&c\Psi s\Psi c^2\Theta - s\Psi c\Psi c^2\Phi + s\Psi c\Psi s^2\Theta s^2\Phi = s\Psi c\Psi s^2\Phi - s\Psi c\Psi s^2\Theta c^2\Phi \\
&s\Psi c\Psi\left(c^2\Theta - c^2\Phi + s^2\Theta s^2\Phi\right) = s\Psi c\Psi\left(s^2\Phi - s^2\Theta c^2\Phi\right) \\
&c^2\Theta - c^2\Phi + s^2\Theta s^2\Phi = s^2\Phi - s^2\Theta c^2\Phi \\
&1 - s^2\Theta - 1 + s^2\Phi + s^2\Theta s^2\Phi = \\
&s^2\Phi + s^2\Theta\left(-1 + s^2\Phi\right) = \\
&s^2\Phi + s^2\Theta\left(-1 + 1 - c^2\Phi\right) = \\
&s^2\Phi - s^2\Theta c^2\Phi = s^2\Phi - s^2\Theta c^2\Phi
\end{aligned}
\tag{98}
$$

$$C_{12}^2 + C_{13}^2 = C_{21}^2 + C_{31}^2$$

$$
\begin{pmatrix} s^2\Psi c^2\Phi - 2s\Psi c\Psi s\Theta s\Phi c\Phi \\ + c^2\Psi s^2\Theta s^2\Phi + s^2\Psi s^2\Phi\ldots \\ + 2s\Psi c\Psi s\Theta s\Phi c\Phi + c^2\Psi s^2\Theta c^2\Phi \end{pmatrix} = s^2\Psi c^2\Theta + s^2\Theta
$$

$$R_x^2\dot{\Phi}^2: \begin{pmatrix} s^2\Psi c^2\Phi + c^2\Psi s^2\Theta s^2\Phi + s^2\Psi s^2\Phi \ldots \\ + c^2\Psi s^2\Theta c^2\Phi \end{pmatrix} = \tag{99}$$

$$s^2\Psi + c^2\Psi s^2\Theta =$$

$$s^2\Psi \cdot 1 + (1 - s^2\Psi)s^2\Theta =$$

$$s^2\Psi s^2\Theta + s^2\Psi c^2\Theta + s^2\Theta - s^2\Psi s^2\Theta =$$

$$s^2\Psi c^2\Theta + s^2\Theta = s^2\Psi c^2\Theta + s^2\Theta$$

$$-2C_{11}C_{12} = 2(C_{21}C_{22} + C_{31}C_{32})$$

$$R_xR_y\dot{\Phi}^2: \begin{aligned} s\Psi c\Psi c\Theta c\Phi - c^2\Psi c\Theta s\Theta s\Phi &= c\Psi s\Psi c\Theta c\Phi + s^2\Psi s\Theta c\Theta s\Phi - s\Theta c\Theta s\Phi \\ -c^2\Psi c\Theta s\Theta s\Phi &= (1 - c^2\Psi - 1)s\Theta c\Theta s\Phi \\ -c^2\Psi c\Theta s\Theta s\Phi &= -c^2\Psi s\Theta c\Theta s\Phi \end{aligned} \tag{100}$$

$$-2C_{11}C_{13} = 2(C_{21}C_{23} + C_{31}C_{33})$$

$$R_xR_z\dot{\Phi}^2: \begin{aligned} -s\Psi c\Psi c\Theta s\Phi - c^2\Psi s\Theta c\Theta c\Phi &= -s\Psi c\Psi c\Theta s\Phi + s^2\Psi s\Theta c\Theta c\Phi - s\Theta c\Theta c\Phi \\ -c^2\Psi c\Theta s\Theta c\Phi &= (1 - c^2\Psi - 1)s\Theta c\Theta c\Phi \\ -c^2\Psi c\Theta s\Theta c\Phi &= -c^2\Psi s\Theta c\Theta c\Phi \end{aligned} \tag{101}$$

$$-2C_{12}C_{13} = 2(C_{22}C_{23} + C_{32}C_{33})$$

$$R_yR_z\dot{\Phi}^2: \begin{aligned} \begin{pmatrix} s^2\Psi s\Phi c\Phi + s\Psi c\Psi s\Theta c^2\Phi \ldots \\ -s\Psi c\Psi s\Theta s^2\Phi - c^2\Psi s^2\Theta s\Phi c\Phi \end{pmatrix} &= \begin{pmatrix} -c^2\Psi s\Phi c\Phi + s\Psi c\Psi s\Theta c^2\Phi \ldots \\ -s\Psi c\Psi s\Theta s^2\Phi + s^2\Psi s^2\Theta s\Phi c\Phi \ldots \\ + c^2\Theta s\Phi c\Phi \end{pmatrix} \\ s\Phi c\Phi(s^2\Psi - c^2\Psi s^2\Theta) &= s\Phi c\Phi(-c^2\Psi + s^2\Psi s^2\Theta + c^2\Theta) \\ s^2\Psi - c^2\Psi s^2\Theta &= -c^2\Psi + s^2\Psi s^2\Theta + c^2\Theta \\ &= -1 + s^2\Psi + s^2\Psi s^2\Theta + 1 - s^2\Theta \\ &= s^2\Psi + s^2\Theta(1 - c^2\Psi - 1) \\ s^2\Psi - c^2\Psi s^2\Theta &= s^2\Psi - s^2\Theta c^2\Psi \end{aligned} \tag{102}$$

$$C_{11}^2 + C_{13}^2 = C_{22}^2 + C_{32}^2$$

$$\begin{pmatrix} c^2\Psi c^2\Theta + s^2\Psi s^2\Phi \ldots \\ + 2s\Psi c\Psi s\Theta s\Phi c\Phi + c^2\Psi s^2\Theta c^2\Phi \end{pmatrix} = \begin{pmatrix} c^2\Psi c^2\Phi + 2s\Psi c\Psi s\Theta s\Phi c\Phi \ldots \\ + s^2\Psi s^2\Theta s^2\Phi + c^2\Theta s^2\Phi \end{pmatrix}$$

$$c^2\Psi c^2\Theta + s^2\Psi s^2\Phi + c^2\Psi s^2\Theta c^2\Phi = c^2\Psi c^2\Phi + s^2\Psi s^2\Theta s^2\Phi + c^2\Theta s^2\Phi$$

$$R_y^2\dot{\Phi}^2: \quad \begin{pmatrix} c^2\Psi c^2\Theta \cdot 1 + s^2\Psi \cdot 1 \cdot s^2\Phi \ldots \\ + c^2\Psi s^2\Theta c^2\Phi \end{pmatrix} = \begin{pmatrix} c^2\Psi \cdot 1 \cdot c^2\Phi + s^2\Psi s^2\Theta s^2\Phi \ldots \\ + 1 \cdot c^2\Theta s^2\Phi \end{pmatrix} \tag{103}$$

$$\begin{pmatrix} c^2\Psi c^2\Theta s^2\Phi_1 + c^2\Psi c^2\Theta c^2\Phi_2 \ldots \\ + s^2\Psi s^2\Theta s^2\Phi_3 + s^2\Psi c^2\Theta s^2\Phi_4 \\ + c^2\Psi s^2\Theta c^2\Phi_5 \end{pmatrix} = \begin{pmatrix} c^2\Psi s^2\Theta c^2\Phi_5 + c^2\Psi c^2\Theta c^2\Phi_2 \ldots \\ + s^2\Psi s^2\Theta s^2\Phi_3 + s^2\Psi c^2\Theta s^2\Phi_4 \ldots \\ + c^2\Psi c^2\Theta s^2\Phi_1 \end{pmatrix}$$

$$C_{11}^2 + C_{12}^2 = C_{23}^2 + C_{33}^2$$

$$\begin{pmatrix} c^2\Psi c^2\Theta + s^2\Psi c^2\Phi \ldots \\ - 2s\Psi c\Psi s\Theta s\Phi c\Phi + c^2\Psi s^2\Theta s^2\Phi \end{pmatrix} = \begin{pmatrix} c^2\Psi s^2\Phi - 2s\Psi c\Psi s\Theta s\Phi c\Phi \ldots \\ + s^2\Psi s^2\Theta c^2\Phi + c^2\Theta c^2\Phi \end{pmatrix}$$

$$R_z^2\dot{\Phi}^2: \quad c^2\Psi c^2\Theta + s^2\Psi c^2\Phi + c^2\Psi s^2\Theta s^2\Phi = c^2\Psi s^2\Phi + s^2\Psi s^2\Theta c^2\Phi + c^2\Theta c^2\Phi$$

$$\begin{pmatrix} c^2\Psi c^2\Theta \cdot 1 + s^2\Psi \cdot 1 \cdot c^2\Phi \ldots \\ + c^2\Psi s^2\Theta s^2\Phi \end{pmatrix} = \begin{pmatrix} c^2\Psi \cdot 1 \cdot s^2\Phi + s^2\Psi s^2\Theta c^2\Phi \ldots \\ + 1 \cdot c^2\Theta c^2\Phi \end{pmatrix}$$

$$\begin{pmatrix} c^2\Psi c^2\Theta s^2\Phi_1 + c^2\Psi c^2\Theta c^2\Phi_2 \ldots \\ + s^2\Psi s^2\Theta c^2\Phi_3 + s^2\Psi c^2\Theta c^2\Phi_4 \ldots \\ + c^2\Psi s^2\Theta s^2\Phi_5 \end{pmatrix} = \begin{pmatrix} c^2\Psi s^2\Theta s^2\Phi_5 + c^2\Psi c^2\Theta s^2\Phi_1 \ldots \\ + s^2\Psi s^2\Theta c^2\Phi_3 + s^2\Psi c^2\Theta c^2\Phi_4 \ldots \\ + c^2\Psi c^2\Theta c^2\Phi_2 \end{pmatrix} \tag{104}$$

The previous calculation demonstrated the validity of Eq. (35).

8. CONCISE SUMMARY

This study introduces the technique of inertial value transformation specifically designed for marine applications. Firstly, this text provides an introduction to irregular seas and an outline of the degrees of freedom of ships. The subsequent part introduces the conventional Kirchhoff motion equations in the body-fixed reference frame. These equations describe the motion of a hydrodynamically influenced Eulerian gyro that is attached to a Newtonian body within the body-fixed perspective. The primary part of the document presents the formal derivation of motion equations for a free-floating body in an inertial coordinate system. The conversion of the equations into the body-fixed system results in the widely recognized Kirchhoff motion equations. The 6DOF simulations of a crew transfer vessel in head seas and beam seas demonstrate the comparability of motions in both reference systems. The temporal integration of the motion equations in an inertial system results in unstable and chaotic ship movements. The present work demonstrates the general applicability and necessity of introducing an additional and opposite directed transformation of rotational accelerations in the Eulerian gyro. This is done within the project SOOP and builds upon the previous work by Korte et al. The findings are particularly relevant for free-moving bodies with six degrees of freedom within an inertial frame. Through consistent utilization, the ship's motion can be stabilized over extended durations. A proof demonstrates the conservation of energy during the changing of inertia values in a rotating body. Ultimately, the contribution has demonstrated a deficiency in the conventional methodology for calculating the motion of vehicles.

The purpose of motion equations in an inertial reference system is to model the behavior of mechanically connected multibody systems in the ocean. In order to analyze the interaction effects, it is necessary to establish the forces and moments of all bodies involved within a common reference system. This necessitates a conversion of the equations of motion, which can be achieved using the proposed technique of transforming inertia values. In a multibody system, a CTV (Cable Tensioning Vessel) is securely attached to the front of an offshore wind turbine tower and is capable of performing perfect rotations. The inclusion of an extra transformation ensures the stability of the system. The scenario entails the interaction between an immobile object and a buoyant object. Future plans include conducting additional research in the subject of multibody dynamics. Currently, a scenario is being built where two ships are mechanically attached in tandem and rotate optimally. The primary difficulty is in the interaction between two buoyant entities, as opposed to a single multibody system. Additional uses of the technology, which the authors intend to explore in future research, include simulating offshore crane operations and doing 3D simulations of a remotely operated vehicle (ROV) that is connected to a mother ship through an umbilical. The research is currently focused on the real-time application of controller parameterization.

REFERENCES

1. Fossen TI. Handbook of Marine Craft Hydrodynamics and Motion Control. 1st ed. Chichester UK: J. Wiley & Sons; 1994. DOI: ISBN 0 471 94113 1
2. Korte H, Takagi T. Dynamic motion calculation of a flexible structure using inertia transformation algorithm. In: Workshop on Fishing and Marine Production Technology; 9-10. August; Trondheim. NTNU. 2004. pp. 1-11. DOI: 10.13140/RG.2.1.4006.4724
3. Kirchhoff G. Über die Bewegung eines Rotationskörpers in einer Flüssigkeit. Crelles Journal. 1869;71:237-273
4. Wendel K. Hydrodynamische Massen und hydrodynamische Massenträgheitsmomente. Jahrbuch der Schiffbautechnischen Gesellschaft. 1950;207-255

5. Korte H, Takagi T. Transformation von Trägheitsgrößen zur Lösung der Mehrkörperdynamik in der Meerestechnik. In: 101. Jahrbuch der Schiffbautechnischen Gesellschaft. Hamburg: Schiffahrts-Verlag "Hansa"; 2007. pp. 313-325

6. Korte H. Track control of a towed underwater sensor carrier. In: IFAC Control in Transportation Systems. Braunschweig: IFAC; 2000. pp. 89-94

7. Paschen M. Contribution to Predict Trajectories of Pelagic Fishing Gears Affected by Ship Manoeuvres (in German) [Dissertation]. Rostock: University of Rostock; 1982

8. Suzuki K, Takagi T, Shimizu T, Hiraishi T, Yamamoto K, Nashimoto K. Validity and visualization of a numerical model used to determine dynamic configuration of fishing net. Fisheries Science. 2003;69(4):1-18. DOI: 10.1046/j.1444-2906.2003.00676.x · Source: OAI

9. Aamo O, Fossen TI. Finite element modelling of moored vessels. Mathematical and Computer Modelling of Dynamical Systems. 2001;7(1):47-75

10. Abreu P, Morishita H, Pascoal A, Ribero J, Silva H. Marine vehicles with streamers for geotechnical surveys: Modeling, positioning and control. In: 10th IFAC CAMS; September; Trondheim. IFAC-PapersOnLine 49-23; 2016. p. 458-464

11. Berntsen P, Aamo O, Leira B, Sørensen A. Structural reliability - based control of moored interconnected structures. Control Engineering Practice. 2008;16(4):495-504

12. Korte H. Motion modelling of a manoeuvring ship at anchor in flowing waters. In: Yamane T, editor. Contributions on the Theory of Fishing Gears and Related Marine Systems; 5.-7. November; University of Kinki, Nara (Jp). Nara:2009. pp. 207-216. DOI: ISBN 4-946-421-13-6

13. Yang K, Wang X, Wu C. Simulation platform for the underwater snakelike robot swimming based on Kane's dynamic model and central pattern generator. Journal Shanghai Jiaotong University. 2014;19(3):294-301

14. Kane TR, Levison DA. Dynamics: Theory and Applications. 1st ed. New York: McGraw-Hill; 1985

15. Takagi T, Miyata S, Fusejima I, Oshima T, Uehara T, Suzuki K, Nomura Y, Kanechiku M, Torisawa S. Potential of computer simulation for buoy-line type of purse seine fishing. In: Paschen M, editors. Contributions on the Theory of Fishing Gears and Related Marine Systems; 9-12. Ocktober; Rostock. Aachen: Shaker Verlag; 2013. p. 55-62. DOI: ISBN 978-3-8440-2251-3

16. Korte H, Stuppe S, Wesuls J-H, Takagi T. Application of the inertia value transformation to simulate the transfer scenario of service staff from a crew transfer vessel to offshore wind turbines. In: Paschen M, O'Neil FG, Ivanovic A, editors. Contributions on the theory of fishing gears and related marine systems; 27-29. October; Aberdeen. Aachen: Shaker Verlag; 2015. p. 227-237. DOI: ISBN 978-3-8440-3955-9

17. Korte H, Wesuls J-H, Stuppe S, Takagi T. Towards an approach to 6 DOF inertial kinematics. In: 10th IFAC CAMS; September; Trondheim. IFAC-PapersOnLine 49; 2016. p. 440-445

18. Piersson WJ, Moskowitz L. A proposed spectral form for fully developed wind seas based on the similarity of S.A. Kitagorodskii. Journal of Geophysical Research. 1964;96(2)

19. ITTC. Report of the Committee on Waves. Final Report and Recommendations to the 23rd International Towing Tank Conference. 2002;505-551

20. Hasselmann K, Barnet TP, Bouws E, Carlson H, Cartwright DE, Enke K, Ewing JA, Gienapp H, Hasselmann DE, Krusemann P, Meerburg A, Müller P, Olbers DJ, Richter K, Seil W, Walden H. Measurements of wind-wave growth and swell decay during the joint North Sea wave project (JONSWAP). Ergänzungsheft zur Deutschen Hydrographischen Zeitschrift. 1973;Reihe A(8)(12):95

21. Ginsberg JH. Advanced Engineering Dynamics. 2nd ed. New York: Cambridge University Press; 2004

22. Woernle C. Dynamic of Multi-Body Systems. Rostock: University of Rostock, Lesson script; 2001

23. Schiehlen W, Eberhard P. Technische Dynamik–Rechnergestützte Modellierung mechanischer Systeme im Maschinen- und Fahrzeugbau. 4th ed. Springer Vieweg; 2014. pp. 24-28. DOI: ISBN 978-3-658-06184-5

24. Korte H, Ihmels I, Richter J, Zerhusen B, Hahn A. Offshore training simulations. In: 9th IFAC Manoeuvring and Control of Marine Crafts; 19–21 September 2012; Arenzano (IT). IFAC; 2012. pp. 37-42. DOI: ISSN 1474-6670

25. Blass A, Gurevich Y. Matrix transformation is complete for the average case. SIAM Journal on Computing. 1995;24(1):3-29

3

Chapter

LOCAL-LEVEL FRAME PATH PLANNING FOR SMALL UNMANNED AIRCRAFT SYSTEMS

1. INTRODUCTION: AN OVERVIEW

The exponential expansion of the unmanned aircraft systems (UASs) business drives the growing need to include UAS into the U.S. national airspace system (NAS). The majority of the efforts have been directed towards the integration of medium or larger Unmanned Aircraft Systems (UAS) into the controlled airspace. Nevertheless, small Unmanned Aircraft Systems (UASs) with a weight below 55 pounds are highly appealing and are expected to experience rapid growth in civil and commercial operations. This is due to their adaptability, as well as their comparatively low starting cost and operational expenses.

At present, Unmanned Aircraft Systems (UASs) are restricted in their ability to enter the National Airspace System (NAS) due to their lack of collision avoidance capabilities with other aircraft [1]. As a result, the Federal Aviation Administration (FAA) has specified that Unmanned Aircraft Systems (UASs) must have the same level of safety as human aircraft in terms of being able to observe and avoid other aircraft [2, 3]. The sense-and-avoid (SAA) obligation is akin to a pilot's capacity to visually survey the surrounding airspace for potential invading aircraft and respond accordingly to prevent a potential accident.

A comprehensive functional sense-and-avoid system typically consists of sensors and their corresponding trackers, collision detection capabilities, and collision avoidance algorithms. The primary emphasis of this chapter is on the avoidance of collisions and the planning of paths. Collision avoidance is a crucial aspect of path planning that entails calculating a path that is free from collisions, starting at a certain place and ending at a desired point, while also optimizing an objective function or performance metric. A comprehensive collision avoidance algorithm takes into account the limitations on the movement of the host vehicle, the behavior of the intruder's motion, and the lack of accuracy in estimating the intruder's position and velocity. The topic of path planning is extensive, with a specific emphasis on collision avoidance, which has received substantial attention in the realm of robotics and autonomous systems. Kuchar and Yang [4] conducted an extensive examination of conflict identification and resolution methodologies. Albaker and Rahim [5] conducted an extensive investigation on techniques used to prevent collisions for Unmanned Aerial Systems (UAS). The predominant collision avoidance strategies include geometric-based guidance methods [6–13], potential field methods [14, 15], sampling-based methods [16, 17], cell decomposition techniques, and graph-search algorithms [18–20].

Geometric methods for collision avoidance are simple and easy to understand. These problems are well-suited for quick analytical solutions that rely on the aircraft's kinematics and the geometry of the encounter situation. The approach employs the geometric correlation between the aircraft that come into contact, together with intuitive reasoning [8, 21]. In general, the geometric approach relies on a linear projection to evaluate if an intruder will enter a virtual zone surrounding a specific ownship. Collision avoidance can be accomplished by altering the velocity vector, assuming a model of constant velocity. Geometric techniques usually do not consider the presence of ambiguity in invader flight plans and the presence of noisy sensor inputs.

The potential field method is a commonly employed technique for collision avoidance in the field of robotics. A conventional potential field operates by applying simulated pressures on the aircraft, often an attracting force from the objective and repulsive forces from barriers or surrounding air traffic. In general, the approach is straightforward to describe and straightforward to apply. Nevertheless, the potential field technique encounters several fundamental problems [22]. One of the problems with this approach is that it employs a greedy strategy, which can be influenced by local minima. Additionally, the literature also presents heuristic advancements aimed at avoiding local minima [23]. Another issue is that conventional potential field methods fail to consider the dynamics of obstacles or the uncertainty in observation or control. Bortoff proposes a technique for representing the trajectory of an unmanned aerial system (UAS) in the context of route planning and collision avoidance. This method involves modeling the UAS path as a sequence of point masses that are interconnected by springs and dampers [24]. This program produces a covert route via a collection of hostile radar stations with predetermined positions.

McLain and Beard provide a trajectory planning technique that is appropriate for coordinating the timing of numerous Unmanned Aerial Systems (UAS) [25]. The routes to the desired destination are represented by a physical comparison to a chain. Argyle et al. propose a path planner that utilizes a simulated chain of unit masses positioned in a force field [26]. This planner aims to identify routes that pass through the highest points of a bounded differentiable reward function.

Sampling-based techniques such as probability road maps (PRM) [16] and rapidly exploring random trees (RRTs) [17] have demonstrated significant effectiveness in path planning and obstacle avoidance, particularly for terrestrial robots. Due to the need for extensive computational resources in path replanning, they are not appropriate for reactive avoidance. Nevertheless, new enhancements to the fundamental RRT algorithm, such as chance-constrained RRT* [27] and close-loop RRT [28], demonstrate encouraging outcomes for environments with uncertainty and complex dynamics [28–30]. Cell decomposition is a commonly employed method for path planning. It involves dividing the unobstructed region of the configuration space into smaller cells, which are then linked together to form a graph [20]. In general, cell decomposition approaches are regarded as global path planners that necessitate prior knowledge of the environment. A viable route is discovered from the initial node to the target node by exploring the connectivity graph using search methods such as A* or Dijkstra's algorithm [18].

The approach suggested in this work will analyze encounter scenarios, such as the one shown in Figure 1, when the ownship confronts one or more intruders. The main objective of this study is to provide a collision avoidance framework for unmanned aircraft. The architecture, however, will be specifically customized for small Unmanned Aircraft Systems (UAS). It is assumed that there is a sensor(s) and tracking system that can offer an estimate of the intruder's track.

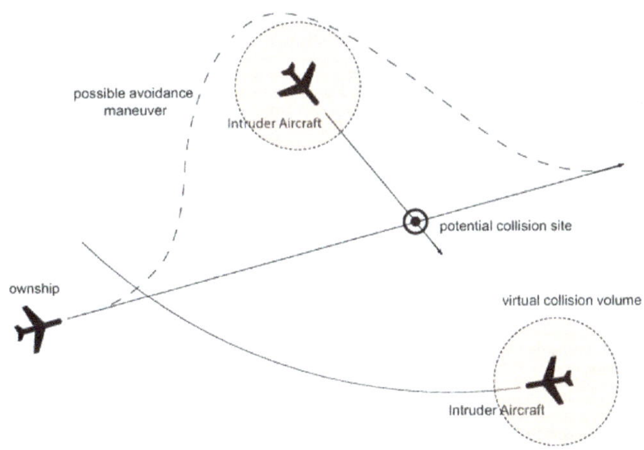

Figure 1. The geometry of an encounter scenario.

2. LOCAL-LEVEL PATH PLANNING

A collision event occurs when two aircraft or more come within the minimum allowed distance between each other. The current manned aviation regulations do not provide an explicit value for the minimum allowed distance. However, it is generally understood that the minimum allowed or safe distance is required to be at least 500 ft. to 0.5 nautical miles (nmi) [21, 31]. For example, the near midair collision (NMAC) is defined as the proximity of less than 500 ft. between two or more aircraft [32]. Similarly and since the potential UAS and intruder aircraft cover a wide range of vehicle sizes, designs, airframes, weights, etc., the choice of a virtual fixed volume boundary around the aircraft is a substitute for the actual dimensions of the intruder.

As shown in Figure 2, the choice for this volume is a *hockey-puck* of radius d_s and height h_s that commonly includes a horizontal distance of 500 ft. and a vertical range of 200 ft. [1, 33, 34]. Accordingly, a collision event is defined as an incident that occurs when two aircraft pass less than 500 ft. horizontally and 100 ft. vertically.

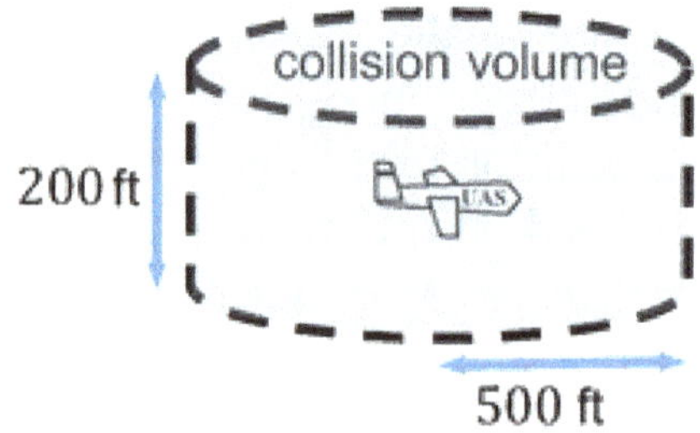

Figure 2. A typical collision volume or protection zone is a virtual fixed volume boundary around the aircraft.

In this work, we develop the path planning logic using a body-centered relative coordinate system. In this body-centered coordinate system, the ownship is fixed at the center of the coordinate system, and the intruder is located at a relative position p_r and moves with a relative velocity v_r with respect to the ownship [35].

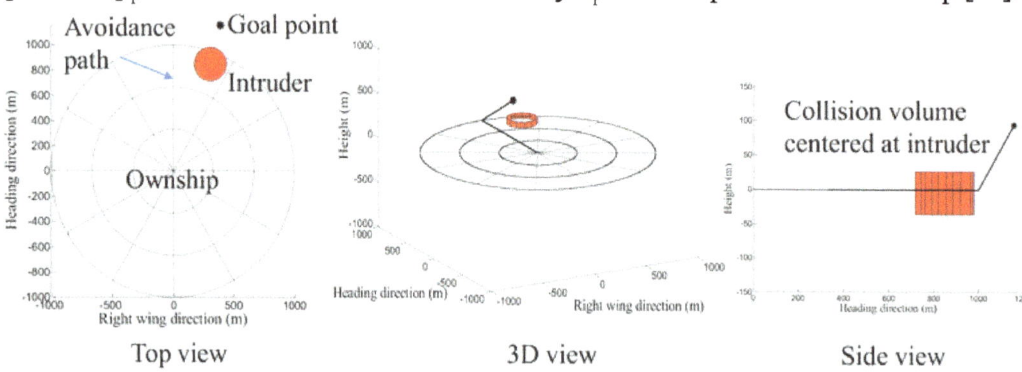

Figure 3. Local-level reference frame.

We call this body-centered coordinate frame the local-level frame because the environment is mapped to the unrolled, unpitched local coordinates, where the ownship is stationary at the center. As depicted in Figure 3, the origin of the local-level reference is the current position of the ownship. In this configuration, the x-axis points out the nose of the unpitched airframe, the y-axis points points out the right wing of the unrolled airframe, and the z-axis points down forming a right-handed coordinate system. In the following discussion, we assume that the collision volume is centered at the current location of the intruder. A collision occurs when the origin of the local-level frame penetrates the collision volume around the intruder.

The detection region is divided into concentric circles that represent maneuvers points at increasing range from the ownship as shown in Figure 4, where the radius of the outmost circle can be thought of as the sensor detection range. Let the region in the space covered by the sensor be called the workspace. Then, this workspace is discretized using a cylindrical grid in which the ownship is commanded to move along the edges of the grid. The result is a directed weighted graph, where the edges represent potential maneuvers, and the associated weights represent the maneuver cost and collision risk. The graph can be described by the tuple $\mathcal{G}(\mathcal{N}, \mathcal{E}, \mathcal{C})$, where $\mathcal{N}$ is a finite nonempty set of nodes, and $\mathcal{E}$ is a collection of ordered pairs of distinct nodes from $\mathcal{N}$ such that each pair of nodes in $\mathcal{E}$ is called a directed edge or link, and $\mathcal{C}$ is the cost associated with traversing each edge.

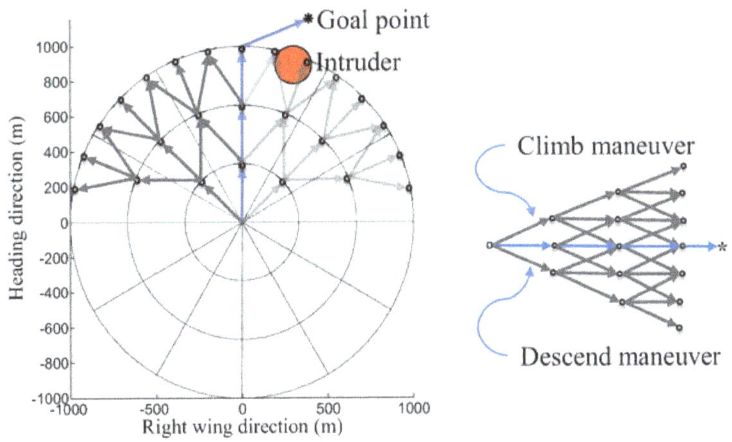

Top view Side view

Figure 4. Discretized local-level reference workspace. The three concentric circles represent three maneuvers points.

The path is then constructed from a sequence of nonrepeated nodes $(\mathbf{n}_1, \mathbf{n}_2, \cdots, \mathbf{n}_N)$ such that each consecutive pair $(\mathbf{n}_i, \mathbf{n}_{i+1})$ is an edges in $\mathcal{G}$. Let the detection range d_r be the radius of the outermost circle, and r be the radius of the innermost circle so that $d_{r=mr}$. As shown in Figure 6, let $\mathcal{L}_l$, $l = 1, 2, \cdots, m$ be the lth level curve of the concentric circles. Assume that the level curves are equally partitioned by a number of points or nodes such that any node on the lth level curve, $\mathcal{L}_l$ connects to a predefined number of nodes k in the next level, that is, in the forward direction along the heading axis as depicted in Figure 4. The nodes on the graph can be thought of as predicted locations of the ownship over a look-ahead time window. Additionally, we assume that only nodes along the forward direction of the heading axis, that is, x=0 connect to nodes in the vertical plane. This assumption allows to command the aircraft to climb or descend by connecting to nodes in the vertical plane as shown in Figure 4. Let the first level curve of the innermost circle be discretized into $|\mathcal{L}_1| = k + 2$ nodes including nodes in the vertical plane. Then, using the notation $|\mathcal{A}|$ to denote the cardinality of the discrete set A, the number of nodes in the lth level curve is given by

$$|\mathcal{L}_l| = \begin{cases} k+2 & \text{if } l = 1, \\ 2|\mathcal{L}_{l-1}| + 2l + 1 & \text{if } l = 2, 3, \cdots, m, \end{cases} \tag{1}$$

where the total number of nodes is $|\mathcal{N}| = \sum_{l=1}^{m} |\mathcal{L}_l|$. For example, assuming that the start node is located at the origin of the reference map and given that k= 3, that is, allowing the ownship to fly straight or maneuver right or left. The total number of nodes in the graph including the start and destination node is given by

$$|\mathcal{N}| = \left(\sum_{l=1}^{m+1} 2^l + 2l - 3 \right) + 1. \tag{2}$$

Figure 5 shows an example of a discretized local-level map. In this example, k= 3 and m= 3, and the total number of nodes in the graph $|\mathcal{N}|$ is 39.

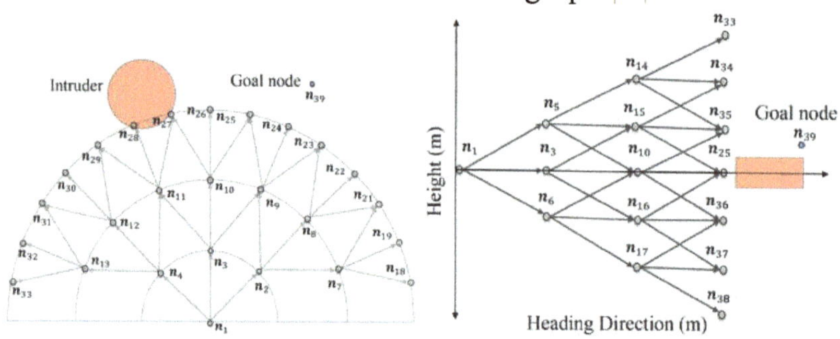

(a) (b)

Figure 5. Example of discretized local-level map. (a) Top view: location and index of nodes and (b) side view: location and index of nodes.

Assuming that the ownship travels between the nodes with constant velocity and climb rate, the location of the ith node at the lth level curve, and $n_{i,l}$ in the horizontal plane of the graph is given by

$$\mathbf{n}_{i,l} = \left[lr \sin \psi_j^{\mathcal{L}_l}, lr \cos \psi_j^{\mathcal{L}_l}, 0 \right]^T, \tag{3}$$

where $\psi_j^l = \frac{j\psi^d}{2^{(l-1)}}$ and $j = \left\{ -\frac{|\mathcal{L}_l|-1}{2}, -\frac{|\mathcal{L}_l|-1}{2} + 1, \ldots, \frac{|\mathcal{L}_l|-1}{2} - 1, \frac{|\mathcal{L}_l|-1}{2} \right\}$ and ψ^d is the allowed heading. In the vertical plane, the location of nodes is $\mathbf{n}_{\bar{j},l} = \left[0, 0, \pm \bar{j} l h^d\right]^T$, where $\bar{j} = \{1, 2, \cdots, l\}$ and h^d are the altitude change at each step as shown in Figure 6.

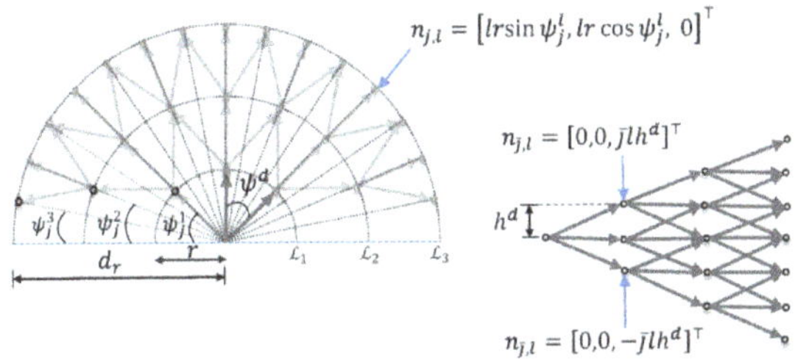

Figure 6. Nodes location in the local-level reference frame.

For example, if $\psi^d = \pi/4$, $h^d = 50$ m, $r = 500$ m, $k = 3$, and $|\mathcal{L}_1| = 5$, then we have $j = \{-1, 0, 1\}$, $\bar{j} = \{-1, 1\}$, $\psi_j^1 = \{-\pi/4, 0, -\pi/4\}$, and the locations of nodes at $\mathcal{L}_1$ in the horizontal plane are $\left\{(-500\sin\pi/4, 500\cos\pi/4, 0)^T,\ (0, 500, 0)^T,\ (500\sin\pi/4, 500\cos\pi/4, 0)^T\right\}$, and in the vertical plane are $\left\{(0, 0, 50)^T, (0, 0, -50)^T\right\}$.

The main priority of the ownship where it is under distress is to maneuver to avoid predicted collisions. This is an important note to consider when assigning a cost of each edge in the resulting graph. The cost associated with traveling along an edge is a function of the edge length and the collision risk. The cost associated with the length of the edge $e_{i,i+1}$ that connects between the consecutive pair nodes $(n_{i,ni+1})$ is simply the Euclidean distance between the nodes n_i and n_{i+1} expressed as

$$C_L(e_{i,i+1}) = \|\mathbf{n}_{i+1} - \mathbf{n}_i\|. \tag{4}$$

The collision cost for traveling along an edge is determined if at any future time instant, the future position of the ownship along that edge is inside the collision volume of the predicted location of an intruder. An exact collision cost computation would involve the integration of collision risk along each edge over the look-ahead time window $\tau \in [t, t + mT]$.

A simpler approach involves calculating the collision risk cost at several locations along each edge, taking into account the projected locations of the intruder over the time horizon τ. Assuming a constant velocity model, a linear extrapolation of the current position and velocity of the detected intruders are computed at evenly spaced time instants over the look-ahead time window. The look-ahead time interval is then divided into several discrete time instants. At each discrete time instant, all candidate locations of the ownship along each edge are checked to determine whether it is or will be colliding with the propagated locations of the intruders. For the simulation results presented in this chapter, the collision risk cost is calculated at three points along each edge in $\mathcal{G}$. If v_o is the speed of the ownship, then the distance along an edge is given by v_oT, where $T=r/v_o$. The three points are computed as

$$\mathbf{p}_1 = \mathbf{n}_i + \mathbf{v}_o T_s \frac{\mathbf{n}_{i+1} - \mathbf{n}_i}{\|\mathbf{n}_{i+1} - \mathbf{n}_i\|}, \tag{5}$$

$$\mathbf{p}_2 = \mathbf{p}_1 + v_o T_s \frac{\mathbf{n}_{i+1} - \mathbf{n}_i}{\|\mathbf{n}_{i+1} - \mathbf{n}_i\|}, \tag{6}$$

$$\mathbf{p}_3 = \mathbf{p}_2 + v_o T_s \frac{\mathbf{n}_{i+1} - \mathbf{n}_i}{\|\mathbf{n}_{i+1} - \mathbf{n}_i\|}, \tag{7}$$

where $T_{s=T/3}$. Let the relative horizontal and vertical position of the intruder with respect to the ownship at the current time t be $p_{r(t)}$ and $p_{rz(t)}$, respectively. Define the collision volume as

$$\mathcal{C}(\mathbf{p}_r(t)) = \left\{ d \in \mathbb{R}^2 : \|\mathbf{p}_r(t)\| - d \le d_s \text{ and } h \in \mathbb{R} : \left| p_{r_z} - h \right| \le h_s/2 \right\}. \tag{8}$$

The predicted locations of each detected intruder over time horizon T at three discrete time samples T_s are

$$\mathbf{p}_{r_{3D}}(t + (1 + 3(l-1))T_s) = \mathbf{p}_{r_{3D}}(t) + \mathbf{v}_{r_{3D}}(t)(1 + 3(l-1))T_s, \tag{9}$$

$$\mathbf{p}_{r_{3D}}(t + (2 + 3(l-1))T_s) = \mathbf{p}_{r_{3D}}(t) + \mathbf{v}_{r_{3D}}(t)(2 + 3(l-1))T_s, \tag{10}$$

$$\mathbf{p}_{r_{3D}}(t + (3 + 3(l-1))T_s) = \mathbf{p}_{r_{3D}}(t) + \mathbf{v}_{r_{3D}}(t)(3 + 3(l-1))T_s, \tag{11}$$

where $\mathbf{p}_{r_{3D}}(t) = \left[\mathbf{p}_r(t), p_{r_z}(t)\right]^T \in \mathbb{R}^3$ and $\mathbf{v}_{r_{3D}}(t) = [v_r(t), \mathbf{v}_{r_z}(t)]^T \in \mathbb{R}^3$ be the 3D relative position and velocity of the intruder with respect to the ownship in the relative coordinate system, where $v_{r(t)}$ and $v_{rz(t)}$ are the relative horizontal velocity and vertical speed at the current time t.

In Eqs. (9)–(11), if $e_{i,i+1}$ is the current edge being evaluated, then the node n_{i+1} determines the value of l. In other words, $\mathbf{n}_{i+1} \in \mathcal{L}_1$, then l=1. For example, if we are to compute the three points along the edge $e_{1,2}$ in Eqs (5)–(7), then $n_{2\in L1}$. Using the definition of the binary cost function, the collision risk cost associated with the $e_{i,i+1}$ edge with respect to each detected intruder is given by the expression

$$C_{col}(\text{int}, e_{i,i+1}) = \begin{cases} \infty & \text{if any of } \mathbf{p}_1, \mathbf{p}_2, \text{ or } \mathbf{p}_3 \in \mathcal{C}\left(\mathbf{p}_{r_{3D}}(t + (\ell + 3(l-1))T_s)\right), \\ 0 & \text{otherwise,} \end{cases} \tag{12}$$

where ℓ={1,2,3}ℓ=123. In Eq. (12), the ∞ or the maximum allowable cost is assigned to any edge that leads to a collision, basically eliminating that edge and the path passing through it. The total collision risk associated with the ith edge is given by

$$C_{col}(e_{i,i+1}) = \sum_{\text{int}=1}^{M} C_{col}(\text{int}, e_{i,i+1}), \tag{13}$$

where M is the number of detected intruders.

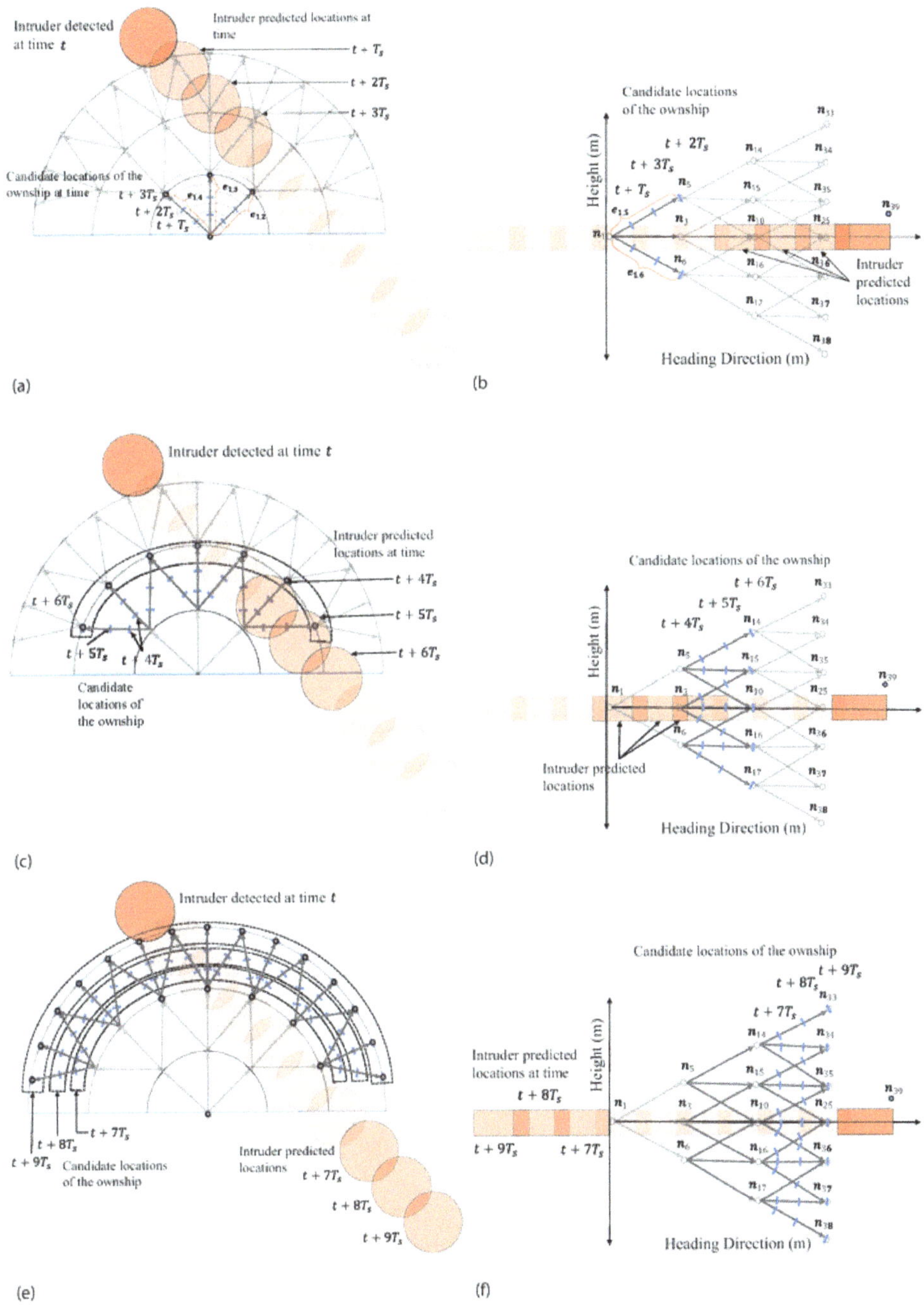

Figure 7. Example illustrating the steps to compute the collision risk. In this example, we have k= 3 *and* **m=** 3. ***(a) Top view: predicted locations of intruder (less transparent circles), and candidate locations of*** **ownship; (b) side view: predicted locations of intruder (less transparent rectangles), and candidate locations of ownship; (c) predicted locations of intruder and candidate locations of ownship over time window (t + 4Ts, t + 6Ts); (d) time window (t + 4Ts, t + 6Ts); (e) time window (t + 7Ts, t + 9Ts); (f) time window (t + 7Ts, t + 9Ts).**

A visual illustration of the collision risk computation is shown in Figure 7. The propagated collision volume of a detected intruder and the candidate locations of the ownship over the first-time interval $[t+T_s, t+3T_s]$ both in the horizontal and vertical plane is depicted in Figure 7a and b. Clearly, there is no intersection between these candidate points the ownship may occupy and the propagated locations of the collision volume over the same interval. Then, according to Eq. (13), the cost assigned to these edges is zero. Next, all candidate locations of the ownship along each edge over the second time interval $[t+4T_s, t+6T_s]$ are investigated. As shown in Figure 7c, edges $e_{2,7}$, $e_{2,8}$, and $e_{2,9}$ intersect with the predicted intruder location at time t+4T_S and t+5T_S, respectively. Similarly, edges $e_{3,15}$ and $e_{3,16}$ in the horizontal plane intersect with the predicted intruder location at time t+4T_S as shown in Figure 7d. Accordingly, the maximum allowable costs will be assigned to these edges, which eliminate these edges and the path passing through them. All the candidate locations of the ownship over the time interval $[t+7T_s, t+9T_s]$ do not intersect with the predicted locations of the intruder as shown in Figure 7e and f. Therefore, by the time, the ownship will reach these edges the detected intruder will be leaving the map, and consequently, a cost of zero is assigned to edges belonging to the third level curve $\mathcal{L}_3$.

To provide an increased level of robustness, an additional threat cost is added to penalize edges close to the propagated locations of the intruder even if they are not within the collision volume. At each discrete time instant, we compute the distances from the candidate locations of the ownship to all the propagated locations of the intruders at that time instant. The cost of collision threat along each edge is then given by the sum of the reciprocal of the associated distances to each intruder

$$C_{th}(\text{int}, e_{i,i+1}) = \frac{1}{d_1} + \frac{1}{d_2} + \frac{1}{d_3}. \tag{14}$$

where d_1, d_2, and d_3 are given by

$$d_1 = \left\| p_1 - \mathrm{p}_{r_{3D}}(t + (1 + 3(l-1))T_s) \right\|,$$

$$d_2 = \left\| p_2 - \mathrm{p}_{r_{3D}}(t + (2 + 3(l-1))T_s) \right\|,$$

$$d_3 = \left\| p_3 - \mathrm{p}_{r_{3D}}(t + (3 + 3(l-1))T_s) \right\|,$$

and the total collision risk cost associated with the ith edge with regard to all intruders is given by

$$C_{th}(e_{i,i+1}) = \sum_{\text{int}=1}^{M} C_{th}(\text{int}, e_{i,i+1}). \tag{15}$$

For example, the edges $e_{1,2}$, $e_{1,3}$, $e_{1,4}$, $e_{1,5}$ and $e_{1,6}$ shown in Figure 7a are not intersecting with the propagated collision volume locations over the first-time interval, yet they will be penalized based on their distances to the predicated locations of the intruder according to Eq. (15). Note that edge $e_{1,2}$ will have greater cost as it is the closest to the intruder among other candidate edges.

Another objective of a path planning algorithm is to minimize the deviation from the original path, that is, the path the ownship was following before it detected a collision. Generally, the path is defined as an ordered sequence of waypoints $\mathcal{W} = \mathbf{w}_1, \mathbf{w}_2, \cdots . \mathbf{w}_f$, where $\mathbf{w}_i = (w_{n,i}, w_{e,i}, w_{d,i})^T \in \mathbb{R}^3$ is the north-east-down location of the ith waypoint in a globally known NED reference frame. The transformation from the global frame to the local-level frame is given by

$$\mathbf{w}_i^b = \mathcal{R}_g^b(\psi_o)\mathbf{w}_i, \quad (16)$$

where

$$\mathbf{R}_g^b(\psi_o) = \begin{pmatrix} \cos\psi_o & \sin\psi_o & 0 \\ -\sin\psi_o & \cos\psi_o & 0 \\ 0 & 0 & 1 \end{pmatrix}$$

where ψ_o is the heading angle of the ownship. Let w_s be the location waypoint of the ownship at the current time instant t and wf∈W be the next waypoint the ownship is required to follow. Assuming a straight-line segment between the waypoints w_s and wf, then any point on this segment can be described $\mathcal{L}(\varrho) = (1-\varrho)\mathbf{w}_s + \rho w_f$ where ϱ∈[0,1]ϱ∈01, and the minimum distance between an arbitrary node n_i in $\mathcal{G}$ can be expressed by [36]

$$D(\mathbf{w}_s, \mathbf{w}_f, n_i) \triangleq \begin{cases} D(\varrho^*), & \text{if } \varrho^* \in [0,1], \\ \|\mathbf{n}_i - \mathbf{w}_s\|, & \text{if } \varrho^* < 0, \\ \|\mathbf{n}_i - \mathbf{w}_f\|, & \text{if } \varrho^* > 1, \end{cases} \quad (17)$$

where

$$D(\varrho^*) = \sqrt{\|\mathbf{n}_i - \mathbf{w}_s\|^2 - \frac{\left((\mathbf{w}_s - \mathbf{n}_i)^T(\mathbf{w}_s - \mathbf{w}_f)\right)^2}{\|\mathbf{w}_s - \mathbf{w}_f\|^2}},$$

and

$$\varrho^* = \frac{(\mathbf{w}_s - \mathbf{n}_i)^T(\mathbf{w}_s - \mathbf{w}_f)}{\|\mathbf{w}_s - \mathbf{w}_f\|^2}.$$

Then, the cost that penalizes the deviation of an edge in $\mathcal{G}$ from the nominal path is given by

$$C_{dev}(e_{i,i+1}) = D(\mathbf{w}_s, \mathbf{w}_f, n_i). \quad (18)$$

If small UASs are to be integrated seamlessly alongside manned aircraft, they may require to follow right-of-way rules. Therefore, an additional cost can be also added to penalize edges that violate right-of-way rules. In addition, this cost can be used to favor edges in the horizontal plane over those in the vertical plane. Since the positive direction of the y-axis in the local-level frame is the right-wing direction, it is convenient to define right and left maneuvers as the positive and the negative directions along the right-wing axis, respectively. Let $\vec{\mathbf{e}}_i \triangleq \mathbf{n}_{i+1} - \mathbf{n}_i$ be the direction vector associated with the edge $e_{i,i+1}$ in $\mathcal{G}$, where $\mathbf{n}_i \triangleq (x_i, y_i, z_i)^T \in \mathbb{R}^3$ is the location of ith node in the local-level reference frame. Let the direction vector $\vec{e}_i$ be expressed as $\vec{\mathbf{e}}_i = (e_{i_x}, e_{i_y} e_{i_z})^T \in \mathbb{R}^3$. We define $\mathrm{E} \triangleq (e_{i_x}, L, R, e_{i_z})^T \in \mathbb{R}^4$, where e_{ix} and e_{iz} are the x and the z components of e→i. The y-component of e→i is decomposed into two components: left L and right R, that are defined by

$$L, R \triangleq \begin{cases} L = e_{i_y}, R = 0 & \text{if } e_{i_y} \leq 0, \\ L = 0, R = e_{i_y} & \text{if } e_{i_y} > 0. \end{cases} \quad (19)$$

If we define the maneuvering design matrix to be $\mathrm{J} = \mathrm{diag}([0, c_L, c_R, c_z])$, then the maneuvering cost associated with each edge is given by

$$C_m(e_{i,i+1}) = \sqrt{\mathbf{E}^T \mathbf{J} \mathbf{E}}, \quad (20)$$

The costs c_L and c_R allow the designer to place more or less cost on the left or right edges. Similarly, c_z allows the designer to penalize vertical maneuvers. Multiple values of these cost parameters may be saved in a look-up table, and the collision avoidance algorithm choses the appropriate value based on the geometry of the encounter.

The overall cost for traveling along an edge comes from the weighted sum of all costs given as [35]

$$C(e_{i,i+1}) = C_L(e_{i,i+1}) + C_{col}(e_{i,i+1}) + k_1 C_{th}(e_{i,i+1}) + k_2 C_{dev}(e_{i,i+1}) + k_3 C_m(e_{i,i+1}), \quad (21)$$

where k_1, k_2, and k_3 are positive design parameters that allow the designer to place weight on collision risk or deviation from path or maneuvering preferences depending on the encounter scenario. Once the cost is assigned to each edge in $\mathcal{G}$, then a graph-search method can be used to find the least cost path from a predefined start point to the destination point. In this work, we have used Dijkstra's algorithm.

Dijkstra's algorithm solves the problem of shortest path in a directed graph in polynomial time given that there are not any negative weights assigned to the edges. The main idea in Dijkstra's algorithm is to generate the nodes in the order of increasing value of the cost to reach them. It starts by assigning some initial values for the distances from the start node and to every other node in the graph. It operates in steps, where at each step, the algorithm updates the cost values of the edges. At each step, the least cost from one node to another node is determined and saved such that all nodes that can be reached from the start node are labeled with cost from the start node. The algorithm stops either when the node set is empty or when every node is examined exactly once. A naive implementation of Dijkstra's algorithm runs in a total time complexity of $O(|\mathcal{N}|^2)$. However, with suitable data structure implementation, the overall time complexity can be reduced to $O(|\mathcal{E}| + |\mathcal{N}| \log_2 |\mathcal{N}|)$ [23, 35].

The local-level path planning algorithm generates an ordered sequence of waypoints $\mathcal{W}_c = \mathbf{w}_{c1}, \mathbf{w}_{c2}, \cdots, \mathbf{w}_{ci}$. Then, these waypoints are transformed from the relative reference frame to the global coordinate frame and added to the original waypoints path W. When the ownship is avoiding a potential collision, the avoidance waypoints overwrite some or all of the original waypoints. Next, a path manager is required to follow the waypoints path and a smoother to make the generated path flyable by the ownship. One possible approach to follow waypoints path is to transit when the ownship enters a ball around the waypoint W_i or a better strategy is to use the half-plane switching criteria that is not sensitive to tracking error [36]. Flyable or smoothed transition between the waypoints can be achieved by implementing the fillet maneuver or using Dubins paths. For further analysis on these topics, we refer the interested reader to Ref. [36].

3. SIMULATION RESULTS

To demonstrate the performance of the proposed path planning algorithm, we simulate an encounter scenario similar to the planner geometry shown in Figure 8. The aircraft dynamics are simulated using a simplified model that captures the flight characteristics of an autopilot-controlled UAS. The kinematic guidance model that we considered assumes that the autopilot controls airspeed, v_a, altitude, hh, and heading angle, ψ. Under zero-wind conditions, the corresponding equations of motion are given by

$$\dot{p}_n = v_a \cos\psi, \tag{22}$$

$$\dot{p}_e = v_a \sin\psi, \tag{23}$$

$$\dot{\psi} = \frac{g}{v_a}\phi, \tag{24}$$

$$\dot{v}_a = b_v(v_a^c - v_a) \tag{25}$$

$$\dot{\phi} = b_\phi(\phi^c - \phi) \tag{26}$$

$$\ddot{h} = b_{\dot{h}}\left(\dot{h}^c - \dot{h}\right) + b_h(h^c - h), \tag{27}$$

where p_n, p_e are the north-east position of the aircraft. The inputs are the commanded altitude, h^c, the commanded airspeed, v^c_a, and the commanded roll angle, φ^c. The parameters b_v, b_φ, b_h, and $b_{\dot{h}}$ are positive constants that depend on the implementation of the autopilot and the state estimation scheme. For further analysis on the kinematic and dynamic guidance models for UAS, we refer the interested reader to [36]. In the following simulation, the ownship starts at $(0,0,-200)^T$ in the NED coordinate system, with an initial heading of 0 deg. measured from north and follows a straight-line path at a constant speed of 22 m/s to reach the next waypoint located at $(1500,0,-200)^T$. The encounter geometry includes three intruders flying at different altitudes: the first is approaching head-on, the second is converging from the right, and the third is overtaking from the left. We chose the intruders's speed similar to the known cruise speed of ScanEagle UAS, Cessna SkyHawk 172R, and Raven RQ-11B UAS. The speed of the intruders is 41, 65, and 22 m/s, respectively. In addition, the intruders are assumed to fly at a constant speed the entire simulation period. As shown in Figure 8, the initial locations of intruders in the NED coordinate system are $(-25,1000,-225)^T$, $(500,1000,-180)^T$, and $(25,-500,-200)^T$, respectively, with initial heading of 180, -90, and 0°, respectively.

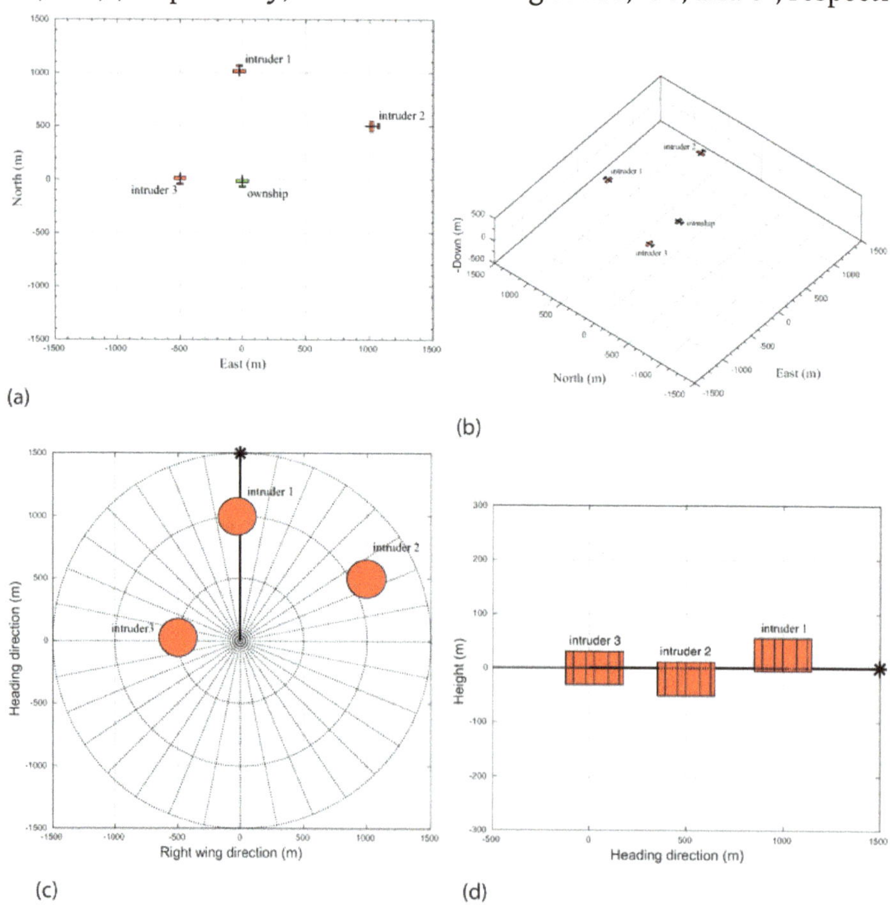

Figure 8. Encounter geometry for the ownship and three intruders at t = *0.1 s. **(a) Overhead view of initial locations of aircraft; (b) 3D view of initial locations of aircraft; (c) overhead view of reference frame; (d) side view of relative reference frame.***

In the following simulation, our choice of the collision volume is a cylinder of radius d_s = 152.4 m (500 ft) and height h_s = 61 m (200 ft) centered on each of the intruders. A collision incident occurs when the horizontal relative range and altitude to the ownship are simultaneously below horizontal and vertical minimum safe distances d_s and $h_{s/2}$. We ssume that there exists a sensor and tracking system that provides the states of the detected intruders. However, not every aircraft that is observed by the sensing system presents a collision threat. Therefore, we implemented a geometric-based collision detection algorithm to determine whether an approaching intruder aircraft is on a collision course. The collision detection approach is beyond the scope of this work, and we refer the interested reader to [37].

At the beginning of simulation, the predicted relative range and altitude at the closest point of approach (CPA) are shown in Table 1. Imminent collisions are expected to occur with the first and second intruders as their relative range and altitude with respect to the ownship are below the defined horizontal and vertical safe distances. The time remaining to the closest point of approach t_{CPA} with respect to the first and second intruders is 15.77 and 16.56 s, respectively. The scenario requires that the ownship plans and executes an avoidance maneuver well before the t_{CPA}. This example demonstrates the need for an efficient and computationally fast avoidance planning algorithm. Table 2 shows the total time required to run the avoidance algorithm, and the maximum and average time required to execute one cycle. The results show that the proposed algorithm takes a significantly reduced time in computation with an average and maximum time to execute one cycle of the code of 20 ms and 0.1326 s, respectively, and a total time of 0.3703 s to resolve the collision conflict.

Table 1. Relative range and altitude, and the time remaining to the closest point of approach.

Intruder	$\|p_r(t_{CPA})\|$ (m)	$\|p_{rz}(t_{CPA})\|$(m)	t_{CPA} (s)
1	24.90	25	15.77
2	141.33	20	16.56
3	437.14	4361.07	1982.25

Table 2. Collision avoidance algorithm run time.

Total run time (s)	Max. run time (one cycle) (s)	Average run time (one cycle) (s)
0.3703	0.1326	0.0206

Figure 9 shows the planned avoidance path by the ownship. These results show that the avoidance path safely maneuvers the ownship without any collisions with the intruders. In addition, the ownship should plan an avoidance maneuver that does not lead to a collision with intruders that were not on a collision course initially such as the case with the third intruder. Initially, the third intruder and the ownship are flying on near parallel courses. The relative range and altitude at CPA with respect to the third intruder are 437.14 and 4361.07 m, respectively, and the time remaining to the CPA is 1982.25 s. Obviously, both aircrafts are not on a collision course. However, the third intruder is descending and changing its heading toward the ownship. The path planner, however, accounts for predicted locations of the detected intruder over the look-ahead time window, allowing the ownship to maintain a safe distance from the third intruder. This example demonstrates that the proposed path planner can handle unanticipated maneuvering intruders. Once collisions are resolved the path planner returns the ownship to the next waypoint of its initial path.

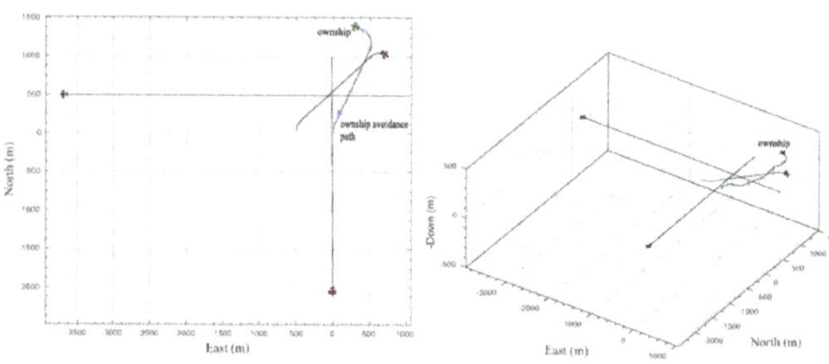

(a) (b)

Figure 9. Avoidance path followed by the ownship and path tracks of the intruders at t = 75 s. (a) Overhead view of avoidance path and (b) 3D view of of avoidance path.

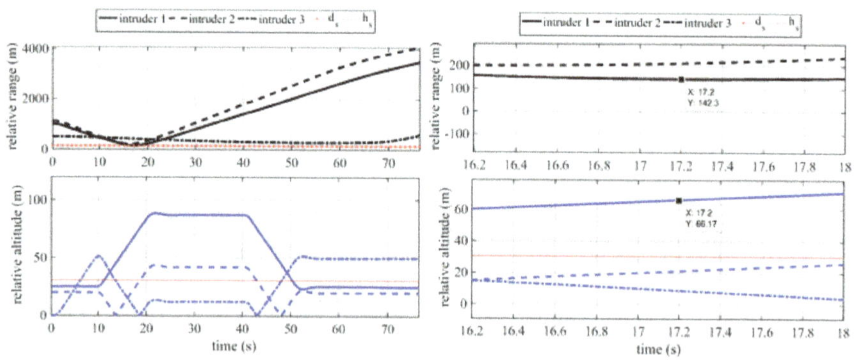

(a) (b)

Figure 10. Relative horizontal range and altitude between the ownship and intruders. (a) Horizontal range and relative altitude to intruders and (b) a close up view of Figure 10a.

The relative range between the ownship and the intruders is shown in Figure 10. The results show that no collisions have occurred, and that the ownship successfully planned an avoidance maneuver. The avoidance planner ensures that when the relative horizontal range is less than d_s, the relative altitude is greater than $h_{s/}$. For example, as shown in Figure 10b, the relative range to the first intruder over time interval [16.2, 18] s is below d_s. However, over the same time interval, the relative altitude is above $h_{s/2}$.

Another important aspect to evaluate the performance of the proposed algorithm is its ability to reduce the length of the avoidance path while avoiding the intruders. This is important because it reduces the amount of deviation from the original path and ultimately the flight time, which is of critical importance for the small UAS with limited power resources. Table 3 shows that the length of the avoidance paths is fairly acceptable compared to the initial path length.

Table 3. Length of the avoidance path.

Scenario number	Initial path length (m)	Avoidance path length (m)
1	1500	1955

4. CONCISE SUMMARY

This chapter has introduced a path planning approach that is well-suited for small unmanned aerial systems (UAS). We have created a collision avoidance algorithm utilizing a coordinate system centered around the ownship. The method constructs a graph of maneuvers in the local-level coordinate system and using Dijkstra's algorithm to identify the path with the lowest cost.

An important characteristic of the suggested method is that the future movement of the ownship is limited to following nodes on the map that are evenly spaced in time. Given that the path is defined by waypoints at specific time intervals, it is rather simple to estimate the approximate location of the ownship at any given time. The time information is utilized to allocate costs to edges in order to optimize path planning and avoid collisions.

One benefit of this method is that collision avoidance is inherently a localized occurrence and can be more accurately described using local coordinates rather than global coordinates. Furthermore, the program takes into consideration the presence of several intruders and unexpected changes in movement in different encounter situations. The algorithm suggested operates with almost no delay in real time using Matlab. Given the brief duration observed in the simulation findings, we anticipate that implementing these algorithms in a compiled language, such as C or C++, will demonstrate the practicality of achieving real-time execution using hardware. This renders the suggested method a feasible solution, especially for small unmanned aerial systems (UAS).

To progress towards a deployable Unmanned Aerial System (UAS), it is crucial to conduct testing and evaluation of the closed-loop system consisting of the sensor, tracker, collision detection, path planning, and collision avoidance components. In practice, the deployment of any Unmanned Aircraft System (UAS) necessitates a thorough and extensive development process, followed by a meticulous certification process. This is then followed by further study, which involves utilizing more accurate models of encounter airspace, doing a sufficient number of simulations, and employing hardware-in-the-loop simulation. Contrary to current collision detection and avoidance systems used in manned aviation, it is not possible to create an encounter model based purely on observed data for unmanned aircraft systems (UASs). This is because UASs are not yet fully integrated into the airspace system and reliable data is not available. A compelling research topic would include creating encounter models that resemble the ones used for assessing and certifying manned aircraft traffic alert and collision avoidance systems (TCAS).

REFERENCES

1. George S. FAA Workshop on Sense and Avoid (SAA) for Unmanned Aircraft Systems (UAS). 2009
2. Hottman SB, Hansen KR, Berry M. Literature review on detect, sense, and avoid technology for Unmanned Aircraft Systems. In: Technical Report. 2009
3. Federal Aviation Administration. Subchapter F-Air Traffic and General Operating Rules. 2015
4. Kuchar JK, Yang LC. A review of conflict detection and resolution modeling methods. IEEE Transactions on Intelligent Transportation Systems. Dec. 2000;1(4):179-189
5. Albaker BM, Rahim NA. A survey of collision avoidance approaches for unmanned aerial vehicles. International Conference for Technical Postgraduates (TECHPOS). 2009:1-7
6. Hyunjin YK. Reactive collision avoidance of unmanned aerial vehicles using a single vision sensor. AIAA Guidance, Control, and Dynamics. 2013;36(4):1234-1240
7. Rajnikant S, Saunders JB, Randal Beard W. Reactive path planning for micro air vehicles using bearing-only measurements. International Robotic Systems. 2012;65(1–4):409-416
8. White BA, Antonios HS. UAV obstacle avoidance using differential geometry concepts. In: 18th IFAC World Congress; Milano, Italy; 2011. Vol. 3. pp. 6325-6330
9. Saunders J, Beard RW. Vision-based reactive multiple obstacle avoidance for micro air vehicles. In: IEEE American Control Conference ACC'09; St. Louis, MO, June 10–12; 2009, pp. 5253-5258
10. George J, Ghose D. A reactive inverse PN algorithm for collision avoidance among multiple unmanned aerial vehicles. In: American Control Conference; St. Louis, MO; June 10–12; IEEE. pp. 3890-3895

11. Bilimoria KD. A geometric optimization approach to aircraft conflict resolution. In: Proceedings of the AIAA Guidance, Navigation and Control Conference and Exhibit; 2010

12. Fiorini P, Shiller Z. Motion planning in dynamic environments using velocity obstacles. The International Journal of Robotics Research. 1998;17(7):760-772

13. Chakravarthy A, Ghose D. Obstical avoidance in a dynamic environment: A collision cone approach. IEEE Transactions on System, Man and Cybernitics, Part A: Systems and Humans. 1998;28(5):562-572

14. Lam TM, Mulder M, Van Paassen M, Mulder JA, Van Der FC. Force-stiffness feedback in uninhabited aerial vehicle teleoperation with time delay. AIAA Guidance, Control, and Dynamics. 2009;32(3):821-835

15. Sahawneh LR, Beard RW, Avadhanamz S, He B. Chain-based collision avoidance for UAS sense and avoid systems. In: AIAA Guidance, Navigation, and Control (GNC) Conference; Boston, MA; 2013

16. Kavraki LE, Svestka P, Latombe JC, Overmars MH. Probabilistic roadmaps for path planning in high-dimensional configuration spaces. IEEE Transactions on Robotics and Automation. 1996;12(4):566-580

17. LaValle SM. Rapidly-exploring random trees: A new tool for path planning. Technical Report TR 98–11. Computer Science Department, Iowa State University; October 1998

18. Dijkstra EW. A note on two problems in connection with graphs. Numerische Mathematik. 1959;1:269-271

19. Dechter R, Pearl J. Generalized best-first search strategies and the optimality of a*. Journal of the ACM (JACM). 1985;32(3):505-536

20. Mirolo C, Pagello E. A cell decomposition approach to motion planning based on collision detection. In: Proceedings of the 1995 International Conference on Advanced Robotics. 1995. pp. 481-488

21. Angelov P. Sense and Avoid in UAS: Research and Applications. Chichester, West Sussex, United Kingdom: John Wiley & Sons, Ltd; 2012

22. Koren Y, Borenstein J. Potential field methods and their inherent limitations for mobile robot navigation. In IEEE International Conference On Robotics And Automation; IEEE. 1991;2:1398-1404

23. LaValle SM. Planning Algorithms. Cambridge University Press; 2006

24. Bortoff SA. Path planning for UAVs. In: Proceedings of the American Control Conference. Chicago, Illinois; June 2000. pp. 364-368

25. McLain TW, Beard RW. Trajectory planning for coordinated rendezvous of unmanned air vehicles. In: Proceedings of the AIAA Guidance, Navigation, and Control Conference. AIAA Reston, VA; 2000. Vol. 4369. pp. 1-8
26. Argyle ME, Chamberlain C, Beard RW. Chain-based path planning for multiple UAVs. In: 50th IEEE Conference on Decision and Control and European Control Conference, Orlando, FL, USA. Dec. 2011
27. Luders BD, Karaman S, How JP. Robust sampling-based motion planning with asymptotic optimality guarantees. In: AIAA Guidance, Navigation, and Control Conference (GNC), Boston, MA. 2013
28. Luders BD, Karaman S, Frazzoli E, How JP. Bounds on tracking error using closed-loop rapidly-exploring random trees. In: American Control Conference (ACC). IEEE; 2010. pp. 5406-5412
29. Luders B, Karaman S, How JP. Robust sampling-based motion planning with asymptotic optimality guarantees. In Guidance, Navigation, and Control (GNC) Conference, Boston, MA. 2013. AIAA
30. Kothari M, Postlethwaite I. A probabilistically robust path planning algorithm for UAVs using rapidly-exploring random trees. International Robotic Systems. 2013;71:231-253
31. Standard Specification for Design and Performance of an Airborne Sense-and-Avoid System. Tech. Rep. TR F2411-07. West Conshohocken, PA: ASTM International; 2007
32. US Department of Transportation and Federal Aviation Adminstration. Aeronautical Information Manual Official Guide to Basic Flight Information and ATC Procedures
33. Lee SM, Park C, Johnson MA, Mueller ER. Investigating effects of well clear definitions on UAS sense-and-avoid operations. In: Aviation Technology, Integration, and Operations Conference, Los Angeles, CA. AIAA. 2013
34. Consiglio M, Chamberlain J, Munoz C, and Hoffler K. Concept of integration for UAS operations in the NAS. In: 28th International Congress of the Aeronautical Sciences (ICAS); Brisbane, Australia; 2012
35. Sahawneh LR, Airborne Collision Detection and Avoidance for Small UAS Sense and Avoid Systems [PhD Thesis] Brigham Young University; 2016
36. Beard RW, McLain TW. Small Unmanned Aircraft: Theory and Practice. New Jersey, USA: Princeton University Press; 2012
37. Sahawneh LR, Argyle ME, Beard RW. 3D path planning for small UAS operating in low-altitude airspace. In International Conference on Unmanned Aircraft Systems (ICUAS). IEEE, 2016. pp. 413-419

4

Chapter

FORWARD AND INVERSE KINEMATICS FOR THE DOBOT ROBOTIC ARM USING THE PSEUDOINVERSE AND TRANSPOSITION METHOD

1. INTRODUCTION: AN OVERVIEW

Robots and manipulators are crucial and potent tools in modern manufacturing. They are performing several diverse duties and operations that do not necessitate comfort, time for relaxation, or payment. However, it requires a significant amount of time and skilled staff to ensure proper functioning of the robot [6].

The motion of a robot can be categorized into two types: forward kinematics and inverse kinematics. Forward kinematics explains the movement of robots based on the angles inputted. There is always a resolution for the forward kinematics of a manipulator. The task of solving inverse kinematics is more challenging than that of forward kinematics. Figure 1 illustrates the connection between forward kinematics and inverse kinematics. Inverse kinematics involves solving the problem in the opposite direction compared to forward kinematics. However, we are always able to develop a solution for the inverse kinematics of a manipulator. There are only a limited number of groups of manipulators, specifically those with Euler wrists, that have a straightforward solution for inverse kinematics [8, 9].

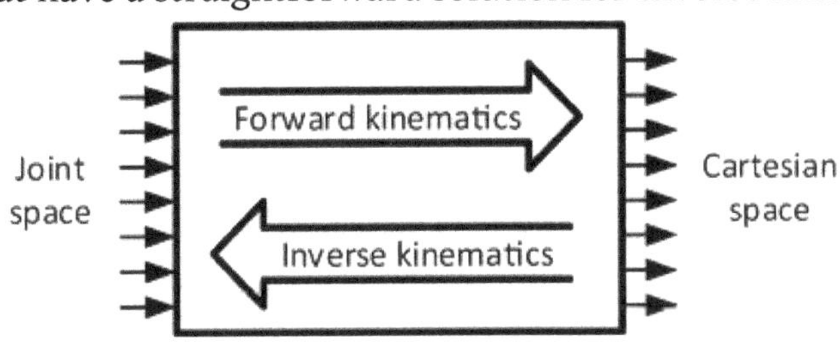

Figure 1. The schematic representation of forward and inverse kinematics.

The two primary approaches for addressing inverse kinematics are analytical and numerical methods. In the first method, the joint variables are solved analytically using the conventional sinus and cosine representation. The second strategy utilizes numerical techniques to explain the joint variables [9].

Chapter will focus only on the DOBOT Magician robot arm, as depicted in Figure 2. Figure 3 displays the fundamental characteristics of the robotic manipulator, while Table 1 presents its motion parameters.

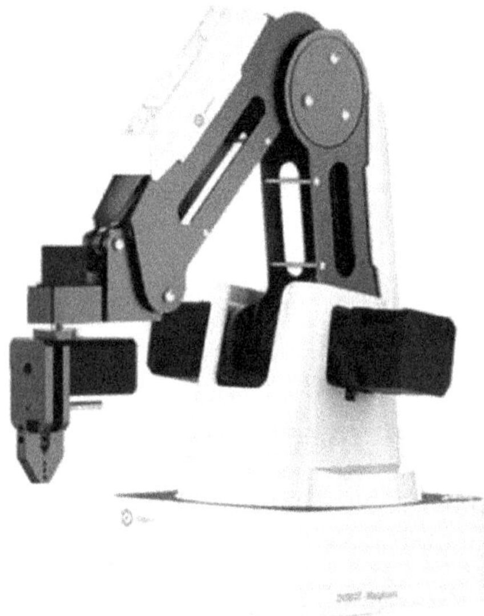

Figure 2. DOBOT Magician [10].

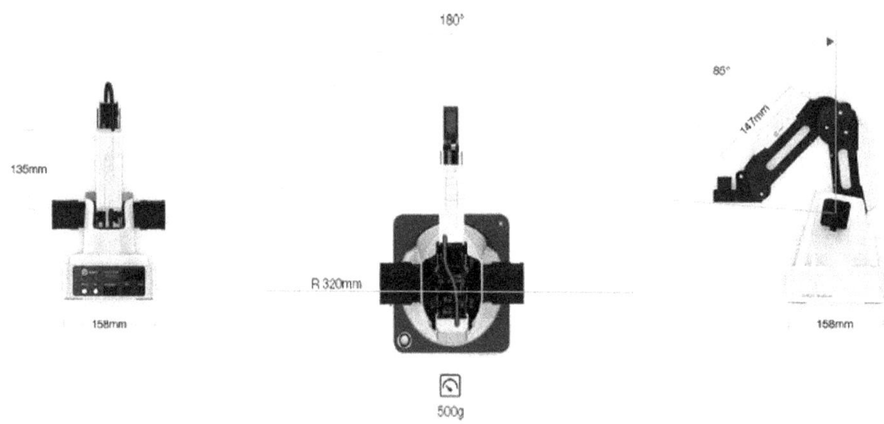

Figure 3. Simple specification of DOBOT [10].

This chapter is structured in the subsequent fashion. In the initial portion, we performed the forward and inverse kinematics conversions for the DOBOT manipulator. Furthermore, we developed the DOBOT Magician simulation within the Matlab framework. Furthermore, we provide a detailed account of the elucidation of Denavit-Hartenberg parameters. Ultimately, we implemented the pseudoinverse and transposition techniques of the Jacobian matrix in the inverse kinematics.

Table 1. Axis movement of DOBOT Magician [10].

Axis	Range	Max speed (250 g workload)
Joint 1 base	-90° to +90°	320°/s
Joint 2 rear arm	0° to +85°	320°/s
Joint 3 fore arm	-10° to +95°	320°/s
Joint 4 rotation servo	+90° to -90°	480°/s

2. KINEMATICS STRUCTURE RRR IN 3D

Kinematic structure of the DOBOT manipulator is shown in Figure 4. It is created from three rotation joints and three links. Joint A rotates about the axis z and joints B and C rotate about the axis x_1.

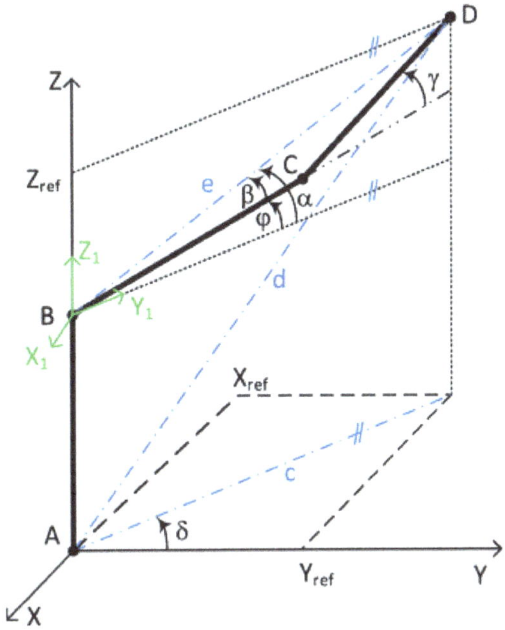

Figure 4. Representation of DOBOT manipulator in 3D view.

Figure 5 shows a view from the direction of axis z and Figure 6 shows a perpendicular view of the plane defined by z axis and line c.

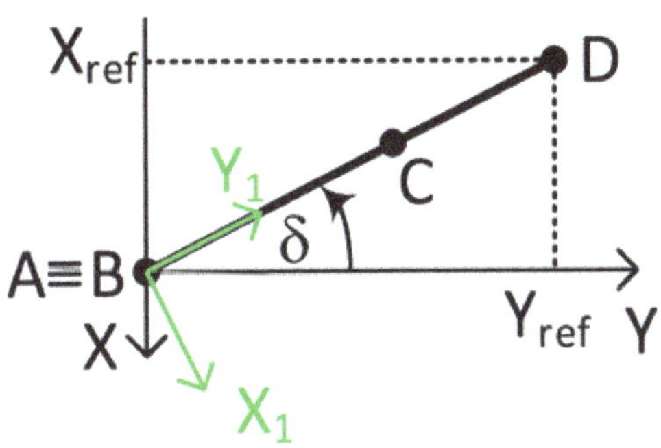

Figure 5. Representation of DOBOT footprint.

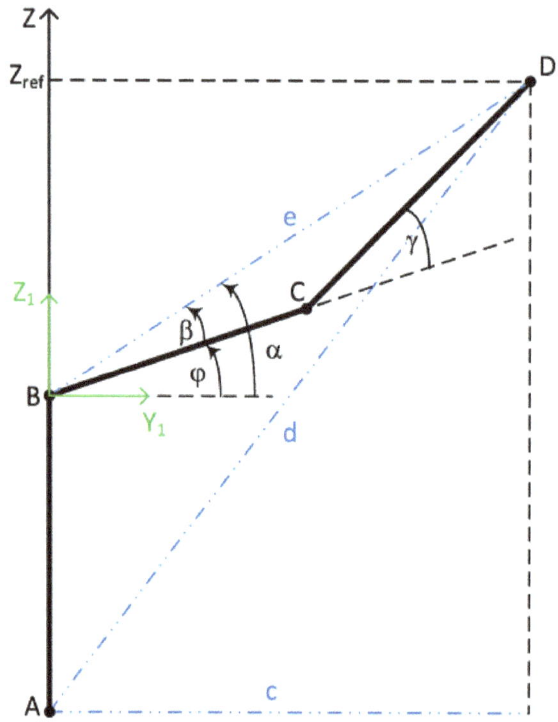

Figure 6. View of the plane defined by z axis and line c.

Kinematic equations of the points B, C, and D, respectively:

$$x_0^B = 0 \tag{1}$$

$$y_0^B = 0 \tag{2}$$

$$z_0^B = l_1 \tag{3}$$

$$x_0^C = -l_2 . \cos\varphi_1 . \sin\delta \tag{4}$$

$$y_0^C = l_2 . \cos\varphi_1 . \cos\delta \tag{5}$$

$$z_0^C = l_1 + l_2 . \sin\varphi_1 \tag{6}$$

$$x_0^D = -(l_2 . \cos\varphi_1 + l_3 . \cos\varphi_2) . \sin\delta \tag{7}$$

$$y_0^D = (l_2 . \cos\varphi_1 + l_3 . \cos\varphi_2) . \cos\delta \tag{8}$$

$$z_0^D = l_1 + l_2 . \sin\varphi_1 + l_3 . \sin\varphi_2 \tag{9}$$

Where $\phi_2 = \phi + \gamma$.

2.1. Forward kinematics

Forward kinematics task is defined by Eq. (10)

$$X = f(Q) \tag{10}$$

where

$$X = \begin{bmatrix} x_0^D \\ y_0^D \\ z_0^D \end{bmatrix} \tag{11}$$

X is position vector of manipulator endpoint coordinates.

$$Q = \begin{bmatrix} \varphi \\ \gamma \\ \delta \end{bmatrix} \tag{12}$$

Q is vector of independent coordinates: $\phi = \phi_1, \gamma, \delta$.

Because the function X = *f*(Q) is nonlinear, it is difficult to solve the inverse task Q = *f*(X) when looking for a vector of independent coordinates (rotation of individual manipulator joints) as a function of the desired manipulator endpoint coordinates. An analytical solution to the inverse task is possible only in the case of a relatively simple kinematic structure of the manipulator (see next chapter).

Therefore, the function X = *f*(Q) linearized using Taylor series, taking into account only the first four (linear) members of the development:

$$x_0^D = x_0^D(\varphi,\gamma,\delta) \approx x_0^D(\varphi_0,\gamma_0,\delta_0) + \left.\frac{\partial x_0^D(\varphi,\gamma,\delta)}{\partial\varphi}\right|_{\varphi_0,\gamma_0,\delta_0}(\varphi-\varphi_0) + \left.\frac{\partial x_0^D(\varphi,\gamma,\delta)}{\partial\gamma}\right|_{\varphi_0,\gamma_0,\delta_0}(\gamma-\gamma_0) + \left.\frac{\partial x_0^D(\varphi,\gamma,\delta)}{\partial\delta}\right|_{\varphi_0,\gamma_0,\delta_0}(\delta-\delta_0) \tag{13}$$

$$y_0^D = y_0^D(\varphi,\gamma,\delta) \approx y_0^D(\varphi_0,\gamma_0,\delta_0) + \left.\frac{\partial y_0^D(\varphi,\gamma,\delta)}{\partial\varphi}\right|_{\varphi_0,\gamma_0,\delta_0}(\varphi-\varphi_0) + \left.\frac{\partial y_0^D(\varphi,\gamma,\delta)}{\partial\gamma}\right|_{\varphi_0,\gamma_0,\delta_0}(\gamma-\gamma_0) + \left.\frac{\partial y_0^D(\varphi,\gamma,\delta)}{\partial\delta}\right|_{\varphi_0,\gamma_0,\delta_0}(\delta-\delta_0) \tag{14}$$

$$z_0^D = z_0^D(\varphi,\gamma,\delta) \approx z_0^D(\varphi_0,\gamma_0,\delta_0) + \left.\frac{\partial z_0^D(\varphi,\gamma,\delta)}{\partial\varphi}\right|_{\varphi_0,\gamma_0,\delta_0}(\varphi-\varphi_0) + \left.\frac{\partial z_0^D(\varphi,\gamma,\delta)}{\partial\gamma}\right|_{\varphi_0,\gamma_0,\delta_0}(\gamma-\gamma_0) + \left.\frac{\partial z_0^D(\varphi,\gamma,\delta)}{\partial\delta}\right|_{\varphi_0,\gamma_0,\delta_0}(\delta-\delta_0)$$

After editing:

$$x_0^D(\varphi,\gamma,\delta) - x_0^D(\varphi_0,\gamma_0,\delta_0) = \left.\frac{\partial x_0^D(\varphi,\gamma,\delta)}{\partial\varphi}\right|_{\varphi_0,\gamma_0,\delta_0}(\varphi-\varphi_0) + \left.\frac{\partial x_0^D(\varphi,\gamma,\delta)}{\partial\gamma}\right|_{\varphi_0,\gamma_0,\delta_0}(\gamma-\gamma_0) + \left.\frac{\partial x_0^D(\varphi,\gamma,\delta)}{\partial\delta}\right|_{\varphi_0,\gamma_0,\delta_0}(\delta-\delta_0) \tag{16}$$

$$y_0^D(\varphi,\gamma,\delta) - y_0^D(\varphi_0,\gamma_0,\delta_0) = \left.\frac{\partial y_0^D(\varphi,\gamma,\delta)}{\partial\varphi}\right|_{\varphi_0,\gamma_0,\delta_0}(\varphi-\varphi_0) + \left.\frac{\partial y_0^D(\varphi,\gamma,\delta)}{\partial\gamma}\right|_{\varphi_0,\gamma_0,\delta_0}(\gamma-\gamma_0) + \left.\frac{\partial y_0^D(\varphi,\gamma,\delta)}{\partial\delta}\right|_{\varphi_0,\gamma_0,\delta_0}(\delta-\delta_0) \tag{17}$$

$$z_0^D(\varphi,\gamma,\delta) - z_0^D(\varphi_0,\gamma_0,\delta_0) = \left.\frac{\partial z_0^D(\varphi,\gamma,\delta)}{\partial\varphi}\right|_{\varphi_0,\gamma_0,\delta_0}(\varphi-\varphi_0) + \left.\frac{\partial z_0^D(\varphi,\gamma,\delta)}{\partial\gamma}\right|_{\varphi_0,\gamma_0,\delta_0}(\gamma-\gamma_0) + \left.\frac{\partial z_0^D(\varphi,\gamma,\delta)}{\partial\delta}\right|_{\varphi_0,\gamma_0,\delta_0}(\delta-\delta_0) \tag{18}$$

We denoted:

$$\Delta x_0^D = x_0^D(\varphi,\gamma,\delta) - x_0^D(\varphi_0,\gamma_0,\delta_0) \tag{19}$$

$$\Delta y_0^D = y_0^D(\varphi,\gamma,\delta) - y_0^D(\varphi_0,\gamma_0,\delta_0) \tag{20}$$

$$\Delta z_0^D = z_0^D(\varphi,\gamma,\delta) - z_0^D(\varphi_0,\gamma_0,\delta_0) \tag{21}$$

$$\Delta\varphi = \varphi - \varphi_0 \tag{22}$$

$$\Delta\gamma = \gamma - \gamma_0 \tag{23}$$

$$\Delta\delta = \delta - \delta_0 \tag{24}$$

Then, we obtained:

$$\Delta x_0^D = \left.\frac{\partial x_0^D(\varphi,\gamma,\delta)}{\partial\varphi}\right|_{\varphi_0,\gamma_0,\delta_0}\Delta\varphi + \left.\frac{\partial x_0^D(\varphi,\gamma,\delta)}{\partial\gamma}\right|_{\varphi_0,\gamma_0,\delta_0}\Delta\gamma + \left.\frac{\partial x_0^D(\varphi,\gamma,\delta)}{\partial\delta}\right|_{\varphi_0,\gamma_0,\delta_0}\Delta\delta \tag{25}$$

$$\Delta y_0^D = \left.\frac{\partial y_0^D(\varphi,\gamma,\delta)}{\partial\varphi}\right|_{\varphi_0,\gamma_0,\delta_0}\Delta\varphi + \left.\frac{\partial y_0^D(\varphi,\gamma,\delta)}{\partial\gamma}\right|_{\varphi_0,\gamma_0,\delta_0}\Delta\gamma + \left.\frac{\partial y_0^D(\varphi,\gamma,\delta)}{\partial\delta}\right|_{\varphi_0,\gamma_0,\delta_0}\Delta\delta \tag{26}$$

$$\Delta z_0^D = \left.\frac{\partial z_0^D(\varphi,\gamma,\delta)}{\partial\varphi}\right|_{\varphi_0,\gamma_0,\delta_0}\Delta\varphi + \left.\frac{\partial z_0^D(\varphi,\gamma,\delta)}{\partial\gamma}\right|_{\varphi_0,\gamma_0,\delta_0}\Delta\gamma + \left.\frac{\partial z_0^D(\varphi,\gamma,\delta)}{\partial\delta}\right|_{\varphi_0,\gamma_0,\delta_0}\Delta\delta \tag{27}$$

In matrix form:

$$\begin{bmatrix} \Delta x_0^D \\ \Delta y_0^D \\ \Delta z_0^D \end{bmatrix} = \begin{bmatrix} \frac{\partial x_0^D(\varphi,\gamma,\delta)}{\partial \varphi} & \frac{\partial x_0^D(\varphi,\gamma,\delta)}{\partial \gamma} & \frac{\partial x_0^D(\varphi,\gamma,\delta)}{\partial \delta} \\ \frac{\partial y_0^D(\varphi,\gamma,\delta)}{\partial \varphi} & \frac{\partial y_0^D(\varphi,\gamma,\delta)}{\partial \gamma} & \frac{\partial y_0^D(\varphi,\gamma,\delta)}{\partial \delta} \\ \frac{\partial z_0^D(\varphi,\gamma,\delta)}{\partial \varphi} & \frac{\partial z_0^D(\varphi,\gamma,\delta)}{\partial \gamma} & \frac{\partial z_0^D(\varphi,\gamma,\delta)}{\partial \delta} \end{bmatrix}_{\varphi_0,\gamma_0,\delta_0} \cdot \begin{bmatrix} \Delta\varphi \\ \Delta\gamma \\ \Delta\delta \end{bmatrix} \tag{28}$$

Where matrix:

$$J = \begin{bmatrix} \frac{\partial x_0^D(\varphi,\gamma,\delta)}{\partial \varphi} & \frac{\partial x_0^D(\varphi,\gamma,\delta)}{\partial \gamma} & \frac{\partial x_0^D(\varphi,\gamma,\delta)}{\partial \delta} \\ \frac{\partial y_0^D(\varphi,\gamma,\delta)}{\partial \varphi} & \frac{\partial y_0^D(\varphi,\gamma,\delta)}{\partial \gamma} & \frac{\partial y_0^D(\varphi,\gamma,\delta)}{\partial \delta} \\ \frac{\partial z_0^D(\varphi,\gamma,\delta)}{\partial \varphi} & \frac{\partial z_0^D(\varphi,\gamma,\delta)}{\partial \gamma} & \frac{\partial z_0^D(\varphi,\gamma,\delta)}{\partial \delta} \end{bmatrix}_{\varphi_0,\gamma_0,\delta_0} \tag{29}$$

is Jacobian matrix. We denoted:

$$\Delta X_0^D = \begin{bmatrix} \Delta x_0^D \\ \Delta y_0^D \\ \Delta z_0^D \end{bmatrix} \tag{30}$$

and

$$\Delta Q = \begin{bmatrix} \Delta\varphi \\ \Delta\gamma \\ \Delta\delta \end{bmatrix} \tag{31}$$

Then, we obtained the matrix equation, which represents linearized forward kinematics in incremental form:

$$\Delta X_0^D = J.\Delta Q \tag{32}$$

After we multiplied the Eq. (32) with inverse matrix J^{-1} from the left, we obtained the equation of inverse kinematics.

$$J^{-1}.\Delta X_0^D = J^{-1}.J.\Delta Q \tag{33}$$

$$J^{-1}.\Delta X_0^D = I.\Delta Q \tag{34}$$

Where I is the identity matrix. After that:

$$\Delta Q = J^{-1}.\Delta X_0^D \tag{35}$$

Derivative of the kinematic equations with respect to the independent coordinates for kinematic structure of DOBOT manipulator:

$$\frac{\partial x_0^D}{\partial \varphi} = [l_2 . \sin\varphi + l_3 . \sin(\varphi + \gamma)] . \sin\delta \tag{36}$$

$$\frac{\partial x_0^D}{\partial \gamma} = l_3 . \sin(\varphi + \gamma) . \sin\delta \tag{37}$$

$$\frac{\partial x_0^D}{\partial \delta} = -[l_2 . \cos\varphi + l_3 . \cos(\varphi + \gamma)] . \cos\delta \tag{38}$$

$$\frac{\partial y_0^D}{\partial \varphi} = -[l_2 . \sin\varphi + l_3 . \sin(\varphi + \gamma)] . \cos\delta \tag{39}$$

$$\frac{\partial y_0^D}{\partial \gamma} = -l_3 . \sin(\varphi + \gamma) . \cos\delta \tag{40}$$

$$\frac{\partial y_0^D}{\partial \delta} = -[l_2 . \cos\varphi + l_3 . \cos(\varphi + \gamma)] . \sin\delta \tag{41}$$

$$\frac{\partial z_0^D}{\partial \varphi} = l_2 . \cos\varphi + l_3 . \cos(\varphi + \gamma) \tag{42}$$

$$\frac{\partial z_0^D}{\partial \gamma} = l_3 . \cos(\varphi + \gamma) \tag{43}$$

$$\frac{\partial z_0^D}{\partial \delta} = 0 \tag{44}$$

Jacobian matrix:

$$J = \begin{bmatrix} \frac{\partial x_0^D}{\partial \varphi} & \frac{\partial x_0^D}{\partial \gamma} & \frac{\partial x_0^D}{\partial \delta} \\ \frac{\partial y_0^D}{\partial \varphi} & \frac{\partial y_0^D}{\partial \gamma} & \frac{\partial y_0^D}{\partial \delta} \\ \frac{\partial z_0^D}{\partial \varphi} & \frac{\partial z_0^D}{\partial \gamma} & \frac{\partial z_0^D}{\partial \delta} \end{bmatrix} \tag{45}$$

2.2. Analytical Solution of the Inverse Kinematics of DOBOT manipulator

The following equations are derived from Figure 4.

$$c^2 = x^2 + y^2 \tag{46}$$

$$d^2 = c^2 + z^2 = x^2 + y^2 + z^2 \tag{47}$$

$$e^2 = l_2^2 + l_3^2 - 2l_2l_3 \cos(\pi - \gamma) \tag{48}$$

$$\gamma = \pm\text{arccos}\left(\frac{e^2 - l_2^2 - l_3^2}{2l_2l_3}\right) \tag{49}$$

$$e^2 = c^2 + (z - l_1)^2 \tag{50}$$

$$\varphi = \alpha - \beta \tag{51}$$

$$\alpha = arctg\frac{z - l_1}{c} = arctg\frac{z - l_1}{\sqrt{x^2 + y^2}} \tag{52}$$

$$l_3^2 = l_2^2 + e^2 - 2l_2.e.\cos\beta \tag{53}$$

$$\beta = \pm\text{arccos}\left(\frac{l_2^2 + e^2 - l_3^2}{2l_2e}\right) \tag{54}$$

$$\varphi = arctg\frac{z - l_1}{\sqrt{x^2 + y^2}} \mp \text{arccos}\left(\frac{l_2^2 + e^2 - l_3^2}{2l_2e}\right) \tag{55}$$

$$\delta = arctg\frac{-x}{y} \tag{56}$$

3. SIMULATION DOBOT MAGICIAN IN MATLAB ENVIRONMENT

For simulation movement of the manipulator, Matlab Simulink environment and SimMechanics toolbox are suitable to use. Some blocks from SimMechanics toolbox are shown in Figure 7, which represents the model of DOBOT manipulator.

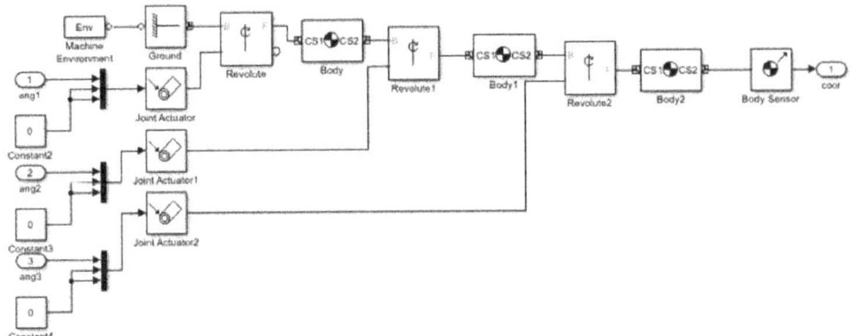

Figure 7. SimMechanics simulation model of DOBOT manipulator.

We used basic block from SimMechanics toolbox in simulation model:

- Joint actuator
- Revolute
- Body
- Body sensor
- Machine environment

The joint actuator block transfers the requested angles to the connected joint. The revolute block defined the rotation of body in space. The body block describes the parameters of body, like dimension, inertia, etc. The body sensor block transfers the coordinates, velocity, and others to simulation, and the last block is the machine environment which defines the parameters of the environment in which the manipulator is located. You can find all necessary data about these blocks in [7].

The simulation model, shown in Figure 8, was designed for considerate results from SimMechanics model, model used D-H parameters and analytical model, which was described in previous chapter. Every result from these models is shown in Figure 9. The fourth part in Figure 9 is the results from analytical simulation model of inverse kinematics. Figure 10 represented the reference angles in the first part of the chart, calculated angles from analytical inverse kinematics model in the second part of chart, and finally the error between both angles. As we can see in Figure 10, the angles are same. This is proof that analytical model of DOBOT manipulator is usable for simulation and implementation to some DSP or microcontroller.

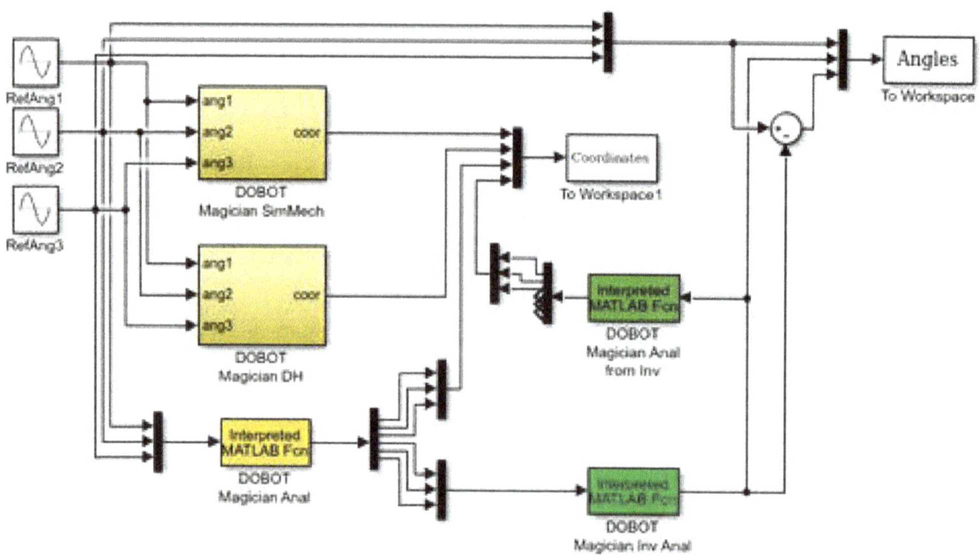

Figure 8. Simulation of DOBOT manipulator in Matlab environment.

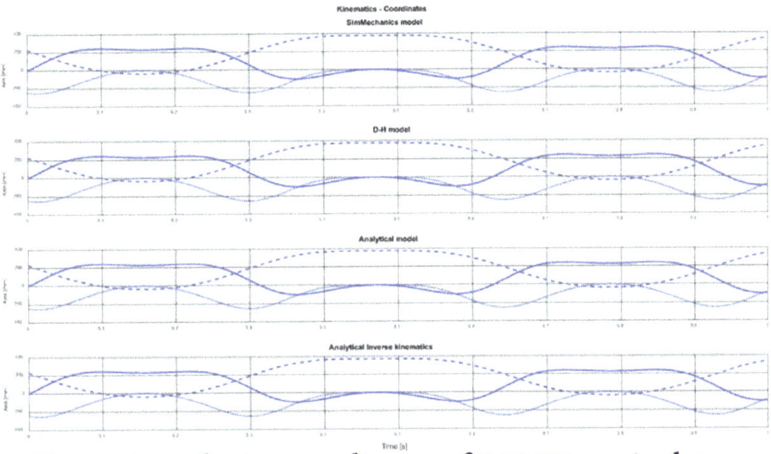

Figure 9. Endpoint coordinates of DOBOT manipulator.

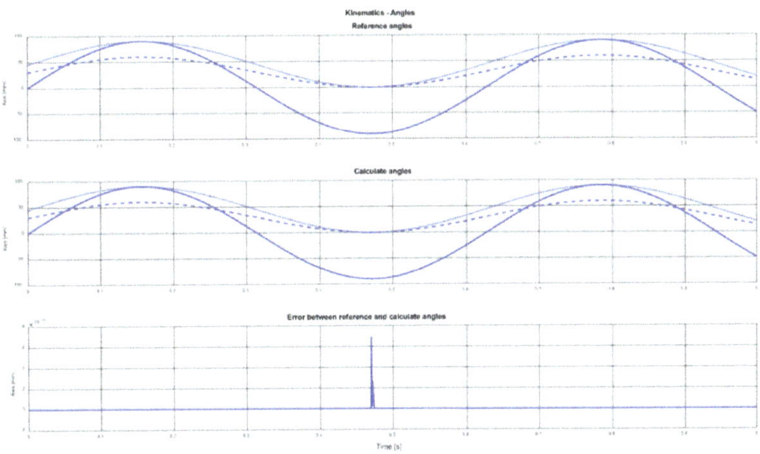

Figure 10. Reference angles, calculated angles, and error between these angles.

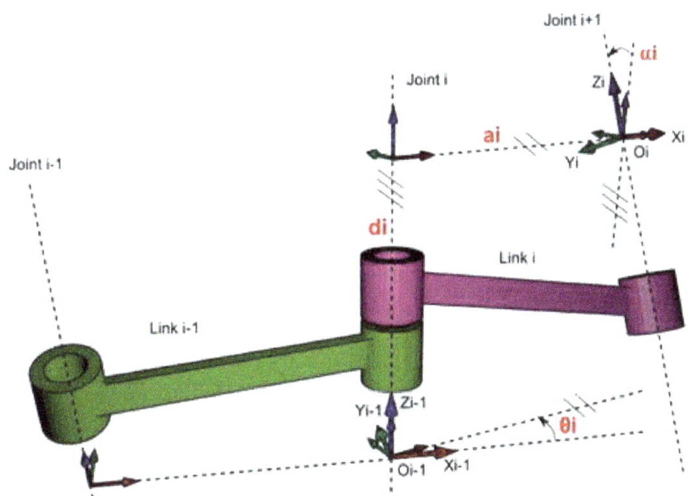

Figure 11. The four parameters of classic DH convention are θ_i, d_i, a_i, α_i *[4].*

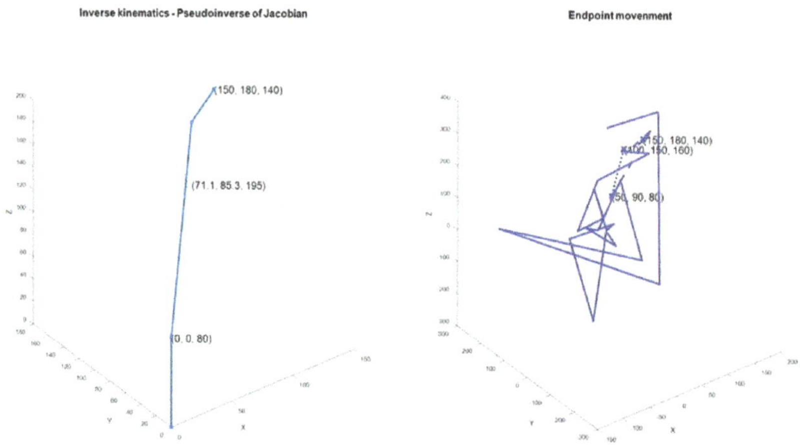

Figure 12. Simulation result of Jacobian matrix pseudoinverse in inverse kinematics model of the DOBOT manipulator.

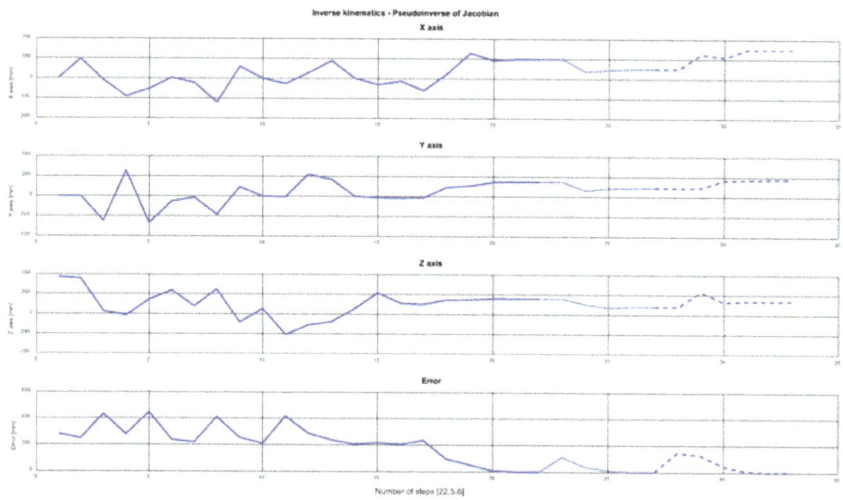

Figure 13. Simulation results of DOBOT axis and total error of coordinates for Pseudoinverse method.

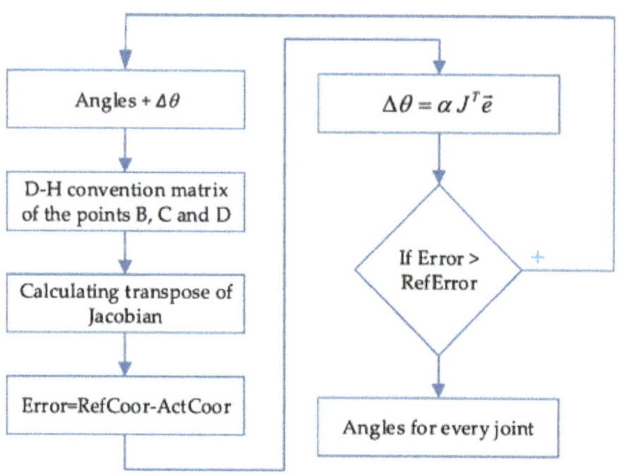

Figure 14. Block diagram of Jacobian matrix transpose method simulation.

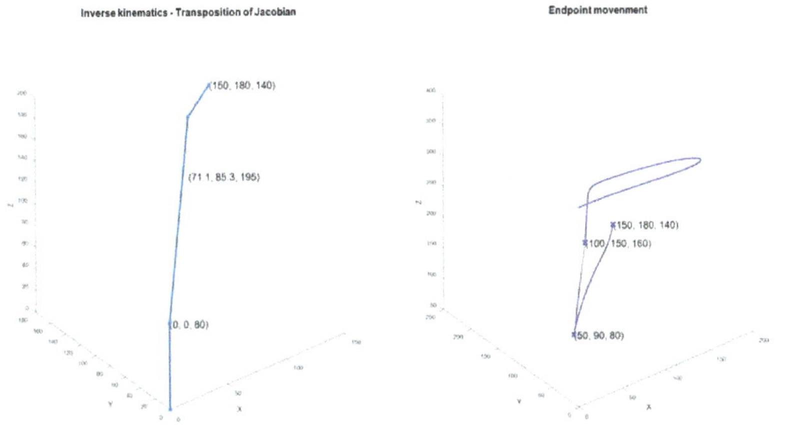

Figure 15. Simulation result of Jacobian matrix transposition in inverse kinematics model of the DOBOT manipulator.

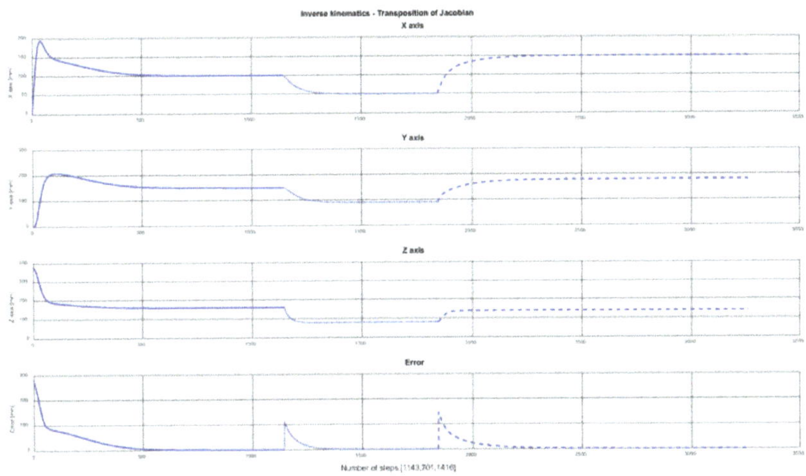

Figure 16. Simulation results of DOBOT axis and total error of coordinates for Transposition method.

4. DENAVIT-HARTENBERG PARAMETERS

The procedure for determining the position using the Denavit-Hartenberg (D-H) convention involves three steps: identifying the D-H parameters, constructing the A matrices, and computing the T matrix using the required coordinate position.

4.1. D-H parameters

D-H notation describes coordinates for different joints of a robotic manipulator in matrix entry. The method includes four parameters:

1. Twist angle α_i
2. Link length a_i
3. Link offset d_i
4. Joint angle θ_i.

The manipulator geometry determines that the twist angle and link length remain constant, but the link offset and joint angle vary based on the kind of joint, which can be either prismatic or revolute. The approach offers a comprehensive set of 10 steps to clearly outline the systematic derivation of the D-H parameters. These steps may be found in references [5] or [6].

4.2. A matrix

The A matrix is a homogenous 4 × 4 transformation matrix. Matrix describes the position of a point on an object and the orientation of the object in a three-dimensional space [6]. The homogenous rotation matrix along an axis is described by the Eq. (57) (Figures 11–13).

$$Rot_{z_{i-1}} = \begin{bmatrix} \cos\theta_i & -\cos\alpha_i \sin\theta_i & \sin\alpha_i \sin\theta_i & 0 \\ \sin\theta_i & \cos\alpha_i \cos\theta_i & -\sin\alpha_i \sin\theta_i & 0 \\ 0 & \sin\alpha_i & \cos\alpha_i & 0 \\ 0 & 0 & 0 & 1 \end{bmatrix} \tag{57}$$

The homogeneous translation matrix is described by Eq. (58).

$$Trans_{z_{i-1}} = \begin{bmatrix} 1 & 0 & 0 & a_i \\ 0 & 1 & 0 & 0 \\ 0 & 0 & 1 & d_i \\ 0 & 0 & 0 & 1 \end{bmatrix} \tag{58}$$

In rotation matrix and translation matrix, we can find the four parameters θ_i, d_i, a_i, and α_i. These parameters derive from specific aspects of the geometric relationship between two coordinate frames. The four parameters are associated with link *i* and joint *i*. In Denavit-Hartenberg convention, each homogeneous transformation matrix A_i is represented as a product of four basic transformations as follows [6]:

$$T_i^{i-1} = Trans_{z_{i-1}}(d_i).Rot_{z_{i-1}}(\theta_i).Trans_{x_i}(r_i).Rot_{x_i}(\alpha_i) \tag{59}$$

D-H convention matrix is given in Eq. (60).

$$T_i^{i-1} = \begin{bmatrix} \cos\theta_i & -\sin\theta_i \cos\alpha_i & \sin\theta_i \sin\alpha_i & r_i \cos\theta_i \\ \sin\theta_i & \cos\theta_i \cos\alpha_i & -\cos\theta_i \sin\alpha_i & r_i \sin\theta_i \\ 0 & \sin\alpha_i & \cos\alpha_i & d_i \\ 0 & 0 & 0 & 1 \end{bmatrix} \tag{60}$$

The previous matrix can be simplified by following equation A_i matrix. The matrix A_i is composed from 3 × 3 rotation matrix R_i, 3 × 1 translation vector P_i, 1 × 3 perspective vector and scaling factor.

$$A_i = \begin{bmatrix} R_{i(3x3)} & P_{i(3x1)} \\ 0_{(1x3)} & 1 \end{bmatrix} \quad (61)$$

4.3. T matrix

The T matrix can be formulated by Eq. (62). The matrix is a sequence of D-H matrices and is used for obtaining end-effector coordinates. The T matrix can be built from several A matrices depending on the number of manipulator joints.

$$T_0^3 = T_0^1 . T_1^2 . T_2^3 \quad (62)$$

Inside the *T* matrix is the translation vector P_i, which includes joint coordinates, where the *X*, *Y*, and *Z* positions are P_1, P_2, and P_3, respectively [6].

5. THE PSEUDOINVERSE METHOD

If the number of independent coordinates *n* (joint parameters) is larger than the number of reference manipulator endpoint coordinates *m* (three in Cartesian coordinate system for the point), it shows that a redundancy problem has occurred. In this case, it can exist in infinite combinations of independent coordinates for the only endpoint position. Jacobian matrix J has a size of *m* rows and *n* columns ($m \neq n$), i.e., J is a non-square matrix. In general, it cannot be computed inverse matrix from non-square matrix.

In order to solve inverse kinematics task for this case, pseudoinverse of Jacobian matrix (denotes J^+) is used. This method uses singular value decomposition (SVD) of Jacobian matrix to determine J^+. Every matrix J can be decomposed with the usage of SVD to three matrices Eq. (63):

$$J = U \Sigma V^T \quad (63)$$

Where

J is $m \times n$ matrix.

U is $m \times m$ orthogonal matrix, i.e. $U^{-1} = U^T$.

V is $n \times n$ orthogonal matrix, i.e. $V^{-1} = V^T$.

Σ is $m \times n$ diagonal matrix, which contains singular values of matrix J on its major diagonal.

$$\begin{bmatrix} j_{11} & j_{12} & \cdots & j_{1n} \\ j_{21} & j_{22} & \cdots & j_{2n} \\ \vdots & \vdots & \ddots & \vdots \\ j_{m1} & j_{m2} & \cdots & j_{mn} \end{bmatrix} = \begin{bmatrix} u_{11} & u_{12} & \cdots & u_{1m} \\ u_{21} & u_{22} & \cdots & u_{2m} \\ \vdots & \vdots & \ddots & \vdots \\ u_{m1} & u_{m2} & \cdots & u_{mm} \end{bmatrix} \cdot \begin{bmatrix} \sigma_1 & 0 & \cdots & 0 \\ 0 & \sigma_2 & \cdots & 0 \\ \vdots & \vdots & \ddots & \vdots \\ 0 & 0 & \cdots & \sigma_d \end{bmatrix} \cdot \begin{bmatrix} v_{11} & v_{21} & \cdots & v_{n1} \\ v_{12} & v_{22} & \cdots & v_{n2} \\ \vdots & \vdots & \ddots & \vdots \\ v_{1n} & v_{2n} & \cdots & v_{nn} \end{bmatrix} \quad (64)$$

Where $d = m$ for $m < n$ and $d = n$ for $m > n$, because Σ is a non-square matrix.

To determine matrices U and Σ, we multiply matrix J by its transpose matrix J^T from the right:

$$J.J^T = (U\Sigma V^T).(U\Sigma V^T)^T \tag{65}$$

$$J.J^T = U\Sigma V^T.V\Sigma^T U^T \tag{66}$$

$$J.J^T = U\Sigma\Sigma^T U^T \tag{67}$$

We multiply the above Eq. (67) by matrix U from the right:

$$JJ^T.U = U.\Sigma\Sigma^T.U^T U \tag{68}$$

$$JJ^T.U = U.\Sigma\Sigma^T \tag{69}$$

It leads to eigenvalue problem for JJ^T matrix. U is $m \times m$ square matrix, which contains eigenvectors of JJ^T matrix in its columns and $\Sigma\Sigma^T$ is diagonal matrix of eigenvalues $\lambda_1, ..., \lambda_m$.

To determine matrices V and Σ, we multiply matrix J by its transpose matrix J^T from the left:

$$J^T.J = (U\Sigma V^T)^T.(U\Sigma V^T) \tag{70}$$

$$J^T.J = V\Sigma^T U^T.U\Sigma V^T \tag{71}$$

$$J^T.J = V\Sigma^T\Sigma V^T \tag{72}$$

We multiply the above Eq. (72) by matrix V from the right:

$$J^T J.V = V.\Sigma^T\Sigma.V^T.V \tag{73}$$

$$J^T J.V = V.\Sigma^T\Sigma \tag{74}$$

It leads to eigenvalue problem for J^TJ matrix. V is $n \times n$ square matrix, which contains eigenvectors of J^TJ matrix in its columns and $\Sigma^T\Sigma$ is diagonal matrix of eigenvalues $\lambda_1, ..., \lambda_n$.

Matrices JJ^T and J^TJ are symmetric matrices and they have the same nonzero eigenvalues. Eigenvalues and eigenvectors of the real symmetric matrices are always real numbers and real vectors.

The eigenvalues are equal to square of the singular values: $\lambda i=\sigma^2_i$, where $i = 1, ..., d$. The number of nonzero eigenvalues is $d = m$ for $m < n$ and $d = n$ for $m > n$. The number of zero eigenvalues is $|m - n|$.

When values of matrices U, Σ, and V were computed, we can determine pseudoinverse of Jacobian matrix as follows:

$$J^{+} = \left(U.\Sigma.V^{T}\right)^{-1} \tag{75}$$

$$J^{+} = V.\Sigma^{+}.U^{T} \tag{76}$$

Where

$$\Sigma^{+} = \begin{bmatrix} \frac{1}{\sigma_1} & 0 & \cdots & 0 \\ 0 & \frac{1}{\sigma_2} & \cdots & 0 \\ \vdots & \vdots & \ddots & \vdots \\ 0 & 0 & \cdots & \frac{1}{\sigma_d} \end{bmatrix} \tag{77}$$

Now, we can solve inverse kinematics task for the cases, when Jacobian matrix is non-square:

$$\Delta Q = J^{+}.\Delta X \tag{78}$$

Pseudoinverse J^{+}, also called Moore-Penrose inverse of Jacobian matrix, gives the best possible solution in the sense of least squares [1].

6. THE JACOBIAN MATRIX TRANSPOSE METHOD

We designed Jacobian matrix transpose method simulation [1–3]. The basic idea was written using Eq. (79). We used the transpose of Jacobian matrix, instead of the inverse of Jacobian matrix, in this method. We set Δθ equal to

$$\Delta\theta = \alpha J^{T}\,\vec{e} \tag{79}$$

Where α is:

$$\alpha = \frac{\left\langle \vec{e}, JJ^{T}\,\vec{e} \right\rangle}{\left\langle JJ^{T}\,\vec{e}, JJ^{T}\,\vec{e} \right\rangle} \tag{80}$$

Whole simulation is described by block diagram shown in Figure 14. In the first step, we defined requested error. This error represented difference between reference coordinates and actual coordinates. Error that we consider as unacceptable, we set to 200 μm. This is position repeatability of DOBOT. We calculate the increment of requesting angles Δθ in each iteration. In the first iteration, Δθ is equal to zero.

Figures 15 and 16 depict the simulation outcome of DOBOT's movement, which is identical to the simulation of the pseudoinverse of the Jacobian matrix. The simulation was divided into three segments. The initial segment (shown by the solid line in the chart) depicts the motion from the starting point to the location (x, y, z) = (100, 150, 160) mm. The second part (shown by the dashed line in the chart) represents the movement from the previous location to the position (50, 90, 80). The third part, represented by the dashed line in the chart, corresponds to a movement towards the point (150, 180, 140).

7. CONCISE SUMMARY

As evident from the simulation results presented in earlier subchapters, each method for inverse kinematics has both advantages and disadvantages. Table 2 displays a comparison of both strategies. The pseudoinverse approach is more efficient than the transposition method, but it poses greater implementation challenges in a digital signal processor (DSP) or a microcontroller. The pseudoinverse approach can be simply implemented in the Matlab environment using the pinv() tool. If we desire to streamline the process of inverse kinematics and do not require rapid computation, it is more advantageous to employ the transposition approach. When utilizing the DOBOT manipulator, it is advisable to employ the analytical model. This strategy is not relevant in the event of a more complex manipulator.

Table 2. Comparison of pseudoinverse and transposition method.

Simulation part	Pseudoinverse method (number of iterations)	Transposition method (number of iterations)
Part 1 (solid line)	22	55
Part 2 (dotted line)	5	34
Part 3 (dashed line)	6	68

The comparison of both strategies is presented in Table 2. Table 2 clearly indicates that the primary factor under consideration is the number of iterations. The pseudoinverse technique is superior, albeit solely for simulation purposes. If we are able to employ this technique in a real-time scenario, such as using dSPACE from MathWorks® or implementing it on a DSP, we will not get the same outcomes as those presented in Table 2. The root of this issue is the utilization of singular value decomposition (SVD), which requires significant computational resources. On the other hand, the process of transposing the Jacobian matrix is far simpler to construct and requires less computational power.

In the upcoming study, we will examine the application of an appropriate iterative technique, such as damped least squares. We have developed multiple approaches for implementing Jacobian matrix transposition on DSP platforms such as TMS430 and C2000™. It is crucial to explore other implementation approaches in order to minimize the calculation time as much as possible.

REFERENCES

1. Buss SR. Introduction to inverse kinematics with Jacobian transpose, pseudoinverse and damped least squares methods. IEEE Journal of Robotics and Automation. 2004;17:1-19

2. Balestrino A, De Maria G, Sciavicco L. Robust control of robotic manipulators. In: Proceedings of the 9th IFAC World Congress; 1984;5:2435-2440

3. Wolovich WA, Elliot H. A computational technique for inverse kinematics. In: Proceedings of the 23rd IEEE Conference on Decision and Control; 1984. pp. 1359-1363

4. Denavit–Hartenberg [Internet]. 2017. Available from: https://en.wikipedia.org/wiki/Denavit%E2%80%93Hartenberg_parameters [Accessed: 03-08-2017]

5. Desai JP. D-H Convention, Robot and Automation Handbook. USA: CRC Press; 2005 ISBN: 0-8493-1804-1

6. Rehiara AB. Kinematics of Adept Three Robot Arm. In: Satoru Goto, editor. Robot Arms. 2011. InTech. ISBN: 978-953-307-160-2. Available from: http://www.intechopen.com/books/robot-arms/kinematicsof-adeptthree-robot-arm [Accessed: 01-08-2017]

7. The MathWorks, Inc., SimMechanics 2User's Guide [Internet], 2007. Available from: https://mecanismos2mm7.files.wordpress.com/2011/09/tutorial-sim-mechanics.pdf [Accessed: 08-04-2017]

8. Bingul KS. The inverse kinematics solutions of industrial robot manipulators. In: IEEE Conference on Mechatronics; Istanbul, Turkey; 2004. pp. 274-279

9. Serdar Kucuk, Zafer Bingul. In: Sam Cubero, editor. Robot Kinematics: Forward and Inverse Kinematics, Industrial Robotics: Theory, Modelling and Control. 2006. InTech. ISBN: 3-86611-285-8. Available from: http://www.intechopen.com/books/industrial_robotics_theory_modelling_and_control/robot_kinematics__forward_and_inverse_kinematics [Accessed 01-08-2017]

10. WebPage [Internet], 2017. Available from: http://www.dobot.cc/ [Accessed: April 04, 2017]

5 Chapter

METHODS FOR INCREASING THE OPERATIONAL AREA OF PARALLEL ROBOTS

1. INTRODUCTION: AN OVERVIEW

The parallel robot has exceptional attributes, including elevated velocity, superior accuracy, and enhanced rigidity [1]. Nevertheless, the workspace of the system is limited because to mechanical collisions between limbs and the presence of complex single configurations. This chapter primarily focuses on the techniques used to increase the translational workspace of the parallel robot. The parallel robot exhibits many solutions for both inverse and forward displacement analysis. The parallel robot can increase the range of its translational workspace by modifying its configurations between different solutions. Nevertheless, traditional nonredundant parallel robots experience singularity when transitioning between modes. This paper presents the introduction of singularity-free mode shifts in parallel robots through the use of redundant actuation. Subsequently, techniques for enlarging the rotational range of motion of the parallel robot are demonstrated. To produce a significant rotation, the moving part has been equipped with mechanical mechanisms such as gears, pulleys, and helical joints. An innovative differential screw-nut system is shown to enhance the rotational workspace of the parallel robot.

2. EXPANDING THE TRANSLATIONAL WORKSPACE

2.1. Mode change of the parallel robot

Figure 1 depicts a schematic representation of a planar serial robot with two degrees of freedom (dof), whereas Figure 2 shows a schematic model of a planar parallel robot with two degrees of freedom. The letters R and P are used to indicate the rotational pair and prismatic (slide) pair, respectively. The presence of an underline signifies the presence of an actuator at the corresponding pair. Specifically, the letters R and P denote active rotational and active prismatic pairs, while R and P represent passive pairs.

The serial robot in Figure 1 is represented as *RR* mechanism, and the parallel robot in Figure 2 is represented as *2PRR* mechanism. In *2PRR*, "2" means that the two *PRR* mechanisms are connected to the output link (end effector or hand) in parallel. In order to control the position of the hand, forward displacement analysis gives the position and orientation of the hand from the displacements of the active pairs (joints), and inverse displacement analysis gives the displacements of the active joints from the position and orientation of the hand.

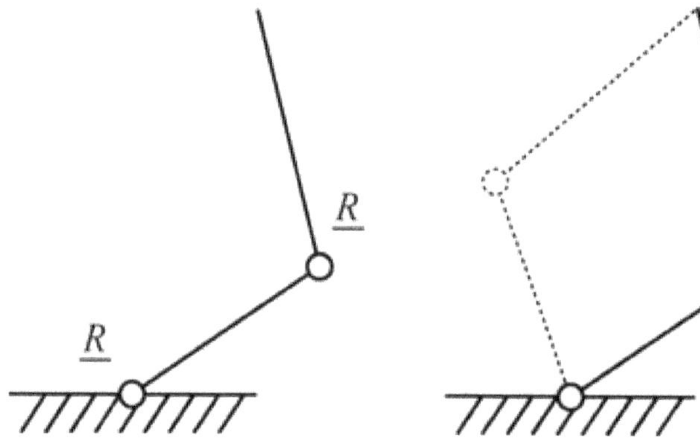

Figure 1. Working mode change of RR serial robot.

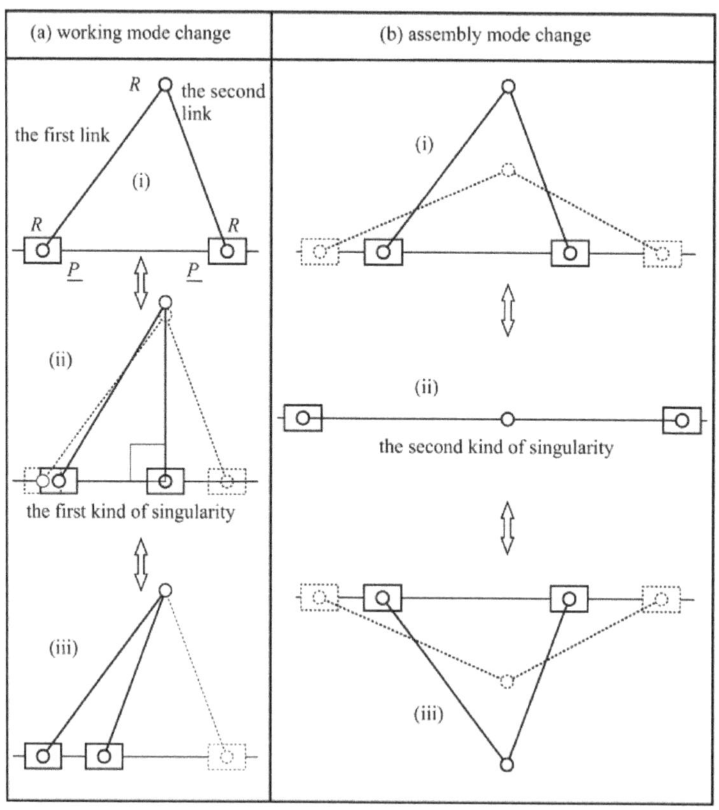

Figure 2. Assembly and working mode change of 2PRR parallel robot.

As shown in Figures 1 and 2(a), different solutions for inverse displacement analysis and different position of the joints for a position and orientation of the hand exist. In the robotics, the solution of the inverse displacement analysis is called as the working mode. Changing the positions of the joints between the different working modes, at that time the robot changes its configuration, is named as the working mode change. In the case of the serial robot, there exists only one solution for the forward displacement analysis; namely, when the displacement of the active joints is given, the position and orientation of the robot are fixed. However, in the case of the parallel robot, there exist different solutions for the forward displacement analysis as shown in Figure 2(b). The solution of the forward displacement analysis is named as the assembly mode. Parallel robot can change the position and orientation of the hand with the same displacements of the actuated joints. Note here that the parallel robot in Figure 2 is designed as free of the mechanical interferences such as limb collisions. Parallel robot can expand the workspace of the hand by the assembly mode change as shown in Figure 2(b).

2.2. Kinematics of the parallel robot

As mentioned before, the parallel robot can expand its workspace by the mode change. However, parallel robot encounters the singular configuration between the mode changes. In this section, kinematics and singular configuration of a *2PRR* parallel robot are shown in Figure 3. Coordinate frames Σ_0 and Σ_i $(i=1,2)$ are defined for the fixed coordinate and position of each actuator, respectively. Vector bi is defined for the position of the origin Σ_i with respect to the coordinate Σ_0. In the case of the *2PRR* parallel robot in Figure 3, $\mathbf{b}_i = 0$ $(i=1,2)$ because each coordinate Σ_i is coincident with Σ_0. Unit direction vector and displacement of the linear actuator are defined as $\mathbf{u}_i$ and $\mathbf{q}_i$, respectively. Unit direction vector and length of the rod are defined as $\mathbf{w}_i$ and l_i, respectively. Position of the hand (output link) is given as

$$\mathbf{x} = \mathbf{b}_i + q_i u_i + l_i w_i \tag{1}$$

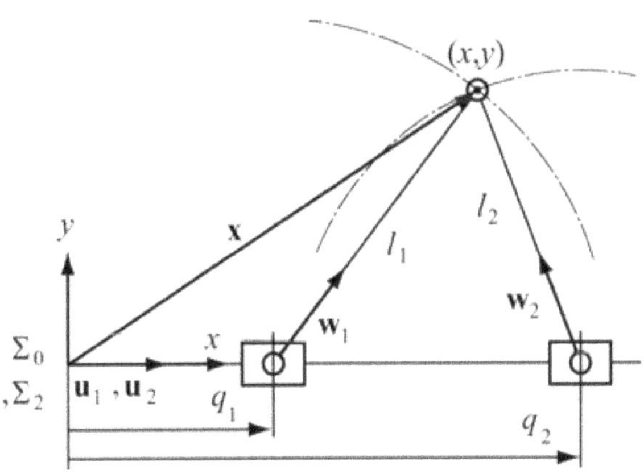

Figure 3. Kinematic model of 2PRR parallel robot.

Equation (1) is the loop closure equation of the parallel robot. Next, the forward displacement analysis and inverse displacement analysis of the parallel robot are going to be derived.

Eq. (1) is deformed as

$$l_i w_i = \mathbf{x} - \mathbf{b}_i - q_i u_i \tag{2}$$

The unit direction vector $\mathbf{w}_i$ is eliminated by raising the both sides of Eq. (2) to the second power as,

$$l_i^2 = (x - b_i - q_i u_i)^T(\mathbf{x} - \mathbf{b}_i - q_i u_i) \tag{3}$$

As shown in Figure 3, $\mathbf{u}_i$ and $\mathbf{b}_i$ are given as,

$$\mathbf{u}_i = \begin{bmatrix}1\\0\end{bmatrix}, \mathbf{b}_i = \begin{bmatrix}0\\0\end{bmatrix} \quad (i = 1, 2) \tag{4}$$

Substituting Eq. (4) into Eq. (3), one obtains a quadratic equation about the displacement of the actuator as,

$$q_i^2 - 2q_i x + (y^2 - x^2 - l_i^2) = 0 \tag{5}$$

Solutions of the inverse displacement analysis are given by the solution of Eq. (5) as,

$$q_i = x \pm \sqrt{l_i^2 - y^2} \tag{6}$$

Eq. (6) represents that there are two solutions for the inverse displacement analysis. It means that the parallel robot has two working modes. Figure 2(a) (i) and (iii) represents the configurations of the parallel robot of the two working modes.

In the forward displacement analysis, positions x and y of the hand are solved by the simultaneous equations about Eq. (1) of each arm as,

$$\begin{cases} (x - q_1)^2 + y^2 = l_1^2 \\ (x - q_2)^2 + y^2 = l_2^2 \end{cases} \tag{7}$$

Each equation of (7) represents that a circle of the central position is $(q_i, 0)$ and the radius is l_i. The solution of Eq. (7) is given as the crossing point of the two circles. There exist two crossing points of the circles; namely, the parallel robot has two assembly modes. Figure 2(b) (i) and (iii) represents each configuration of the assembly mode of the parallel robot.

When $y = l_i$, at that time, the value in the square root in Eq. (6) becomes zero and the inverse displacement analysis has a duplication solution. When $y > l_i$, at that time, value in the square root in Eq. (6) becomes negative and the inverse displacement analysis has no solution. When the distance between the two circles of Eq. (7) equals $(l_1 + l_2)$, the forward displacement analysis has a duplication solution. When the distance between the two circles of Eq. (7) becomes larger than (l_i+l_2), the forward displacement analysis has no solution.

2.3. Singularity analysis of the parallel robot

2.3.1. The Jacobian of the parallel robot

Differentiating the both sides of Eq. (1) with respect to time, one obtains

$$\dot{\mathbf{x}} = \dot{q}_i u_i + l_i \dot{\mathbf{w}}_i \quad (8)$$

In Eq. (1), $\mathbf{b}_i$, $\mathbf{u}_i$, and l_i are constant values and their time derivatives become zero. Because $\mathbf{w}_i$ is unit direction vector, one obtains

$$\mathbf{w}_i^T \mathbf{w}_i = 1 \quad (9)$$

Differentiating the both sides of Eq. (9) with respect to time, one obtains

$$w_i^T \dot{\mathbf{w}}_i = 1 \quad (10)$$

Multiplying the both sides of Eq. (8) by $\mathbf{w}_i^T$ and taking into consideration about Eq. (9), one obtains

$$\mathbf{w}_i^T \dot{\mathbf{x}} = \mathbf{w}_i^T u_i \dot{q}_i \quad (11)$$

The displacements of all actuators q_1 and q_2 are defined by vector form as $\mathbf{q}=[q_1 \;\; q_2]^T$; then, Eq. (10) is expressed by matrix form as

$$\mathrm{J}_x \dot{x} = \mathrm{J}_q \dot{q} \quad (12)$$

$$\mathrm{x} = [\dot{x} \;\; \dot{y}]^T, \mathbf{q} = [\dot{q}_1 \;\; \dot{q}_2]^T$$

$$\mathrm{J}_x = \begin{bmatrix} w_1^T \\ w_2^T \end{bmatrix}, \mathrm{J}_q = \begin{bmatrix} w_1^T u_1 & 0 \\ 0 & w_2^T u_2 \end{bmatrix}$$

$\mathbf{J}_x$ and $\mathbf{J}_q$ are the Jacobian of the parallel robot.

2.3.2. Second kind of singularity

When the Jacobian matrix $\mathbf{J}_x$ becomes singular, namely the determinant of the Jacobian becomes zero as

$$\det(J_x) = 0 \quad (13)$$

Then, the parallel robot is in the second kind of singularity [2] or in direct kinematics singularity [3]. The *2PRR* parallel robot occurs in the second kind of singularity when the direction of the two rods $\mathbf{w}_1$ and $\mathbf{w}_2$ becomes identical.

$$\det\left(\begin{bmatrix} w_1^T \\ w_2^T \end{bmatrix}\right) = 0 \quad (14)$$

This case corresponds to that two rods of the parallel robot lay in one line as shown in Figure 2(b) (ii). As shown in Figure 2(b), the parallel robot encounters the second kind of singularity during the assembly mode change.

2.3.3. First kind of singularity

When the Jacobian matrix $\mathbf{J}_q$ is singular

$$\det\left(J_q\right) = 0 \tag{15}$$

Then, the parallel robot is in the first kind of singularity [2] or in inverse kinematics singularity [3]. The *2PRR* parallel robot occurs in the first kind of singularity when the direction of at least one rod is perpendicular to the direction of the actuator as shown in Figure 2(a) (ii).

$$\det\left(\begin{bmatrix} w_1^T u_1 & 0 \\ 0 & w_2^T u_2 \end{bmatrix}\right) = 0 \tag{16}$$

As shown if Figure 2(a), the parallel robot encounters the first kind of singularity during the working mode change.

2.3.4. The third kind of singularity

The third kind of singularity [2] occurs when both $\mathbf{J}_q$ and $\mathbf{J}_x$ are simultaneously singular. It requires certain conditions of the linkage parameter. In the case of the *2PRR* parallel robot, when the lengths of the rods are identical, the third kind of singularity occurs as shown in Figure 4. The third kind of singularity is referred as the combined singularity [3].

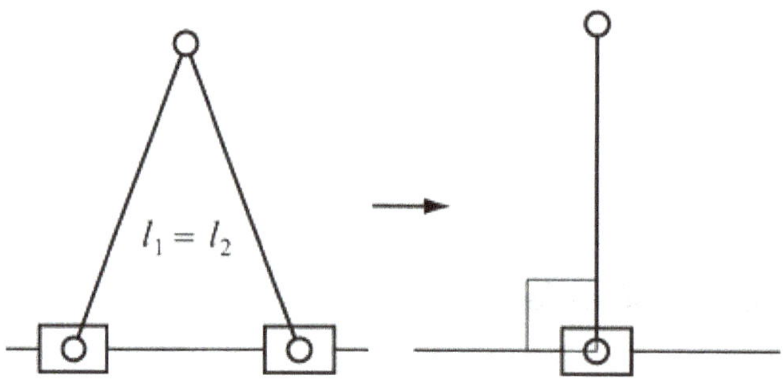

Figure 4. The third kind of singularity of the 2PRR parallel robot.

2.4. Passing the singular configuration during the mode changes

2.4.1. Passing the second kind of singularity during the assembly mode change

2.4.1.1. Using the inertia

As shown in Figure 5(a), if the parallel robot has some speed just before the singular configuration, the robot keeps moving according to the law of inertia; then, the parallel robot passes the second kind of singularity. Note here if the parallel robot stops at the singular configuration, it is impossible to pass the singular configuration by the inertia.

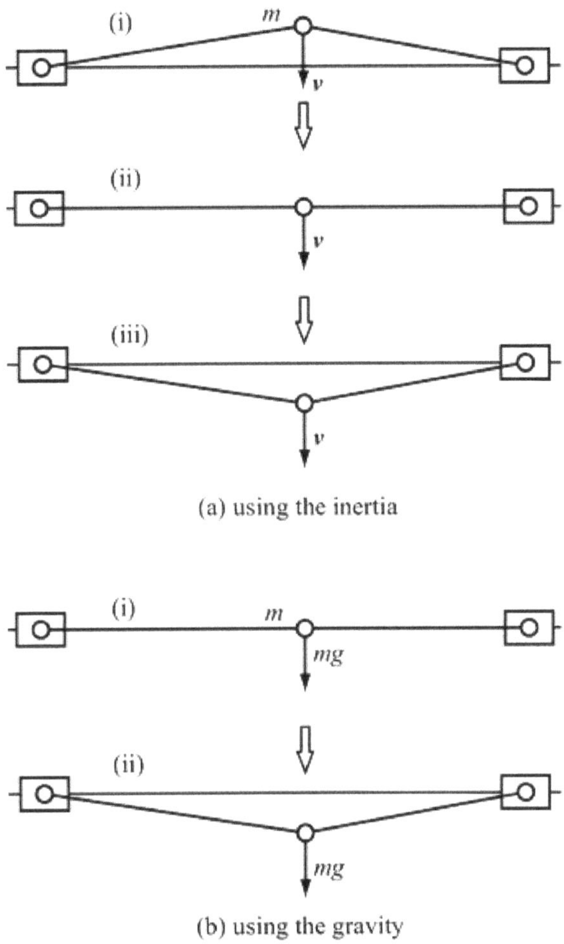

Figure 5. Assembly mode change of nonredundant 2PRR parallel robot.

2.4.1.2. Using the gravity

As shown in Figure 5(b), when the gravity force acts to the lower direction, the parallel robot can pass the singular configuration even if the parallel robot stops at the singular configuration. However, the robot cannot move to the upper direction against the gravity. When the robot moves on the horizontal plane where the gravity force does not act on the robot, the parallel robot cannot pass the singular configuration.

2.4.1.3. Using the redundancy

A robot has actuation redundancy when it is driven by number of actuators greater than the degree of freedom. The actuation redundancy may increase the cost of the robot and complexity of control. However, the actuation redundancy is one of the most effective methods for avoiding the singularity during the mode change.

Figure 6 represents 3*PRR* parallel robot, a planar two-dof parallel robot redundantly actuated three actuators. The Jacobian of the parallel robot is given as

$$J_x \dot{x} = J_q \dot{q} \tag{17}$$

$$x = [\dot{x} \quad \dot{y}]^T, q = [q_1 \quad q_2 \quad q_3]^T$$

$$J_x = \begin{bmatrix} w_1^T \\ w_2^T \\ w_3^T \end{bmatrix}, J_q = \begin{bmatrix} w_1^T u_1 & 0 & 0 \\ 0 & w_2^T u_2 & 0 \\ 0 & 0 & w_2^T u_2 \end{bmatrix}$$

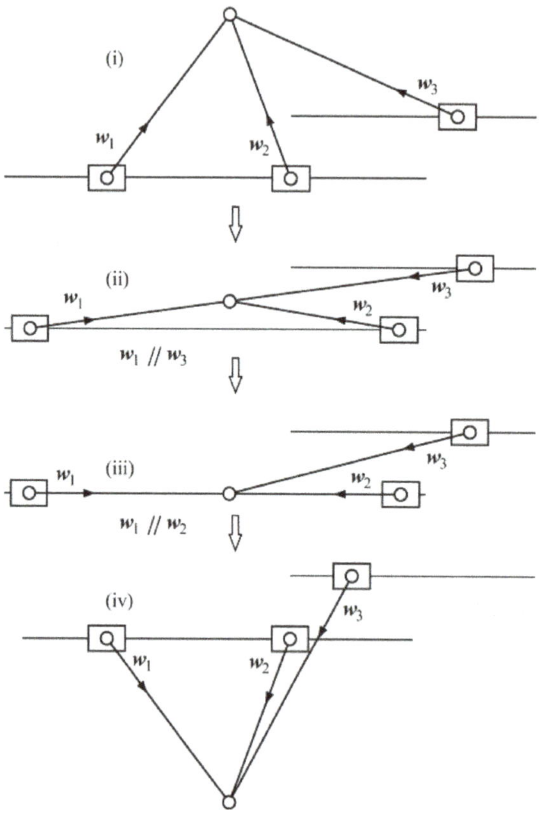

Figure 6. Assembly mode change of redundant 3PRR parallel robot.

where $\mathbf{J}_x$ is 3×2 matrix of its full rank that equals two. For convenience, singularity analysis is applied to the 3×2 transposed Jacobian matrix of $\mathbf{J}_x^T$. When at least one 2×2 minor of $\mathbf{J}_x^T$ is nonsingular, the rank of the $\mathbf{J}_x$ equals two; namely, the $\mathbf{J}_x$ has full rank. At this time, the robot still works as two-dof parallel robot. For example, when the first rod and the second rod are collinear as shown in Figure 6 (iii), 2×2 minors of the $\mathbf{J}_x^T$ become

$$\begin{cases} \det([w_1 \quad w_2]) = 0 \\ \det([w_1 \quad w_3]) \neq 0 \\ \det([w_2 \quad w_3]) \neq 0 \end{cases} \tag{18}$$

As shown in Eq. (18), one minor is singular and the other two minors are nonsingular; namely, the $\mathbf{J}_x$ still has the full rank.

Now, the parallel robot loses the redundancy but keeps the nonsingularity. The parallel robot can pass the singular configuration of Figure 6 (iii). In the same way, the parallel robot can pass the singular configuration of Figure 6 (ii). Thus, the parallel robot achieves the assembly mode change from Figure 6 (i) to (iv).

2.4.2. Passing the first kind of singularity during the working mode change

The parallel robot encounters the first kind of singularity during the working mode change. In the velocity control of a robot, the velocity of the actuator is controlled to trace the desired velocity of the actuator $\dot{q}_r$, which is given from the desired velocity of the end-point $\dot{x}_r$ from Eq. (12) as

$$\dot{q}_r = (J_q^{-1} J_x)\, \dot{x}_r \tag{19}$$

However, $\mathbf{J}_q^{-1}$ cannot be calculated when $\mathbf{J}_q$ is singular at the first kind of singularity. In this case, the desired velocity of the actuator is directly given instead of being indirectly given from Eq. (12).

2.5. Researches on mode change of the parallel robot

Researches on expanding the workspace by the mode change of the parallel robot have been reported for nonredundant two-dof planar robot [4, 5], nonredundant three-dof spatial translational robot [6], redundantly driven three-dof translational and rotational planar robot [7], and redundantly driven three-dof spatial translational robot [8]. Redundant actuation is the actuation in one of the most effective methods for avoiding the singularity. However, the redundancy is not always the answer to avoiding the singularity. Additional ingenuities, for example, path planning for mode change, or asymmetrical design for the robot, are required.

3. EXPANDING THE ROTATIONAL WORKSPACE

3.1. Conventional methods

In this section, methods for expanding the rotational workspace of the parallel robot are introduced. Figure 7(a) represents the Stewart Platform [9], six-dof parallel robot with three-dof translations and three-dof rotations. The moving plate is driven by linear actuator embedded six limbs. Each limb is paired by a passive universal joint (*U*) with the base plate and paired by a passive spherical joint (*S*) with the moving plate.

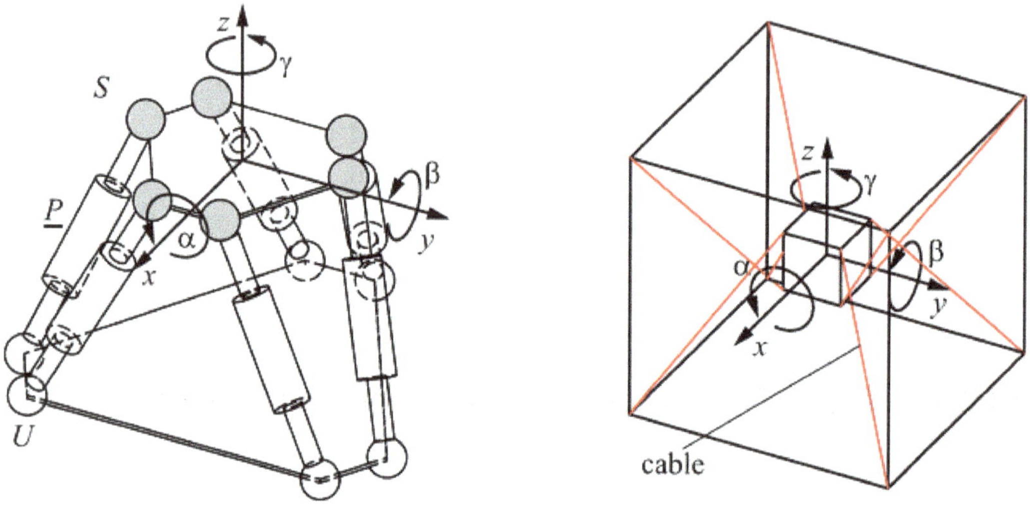

(a) Stewart Platform (b) Cable driven parallel robot

Figure 7. Conventional six-dof parallel robot.

In order to cope with the problem, Clavel invented a novel parallel robot DELTA [11] as shown in Figure 8. The DELTA generates three-dof translation and one-dof rotation around the z-axis.

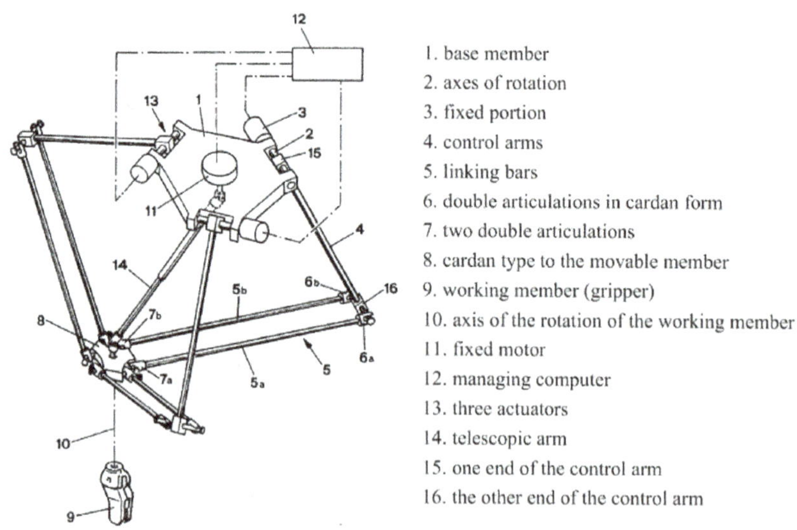

Figure 8. The DELTA robot [11].

The Stewart Platform is categorized into 6*UPS* parallel mechanism. The Stewart Platform generates high power with a hydraulic linear actuator. Flight simulator and driving simulator for carrying heavy cockpit of aircraft are typical applications of the Stewart Platform. However, the robot has a drawback of small rotation around z axis because of the mechanical interference between the limbs. Figure 7(b) represents a six-dof cable-driven parallel robot [10]. This robot also has the same drawback of small rotation around z axis because of the cable interference. Small rotation of parallel robots puts restriction to their applications.

Three control arms are rotated by three motors located on the base plate. Each arm is connected to the moving part by a parallelogram rod with spherical joints. The gripper on the moving part is connected to the motor on the base plate with a telescopic arm via universal joints (Figure 9).

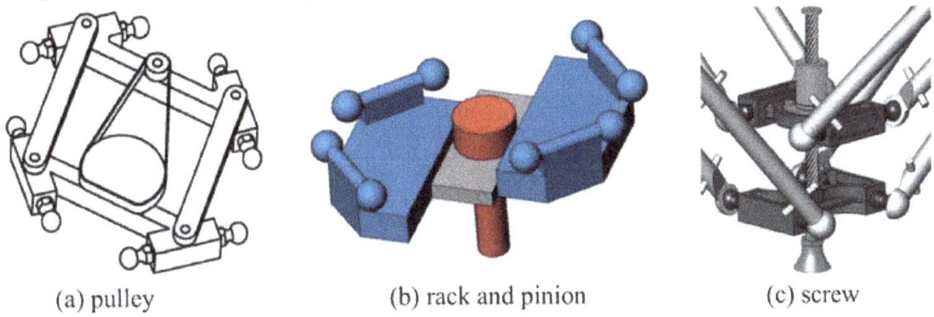

Figure 9. Mechanical gimmicks enlarge the rotational workspace of parallel robot [12].

Researches on expanding the rotational workspace of the DELTA like parallel robot have been reported from research group at the Montpelier University. The basic idea is to convert the translational motion to the rotational motion by the mechanical gimmicks such as pulley, rack and pinion, and screw [12]. Crank embedded moving part has been proposed by McGill University for two-limb parallel robot [13] and by Fraunhofer IPA for cable-driven parallel robot [14].

3.2. Differential screw drive system

Recently, differential screw drive systems composed of two screw-nut pairs of different leads have been applied to robot systems [15, 16]. There are four driving methods of the differential drive systems;

- Rotations of the two nuts are converted to the rotation and translation of the screw pair, which is coaxially arranged with one end interconnected with each other.
- Translations of the two nuts are converted to the rotation and translation of the screw pair, which is coaxially arranged with one end interconnected each other.
- Rotations of the two screws are converted to the rotation and translation of the nut pair, which is coaxially arranged with one end interconnected each other.

- Translations of the two screws are converted to the rotation and translation of the nut pair, which is coaxially arranged with one end interconnected each other.

In this section, kinematics of the differential drive system (b) as shown in Figure 10 is discussed for enlargement of the rotational workspace of the parallel robot.

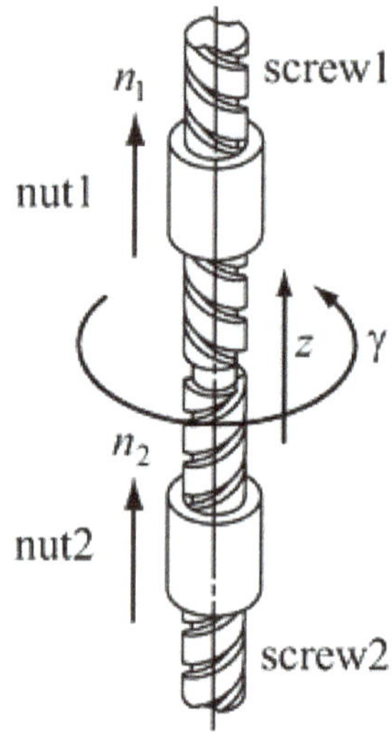

Figure 10. Differential screw drive system (translations of the two nuts).

Let l_i $(i=1,2)$ be the lead of the ith screw nut. n_i $(i=1,2)$, z, and γ represent the position of the ith nut, position, and angle of the screw pair, respectively. Relation of these parameters is given as

$$n_i = z - \frac{l_i}{2\pi}\gamma \tag{20}$$

Equation (20) is given in the matrix form as

$$\begin{bmatrix} n_1 \\ n_2 \end{bmatrix} = \begin{bmatrix} 1 & -\frac{l_1}{2\pi} \\ 1 & -\frac{l_2}{2\pi} \end{bmatrix} \begin{bmatrix} z \\ \gamma \end{bmatrix} \tag{21}$$

Equation (21) gives the inverse kinematics of the differential drive system. The forward kinematics of the system is derived by inverting Eq. (21) as

$$\begin{bmatrix} z \\ \gamma \end{bmatrix} = \frac{1}{l_2 - l_1} \begin{bmatrix} l_2 & -l_1 \\ 2\pi & -2\pi \end{bmatrix} \begin{bmatrix} n_1 \\ n_2 \end{bmatrix} \tag{22}$$

If two leads of the screw-nut pair are identical as then the matrix in Eq. (22) becomes singular. It is necessary that the leads of the two screw nuts are different from the differential drive system.

$$l_1 = l_2 \tag{23}$$

Figure 11 represents a 4-dof parallel robot that the differential drive system (a) embedded in the moving part [15, 16]. *H* represents a passive helical joint composed of screw and nut pair. In Figure 11, two *H* pairs have the same lead, but the different helix hands (right and left). In Figure 9 (c), one *H* pair and one *R* pair are embedded in the moving part. This is included in the differential drive system (a), one of the *H* pair has zero lead.

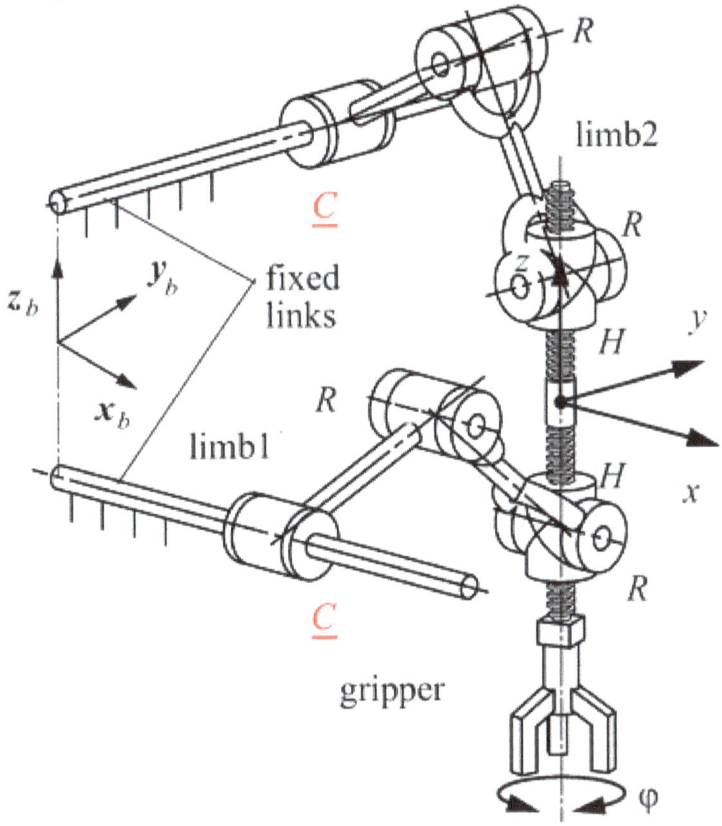

Figure 11. Parallel robot with the differential drive system embedded moving part [15].

4. CONCISE SUMMARY

Following the commercialization of the DALTA robot, other parallel robots have been implemented in machine factories and various other industries. Currently, it is possible to acquire a 3D printer machine kit that utilizes a linear DELTA mechanism for a price below 500 USD. The first-generation parallel robots, such as DELTA and Stewart Platform, are approaching maturity. This chapter introduced the concept of increasing the workspace of the parallel robot. The actuation redundancy was utilized to widen the translational workspace through singularity-free mode change. The differential drive system transforms linear motion into rotational motion, hence increasing the range of rotational movement for the parallel robot.

These parallel robots are anticipated to be the robots of the next generation. The subsequent section of this chapter presents an illustration of a parallel robot that belongs to the next generation. Figure 12(a) depicts a new type of parallel robot with two limbs and six degrees of freedom [17]. Figure 12(b) illustrates the integration of a differential screw mechanism into the output link to expand the gripper's rotating range. The degree of freedom of this parallel robot is increased from four (consisting of three translations and one rotation, as depicted in Figure 11) to six (comprising of three translations and three rotations, as illustrated in Figure 13). We are currently engaged in the kinematic analysis and intricate mechanical design of the parallel robot [18].

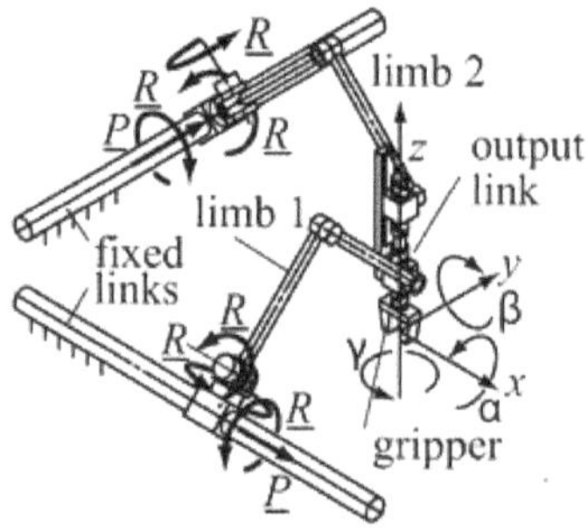

(a) 6-dof parallel robot

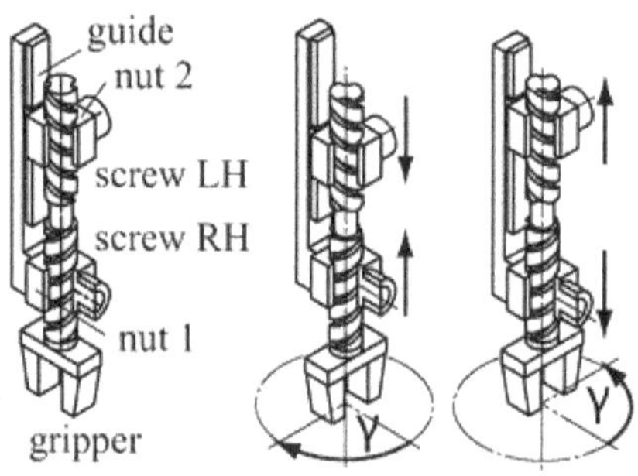

(b) differential screw embedded output link

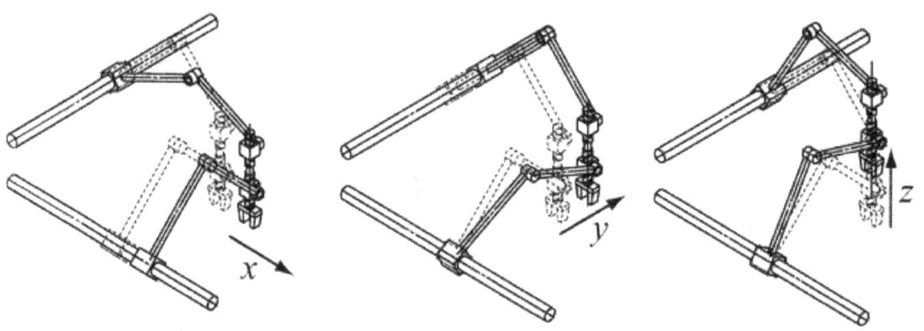

Figure 12. Two-limb six-dof parallel robot with the differential drive embedded output link [18].

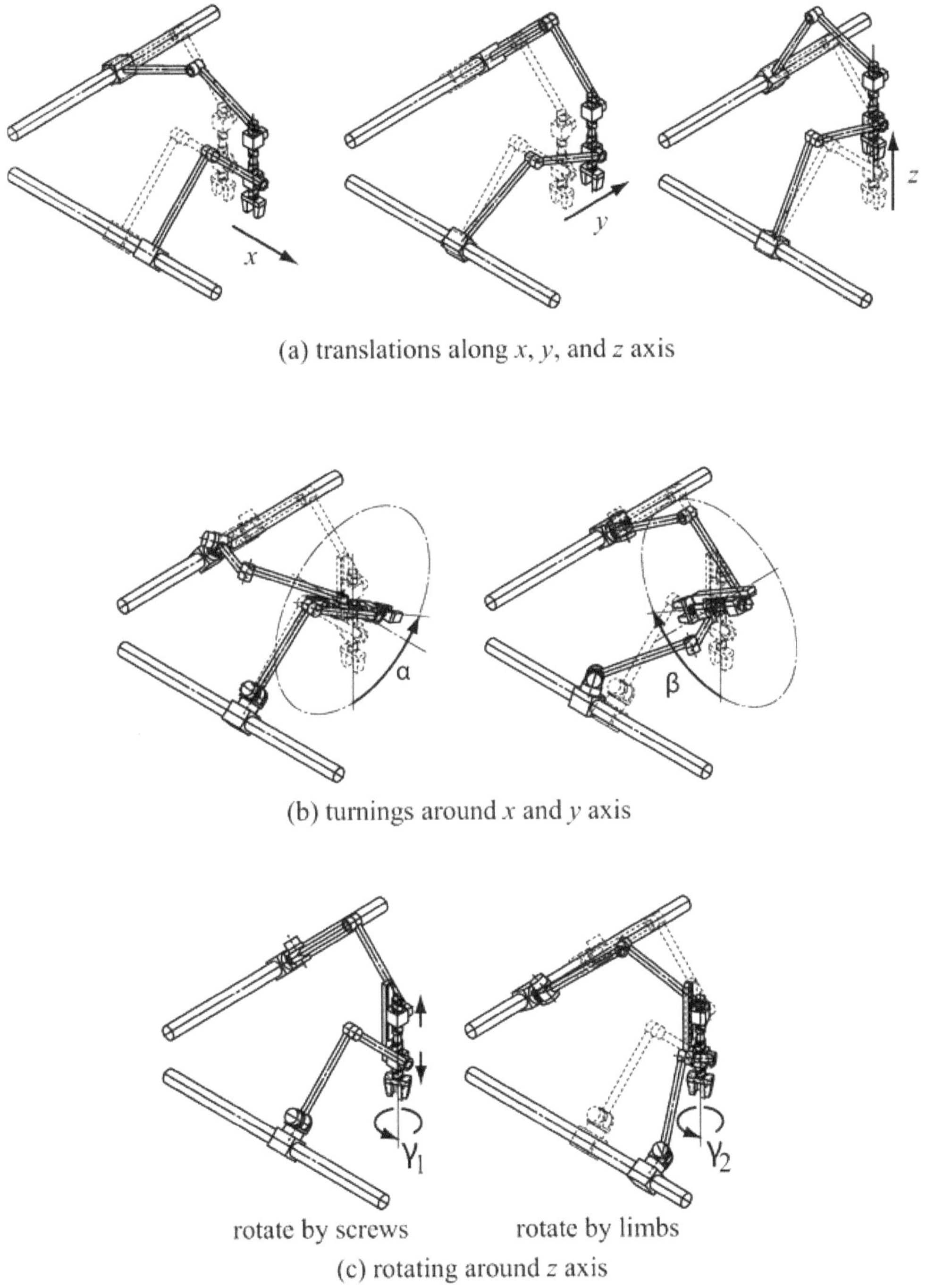

Figure 13. Motion of the two-limb six-dof parallel robot.

REFERENCES

1. Merlet JP. Parallel Robots. 2nd ed. Netherlands: Springer; 2006. 402 p. DOI: 10.1007/1-4020-4133-0

2. Gosselin C, Angeles J. Singularity analysis of closed-loop kinematic chains. IEEE Transactions on Robotics and Automation. 1990;6(3):281-290. DOI: 10.1109/70.56660

3. Tsai LW. Robot Analysis. 1st ed. New York: Wiley-Interscience; 1998. 520 p. ISSN: 978-0471325932

4. Hesselbach J, Woffmeister H, Loohß T, Krefft M, Armbrecht C. Parallel kinematic concept for stationary high performance cutting in wood machining. Production Engineering. 2007;1(2):205-212. DOI: 10.1007/s11740-007-0033-9

5. Campos L, Bourbonnais F, Bonev I and Bigras P. Development of a five-bar parallel robot with large workspace. In: Proceedings of the ASME 2010 International Design Engineering Technical Conferences & Computers and Information in Engineering Conference; 15-18 August 2010; Montreal. 2010. pp. 917-922. DOI: 10.1115/DETC2010-28962

6. Budde C, Last P and Hesselbach J. Development of a triglide-robot with enlarged workspace. In: Proceedings of IEEE International Conference on Robotics and Automation; 10-14 April 2007; Roma. 2007. pp. 543-548. DOI: 10.1109/ROBOT.2007.363043

7. Harada T. Mode changes of a planar 3 DOF redundantly actuated parallel robot. International Journal of Materials, Mechanics and Manufacturing. 2016;4(2):123-126. DOI: 10.7763/IJMMM.2016.V4.238

8. Harada T. Mode changes of redundantly actuated asymmetric parallel mechanism. Journal of Mechanical Engineering Science. 2015;230(3):454-462. DOI: 10.1177/0954406215588479

9. Steward D. A platform with six degrees of freedom. Proceedings of the Institution of Mechanical Engineers. 1965;180(1):371-386. DOI: 10.1243/PIME_PROC_1965_180_029_02

10. Bruckmann T, Pott A, editors. Cable-Driven Parallel Robots. 1st ed. New York: Springer; 2013. 454 p. DOI: 10.1007/978-3-642-31988-4

11. Clavel R. Device for the Movement and Positioning of an Element in Space. US Patent No. 4,976,582. 1989

12. Company O, Pierrot F, Krut S, Nabat V. Simplified dynamic modelling and improvement of a four-degree-of-freedom pick-and-place manipulator with articulated moving platform. Journal of Systems and Control Engineering. 2009;223(1):13-29. DOI: 10.1243/09596518JSCE616

13. Gauthier J, Angeles J, Nokleby SB, Morozov A. The kinetostatic conditioning of two-limb schonflies motion generators. Transactions of the ASME Journal of Mechanisms and Robotics. 2008;1(1):011010. DOI: 10.1115/1.2960544

14. Miermeister P, Pott A. Design of cable-driven parallel robots with multiple platforms and endless rotating axes. In: Kecskeméthy A, Flores FG, editors. Interdisciplinary Applications of Kinematics. 1st ed. Lima: Springer; 2013. p. 21-29. DOI: 10.1007/978-3-319-10723-3_3

15. Harada T, Angeles J. Kinematics and singularity analysis of a CRRHHRRC Schönflies motion generator. Transactions of the Canadian Society for Mechanical Engineering. 2014;38(2):173-183. ISSN: 0315-8977

16. Harada T, Friedlaender T and Angeles J. The development of an innovative two-DOF cylindrical drive: Design, Analysis and preliminary tests. In: Proceedings of IEEE International Conference on Robotics and Automation; 31 May-7 June 2014; Hong Kong. 2014. pp. 6338-6344. DOI: 10.1109/ICRA.2014.6907794

17. Harada T and Makino T. Design of a novel 6 DOF parallel robot for haptic device. In: 12th International Conference on Ubiquitous Healthcare; October 30–November 2; Osaka, Japan. 2015

18. Harada T and Angeles J. From the McGill Pepper-Mill Carrier to the Kindai Atarigi Carrier. In: Proceedings of the 2017 IEEE International Conference on Robotics and Biomimetics; December 5-8; Macau SAR, China. 2017. Printing

6 Chapter

MODELING THE KINEMATICS AND BIODYNAMICS OF THE LONG JUMP TECHNIQUE

1. INTRODUCTION: AN OVERVIEW

The long jump comprises four interrelated phases: approach, take-off, flight, and landing. Based on several existing research [2, 6, 9, 11], it has been found that the approach and take-off are the primary aspects that significantly impact the outcome. The primary issue in long jump, as viewed from a biomechanical perspective, is the conversion of horizontal velocity into a resultant velocity that combines both vertical and horizontal components during the take-off phase. The athlete must prioritize achieving the highest vertical velocity while minimizing any decrease in horizontal velocity. The jumping distance is determined by the take-off distance (L1), flight distance (L2), and landing distance (L3), as stated in the theoretical model proposed by Ballreich and Bruggemann [1]. The CM flight parabola is determined by the following parameters: initial altitude of the CM at take-off, vector sum of the vertical and horizontal velocities at take-off, launch angle, and the effect of air resistance [1, 5, 6, 7]. Prior biomechanical investigations on long jump technique mostly focused on analyzing the kinematic properties using high-speed film or video cameras. Nevertheless, there is a scarcity of research focusing on the dynamic properties of take-off in elite long jumpers, despite its crucial role as the primary determinant of long jump performance. Our study aimed to conduct a comprehensive investigation of both the kinematic model of long jump and the dynamic model of take-off.

2. METHODS AND PROCEDURES

2.1. Kinematic model

An study was conducted on two leaps completed by G.C., one of the world's top long jumpers. G.C. has a body height of 178.5 cm, body mass of 69 kg, and is 24 years old.

Their personal best is 8.40 m. The initial jump measured 7.93 meters, while the subsequent jump measured 8.25 meters.

The kinematic parameters were acquired using a 3-D kinematic system called ARIEL, manufactured by Ariel Dynamics Inc. in the United States. The system consisted of four synchronized cameras, specifically Sony DVCAM DSR 300 PK, which recorded at a frequency of 100 Hz. The initial two cameras, positioned at a 90° angle relative to the object being videotaped, were employed to scrutinize the final two steps of the approach and the take-off. The remaining two cameras were utilized to capture the flight and landing phases (Figure 1). A 15-segment model was digitized for the purpose of kinematic analysis. The model segments correspond to distinct regions of the body, interconnected by point joints. The masses and centers of gravity of the segments, as well as the center of gravity of the entire body (CM), were calculated using the anthropometric model developed by Dempster [1]. The seventh-level Butterworth filter was used to filter all kinematic parameters.

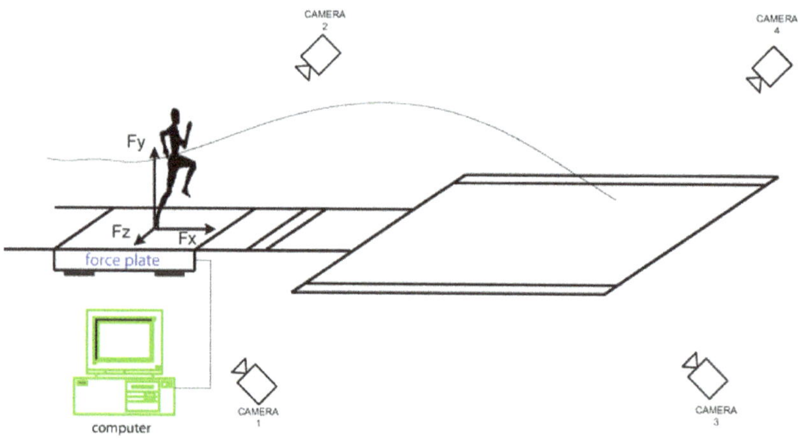

Figure 1. Measurement system for kinematic and dynamic parameters of the take-off in long jump.

2.2. Dynamic model

The dynamic parameters of take-off were registered with a force-platform Kistler 9287, area 900 × 600 mm, covered with a tartan surface and installed before the take-off board on the approach track of long jump. Forces were measured in three directions: X—horizontal, Y—vertical and Z—lateral (Figure 2).

The jump length (L) was measured from the front impression of the foot of the take-off leg on the force-platform to the point of contact of the feet in the sand of the landing pit. The programme package Matlab (Mathworks Inc., USA) was used to analyse the measured forces. The registration frequency on the force-platform was 2000 Hz.

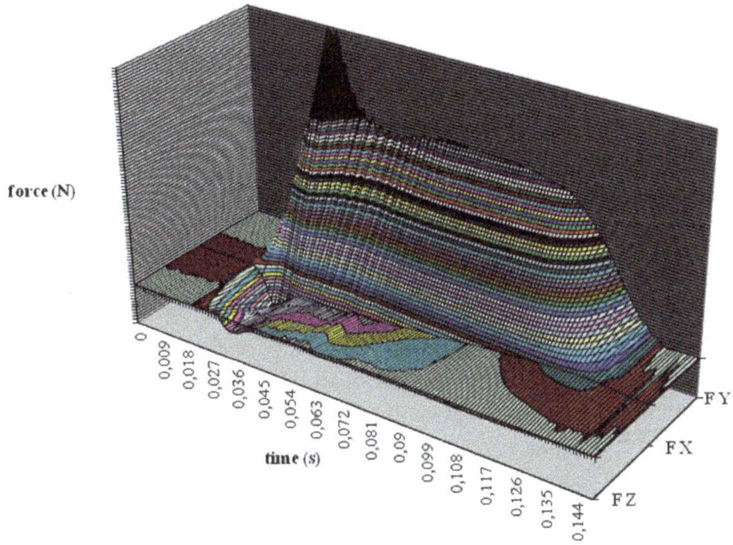

Figure 2. Take-off phase 3D-surface contour plot.

2.3. Optimising angle of projection at take-off action

We ask ourselves how to determine the best angle of projection at take-off for long jumpers. In our case, we model jumpers as mass points and observe their trajectories for a case where there is no air friction. We already know that energy losses have to be taken into consideration because jumper always loses some energy in a take-off phase, when he/she is transforming his/her horizontal velocity before he/she jumps into combination of horizontal and vertical velocity.

Let us first look at the example when height of centre of mass at take-off and height of centre of mass when he/she first touches the ground are the same. Let us denote velocity before jump with v_0 and velocity after take-off with v_1. Angle of projection shall be denoted with ϕ and jumpers mass with m. All mechanical energy before take-off is $W_0 = \frac{1}{2}mv_0{}^2$. This is jumpers kinetic energy. δW_0 denotes the amount of energy that is lost at the time of take-off as heat and sound [16], and of energy before take-off transforms into vertical kinetic energy. We denote this part as $\Delta W = \frac{1}{2}mv_1{}^2(\sin\varphi)^2$. We can therefore write:

$$W_1 = W_0 - \delta W - \Delta W \quad (1)$$

which means

$$\frac{1}{2}mv_1{}^2 = \frac{1}{2}mv_0{}^2 - \frac{\delta}{2}mv_0{}^2 - \frac{1}{2}mv_1{}^2(\sin\varphi)^2. \quad (2)$$

From this we get

$$v_1 = \sqrt{\frac{1-\delta}{1+\sin\varphi^2}}v_0. \quad (3)$$

The length of a jump of a mass point, which in our case is the centre of mass, is

$$D = D(\varphi) = \frac{v_1^2 \sin 2\varphi}{g} = \frac{\sin 2\varphi}{1+(\sin\varphi)^2} p, \tag{4}$$

where

$$p = \frac{(1-\delta) v_0^2}{g}. \tag{5}$$

To find the optimal value of ϕ, we need to find the extreme of function $D = D(\varphi)$. We do so by finding the zeros of the first derivate of function

$$\frac{\partial D}{\partial \varphi} = \frac{3\cos 2\varphi - 1}{(1+(\sin\varphi)^2)^2} p. \tag{6}$$

We must also take a look at the second derivate to establish if angle that we compute is maximum or minimum

$$\frac{\partial^2 D}{\partial \varphi^2} = \frac{4(3\sin 4\varphi + 14\sin 2\varphi)}{(\cos 2\varphi - 3)^2}. \tag{7}$$

Zeros of this function coincide with zeros of $f(\varphi) = 3\cos 2\varphi - 1$. From this equation we get the optimal angle of projection

$$\varphi_{opt} = \frac{1}{2}\cos^{-1}\frac{1}{3} = 35.26°. \tag{8}$$

With this we have established that optimal angle is independent of values v_0 and δ. The fact $\frac{\partial^2 D}{\partial \varphi^2}(\varphi_{opt}) < 0$ also holds true, which brings us to the conclusion that φ_{opt} in fact is the optimal angle for longest jump, that is, is the maximum of function $D = D(\varphi)$.

Of course height of the centre of mass of the jumper at take-off is bigger than the height of the centre of mass when the jumper first touches the ground because he/she pulls his/her legs closer to their bodies when landing (Figure 2—kinematic and biodynamic model of the long jump technique). In this case, the length of the jump is defined as $L = D + d$, where point $(D, -h)$ is defined as an intersection of curves:

$$y(x) = x\tan\varphi - \frac{g x^2}{2h v_1^2 (\cos\varphi)^2} \tag{9}$$

And $y(x) = -h$, where h denotes the height difference of centre of mass before and after the jump. Computing this intersection gives us

$$D(\varphi, h, v_1) = \frac{v_1^2 \sin 2\varphi}{2g}\left(1 + \sqrt{1 + \frac{2gh}{v_1^2 (\sin\varphi)^2}}\right) \tag{10}$$

which means

$$L(\varphi, h, v_1, d) = \frac{v_1{}^2 \sin 2\varphi}{2g}\left(1 + \sqrt{1 + \frac{2gh}{v_1{}^2 (\sin\varphi)^2}}\right) + d. \tag{11}$$

Taking into consideration that $v_1 = \sqrt{\frac{1-\delta}{1+\sin\varphi^2}}\, v_0$ and $p = \frac{(1-\delta)v_0{}^2}{g}$ we get

$$L(\varphi, p, h, d) = \frac{p}{2}\frac{\sin 2\varphi}{(1+(\sin\varphi)^2)^2}\left(1 + \sqrt{1 + 2\frac{1+(\sin\varphi)^2}{(\sin\varphi)^2}\frac{h}{p}}\right) + d. \tag{12}$$

In the same way, as we did before, we can compute the maximum value of function $L = L(\varphi, p, h, d)$ by finding the zeros of the function $\frac{\partial L}{\partial \varphi}$. With a lot of algebra and using Wolfram alpha we can show that

$$\varphi_{opt} = 2\tan^{-1}\left(\sqrt{\frac{16h}{p} + 5 - \frac{2\sqrt{2}\sqrt{3p^2 + 20ph + 32h^2}}{p}}\right). \tag{13}$$

In McFarland's research [10], δ is estimated to be 0.1.

3. MAIN BODY

3.1. Kinematic analysis

Take-off is one of the key generators of a successful long jump. The take-off model is defined by kinematic (Table 1 and Figure 3) and dynamic parameters. These are interrelated, in accordance with biomechanical laws. In this phase, the jumper must optimally transform the horizontal velocity developed in the approach into speed of take-off.

Table 1. Kinematic parameters of take-off action.

Parameter		Unit	Result
Touchdown			
Length of last stride	(S1)	m	2.05
Touchdown distance	(LTD)	m	0.63
Height of CM at TD	(HTD)	m	0.95
Horizontal velocity at TD	(VXTD)	m s^{-1}	9.46
Vertical velocity at TD	(VYTD)	m s^{-1}	-0.26
Speed at TD	(STD)	m s^{-1}	9.47
Maximum knee flexion	(MKF)	deg.	148.0
Height of CM at max. knee flexion	(HMA)	m	1.01
Take-off			

Parameter		Unit	Result
Take-off distance	(L1)	m	0.29
Height of CM at take-off	(HTO)	m	1.23
Horizontal velocity at TO	(VXTO)	m s^{-1}	8.10
Vertical velocity at TO	(VYTO)	m s^{-1}	3.90
Speed at TO	(STO)	m s^{-1}	8.99
Angular velocity of thigh at TO	(AVTO)	deg s^{-1}	790
Angle of projection at TO	(PATO)	deg.	24.1

The take-off speed is the resultant of horizontal velocity at TO and vertical velocity at TO and is one of the most important predictors of an effective jump length. Hay et al. [6], Nixdorf and Bruggemann [14], found that the correlation between the take-off speed and end result is between 0.74 and 0.83. Horizontal velocity at TD for the athlete GC is 9.46 m s^{-1}. In take-off (TD-TO), a reduction of the horizontal velocity of 1.36 m s^{-1} occurs, representing 14.3%. Its decrease is connected with the increase of vertical velocity at TO, amounting to 3.90 m s^{-1} (Figure 3). We can conclude that the loss in horizontal velocity is proportional to the increase in vertical velocity. The vertical velocity at touchdown is directed downwards and has a negative sign (VYTD = -0.26 m s^{-1}).

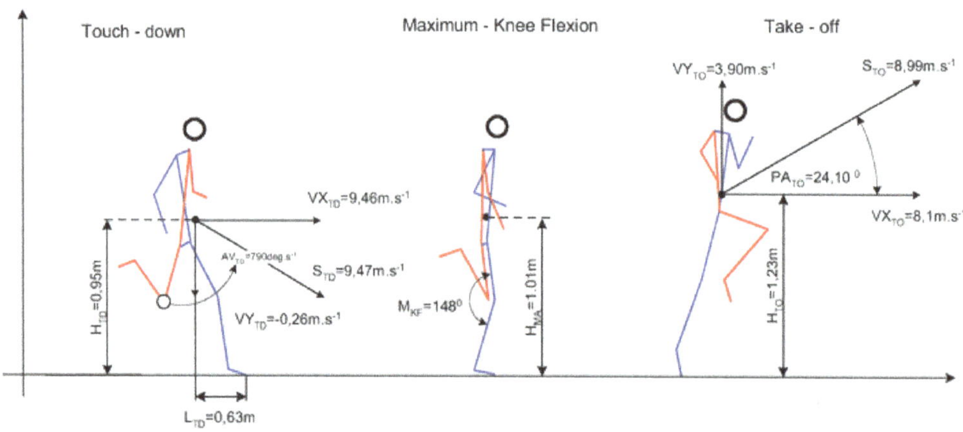

Figure 3. Kinematic model of take-off (G.C., distance 8.25 m).

The realisation of vertical velocity is directly connected with the magnitude of the projection of CM on the surface at touchdown (touchdown distance = 0.63 m). One of the key biomechanical problems of take-off is how to ensure the greatest possible vertical velocity, while keeping the decrease in horizontal velocity to a minimum. The ratio of the take-off velocity components VXTO:VYTO for the athlete GC is 2.08:1. The results of some similar studies [1, 5, 6, 9, 14] show us that this athlete has an optimal vertical velocity at take-off and a horizontal velocity

(VXTO) that is a little too low to allow even longer jumps. The consequence of a relatively high vertical velocity at take-off is also the magnitude of the angle of projection at take-off PATO = 24.10 deg. This magnitude of the angle of projection at take-off later defines the CM parabola of the jumper.

The realisation of a high vertical velocity can be connected with an efficient elevation of CM at take-off. The difference between the lowest position of CM at touchdown and at take-off is 28 cm. This is augmented also with a favourable ratio between the length of the last two strides (2.35 m:2.05 m) and the lowering of CM for 11 cm in the last-but-one stride. The average lowering of CM of the finalists of the 1997 World Championship in Athens was 8 cm [12]. This lowering of CM in the last stride but one increases the vertical acceleration distance in the take-off phase.

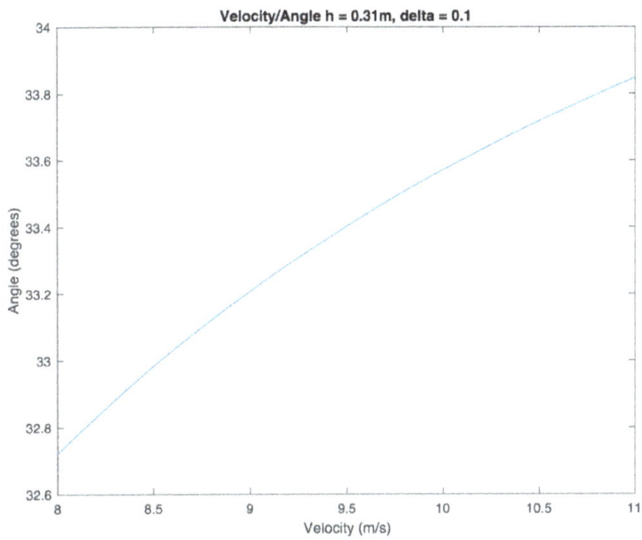

Figure 4. Effect of initial velocity on optimal angle of projection.

A very important factor contributing to the take-off efficiency is the angular velocity of the swing leg (AGTO), 790 deg s^{-1} for the athlete GC. This velocity is not constant, it changes. It is highest in the middle part of the swing amplitude, the lowest at the end of the swing, when the movement of the swing leg is blocked. The consequence of this is the transfer of the force of the mass inertia of this segment (the leg represents, according to the anthropometric model of Dempster, 16.1% of the entire body mass) to the common body centre of gravity of the jumper. Hay et al. [6] states that the contribution of the swing leg to the entire take-off impulse is between 17 and 20%.

Combining this information with the measured data:

$v_0 = 9.46\frac{m}{s}, v_1 = 8.99\frac{m}{s}, h = 0.31m$, and $d = 0.63$ we can compute $\varphi_{opt} = 33.39^\circ$ and $L = L(\varphi, p, h, d) = 6.86m$, while the measured distance of the jump was $8.25m$.

Because, we have all the needed data, it is interesting to see how initial velocity before the take-off v_0 affects the optimal angle of projection φ_{opt} computed above. The values for initial velocity were chosen from the interval [8,11] because most elite long jumpers have their initial velocity from this interval (Figure 4).

Similarly, we can investigate the relation between coefficient of efficiency δ and the length of the jump (Figure 5).

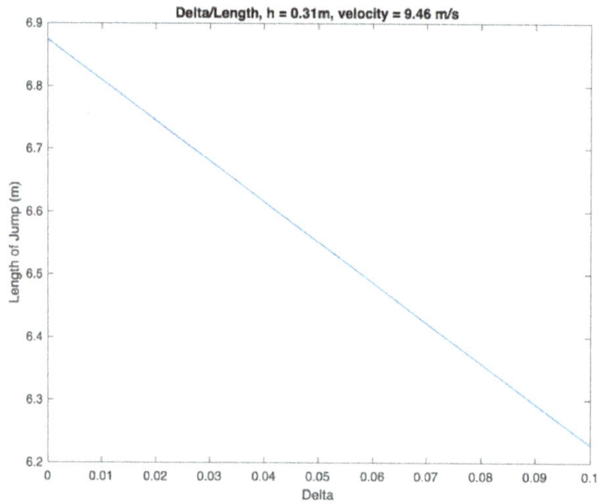

Figure 5. Effect of efficiency coefficient on length of the jump.

Next graph (Figure 6) is in fact actually another graph of correlation between initial velocity before take-off and optimal angle of projection with the difference that this graph shows how the length of jump depends on initial angle of projection for different initial velocities. The optimal angle can be seen as a value of an angle where maximum is achieved.

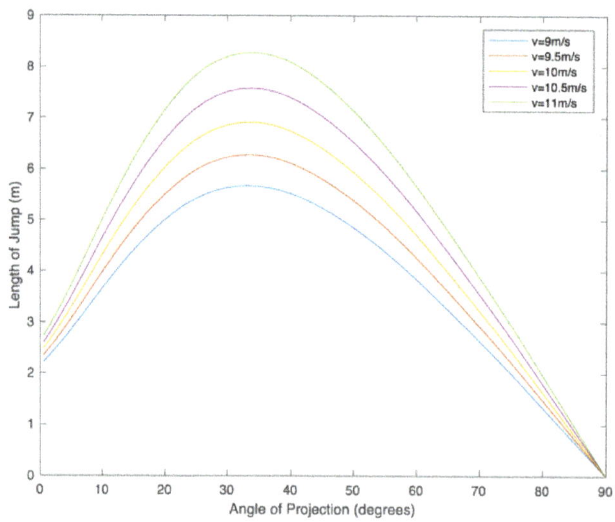

Figure 6. Relation between length of a jump and initial angle of projection.

3.2. Dynamic analysis of take-off action

The dynamic parameters of take-off (Table 2 and Figure 7) were measured directly with a force-platform. In this way, we measured the forces developed by the jumper at take-off in a realistic situation, which is a rarity in studies of this kind. The period of the contact phase was 127 ms. Here, the ratio between the period of compression (84 ms) and the period of lift (43 ms) is important—this was 66%:34% for our athlete, which is a good indicator of an efficient take-off [9, 11], from the point of view of the jump dynamics.

Table 2. Dynamic parameters of take-off action.

Parameter		Unit	Result
Contact time	(TD-TO)	ms	127
Period of compression phase	(TD-MKF)	ms	84
Period of lift phase	(MKF-TO)	ms	43
Max. force in X—horizontal axis	(FXMAX)	N	4581
Max. force in Y—vertical axis	(FYMAX)	N	5132
Max. force in Z—lateral axis	(FZMAX)	N	1396
Force impulse in compression phase	(FIMCP)	N s	-101.9
Force impulse in lift phase	(FIMLP)	N s	8.1
Total force impulse	(TFIMP)	N s	328.8

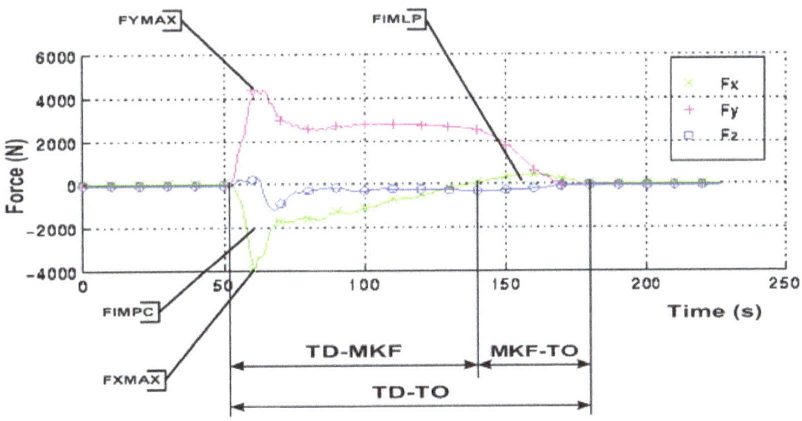

TD-TO	=	0,127	s	MZMAX	=	1395,90	N
TD-MKF	=	0,084	s	TFIMP	=	328,802	N/s
MKF-TO	=	0,043	s	FIMLP	=	8,106	N/s
FXMAX	=	4580,94	N	FIMPC	=	-101,920	N/s
MYMAX	=	5131,57	N				

Figure 7. Force diagram of the take-off phase.

The compression phase by definition lasts from the instant the foot of the take-off leg is placed on the ground till the moment of maximal amortisation in the knee of the take-off leg (TD-MKF). The lift phase then follows till take-off (MFK-TO). In the compression phase, our athlete developed a maximal force of 4581 N in the horizontal direction and 5132 N in the vertical direction. The muscular regime is eccentric in the compression phase period and concentric in the lift phase period. The long jump take-off is a typical example of a two-phase eccentric-concentric muscular contraction, whose efficiency depends mainly on two physiological factors. The first in the switching time—the utilisation of the elastic energy potential saved in the muscles, connecting tissue and tendons. This elastic potential is available only for a certain period of time. This time is defined by the life-time of the cross-bridges between the actin and the myosin in the sarcomeres, lasting from 15 to 100 ms. If the concentric contraction does not follow the eccentric one quickly enough, there is no transfer of the elastic energy from the first to the second phase. Studies have shown that elastic energy is the main force generator in the eccentric (compression) phase and chemical energy of the muscles in the concentric (lift) phase. An efficient integration of elastic and chemical energy in the muscles and tendons can cause a 40% greater resultant force [8, 15, 17].

The second factor of an economic eccentric-concentric contraction is the ability of the muscles to resist rapid extension—stiffness. Stiffness, as a neural mechanism, depends mostly on the pre-activation of the muscles and the action of reflexes: miotatic and Golgi tendon reflex [3, 4, 8, 13]. In light of the biomechanical characteristics, the short-range elastic stiffness is a characteristic for long jump take-off, where it is a matter of an immediate mechanic response of the activated muscle to eccentric contraction. In our experiment, we found the eccentric contraction time to be 84 ms and the athlete develops in this phase a maximal vertical ground reaction force of 5132 N. This force is almost 7.5 times the body mass of the athlete. An important kinematic criterion of the muscles' efficiency while resisting stretching (lowering of CM) is the angle of flexion in the take-off leg's knee—MKF = 148 deg. This angle must be sufficiently large. A marked lowering of CM in the MKF phase results in prolonging the compression phase period and through this an inefficient integration of the eccentric and concentric action of the muscles.

The results in the study of Lees et al. [9] show that force impulse in the compression phase is the primary indicator of vertical velocity in the lift phase. A study by Ballreich and Bruggemann [1] namely showed a much higher correlation between vertical velocity of take-off with effective length of the jump (r = 0.89), than with horizontal velocity of take-off (r = 0.21). The force impulse in the compression phase for the athlete GC was −101.9 N s.

3.3. Neuromuscular mechanisms of explosive power

In many sport disciplines, power is a major biomotor ability used in the prediction of results. To some extent, other biomotor abilities are also associated with power.

It is of little wonder that many kinesiological studies delve into the subject of power, investigating its structure, the training methodology, the application of new methods and diagnostic procedures. In modern kinesiological science, power is undoubtedly one of the most meticulously researched biomotor abilities, yet many questions in this field remain unanswered. The classification of power is based on different criteria. According to the criterion of action, authors [2, 4, 7, 8, 13] distinguish among peak power, explosive power and endurance. Another criterion is that of neuromuscular activity, where power manifests itself in the form of isometric contraction as well as concentric, eccentric or eccentric-concentric contraction. Isometric contraction occurs in conditions where muscular force equals an external force, which is why there is no movement between two points of the muscle. Eccentric contraction occurs in conditions where the external loading is greater than the force of the activated muscles. In real motor situations, an eccentric-concentric muscle contraction is the most common type and manifests itself in take-off power. Take-off power is a special type of explosive power in eccentric-concentric conditions and is most often seen in cyclic, acyclic and combined movement structures. Its main characteristic is the utilisation of elastic energy in the eccentric-concentric cycle of a muscle contraction. The contribution of the elastic properties of the muscle-tendon complex depends on the velocity of the transition. The transition must be as fast as possible and should not exceed an interval of 260 ms [7]. This type of muscle contraction uses less chemical energy to perform mechanical work compared to a concentric contraction alone, thus speeding up the movement. If a concentric phase of contraction follows an eccentric phase quickly enough, the elastic elements release the accumulated energy in kinetic and mechanical work at the beginning of the concentric phase, thereby increasing the muscular force [15].

The movement structures that occur in specific sport situations are associated with different inputs of eccentric and concentric muscle contractions. Due to the inter-segmentary transmission of energy and the optimisation of the take-off action, the thigh muscles' activity is important in vertical jumps in conditions of a concentric or eccentric-concentric muscle contraction [2, 13, 16]. With vertical jumps, the muscles engage in the take-off action following the proximal-distal principle of muscle activation. In the first phase of the jump when the vertical velocity of the body's centre of gravity increases, the extensors of the trunk and hip are the most active muscles. The key role in this phase of the take-off action is played by m. gluteus maximus, which is able to develop great force due to the relatively low angular velocity of the hips. The thigh muscles generate the peak activation at the beginning of the hip extension [4, 6]. Electrical activity in a muscle results from a change in the electrical potential of the muscle membrane. It is measured by means of electrical potential that can be detected on the surface of the skin. For this purpose, surface electrodes are fastened above the muscle whose EMG signal is to be measured. Silver-silverchloride (Ag-AgCl) bipolar surface electrodes are generally used. The skin at the recording site where the electrodes are applied must be carefully prepared, cleansed and treated. Take-off power in conditions where the active muscles first extend (eccentric contraction) and then shorten (concentric contraction) is measured by means of a countermovement vertical jump and a drop jump.

3.4. Kinematic analysis of flight phase

On the basis of the kinematic parameters (Table 3 and Figure 8), we see different contributions of the individual component distances to the final result. For the athlete GC, the flight distance (L2) has the greatest absolute and relative share in the total jump length—88%, then landing distance (L3)—7.7% and the smallest take-off distance (L1)—3.5%. The flight distance is defined as the horizontal distance between the point CM at take-off and the same point at the moment of contact with the sand.

Table 3. Kinematic parameters of the flight phase.

Parameter		Unit	Result
Distance of jump	(L)	m	8.25
Official distance	(OD)	m	8.10
Flight distance	(L2)	m	7.33
Max. height of CM in flight	(HMF)	m	1.88

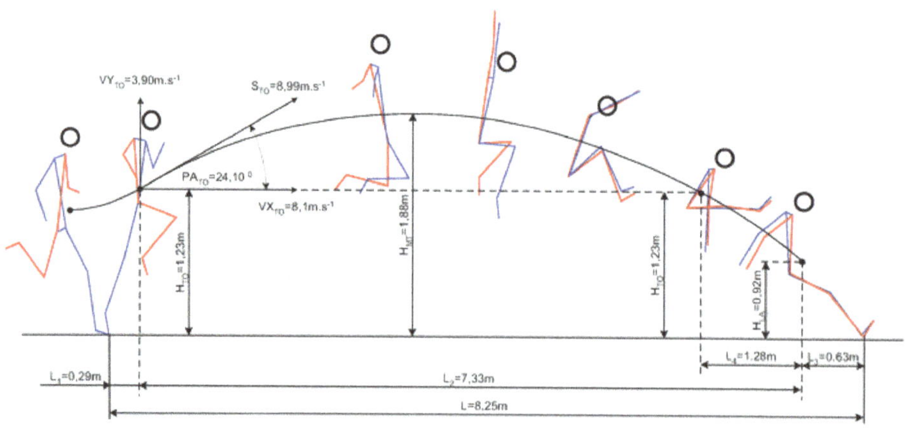

Figure 8. Kinematic model of flight and landing (GC, distance 8.25 m).

The small partial share of the take-off distance of the athlete GC is the consequence of the take-off angle, which amounted to 71.9 deg and the ratio between the vertical (VYTO) and horizontal component of velocity (VXTO) at take-off. The relatively large share of the landing distance (L3) can be ascribed to a very economic technique of landing. In the flight phase, the hang-style long jump is characteristic for the athlete GC, which he utilises very effectively as preparation for landing. A high parabola of CM flight can be seen for this athlete, which is the consequence of a high take-off angle (PATO = 24.1 deg). The highest point of the parabola is 65 cm above the height of CM at take-off. The characteristics of the flight phase are of course also influenced by other factors, especially air resistance, which was not the object of study in this research.

3.5. Kinematic analysis of landing phase

An economic technique of landing is defined by: landing distance (L3), height of CM at the moment of contact of the feet with the sand (HLA) and landing flight distance (L4)—Table 4 and Figure 8.

Table 4. Kinematic parameters of landing.

Parameter		Unit	Result
Landing distance	(L3)	m	0.63
Height of CM at landing	(HLA)	m	0.92
Landing flight distance	(L4)	m	1.28
Fall-back distance	(LFB)	m	0.15

The horizontal landing distance in landing is a crucial variable. It is determined by the projection of point CM and the initial contact with the sand, minus the distance lost by the athlete due to falling backward or breaking the sand at a point closer to the board than the initial point of contact (LFB). The economic index of the approach is calculated using the formula: IR = L3 - LFB. Hay and Nohara [7] discovered that, among a group of 12 top American long jumpers, the average distance they fell back after a jump was 11 cm. In the case of our athlete, this distance is 15 cm. Therefore, we can deduce that his technique is cost-effective and very efficient. Mendoza [11] further determines that landing distances (L3) ranging from 57 to 65 cm indicate a highly efficient landing. The athlete's landing technique plays a crucial role in determining the competitive result. This phase relies on specific motor abilities, particularly the power of the hip flexors and abdominal muscles, as well as the athlete's morphological characteristics and coordination skills. Additionally, the quality of the landing area also affects the outcome.

4. CONCISE SUMMARY

The findings from the kinematic and dynamic analysis indicate that the efficiency of a long jump is primarily determined by specific take-off parameters. These parameters include the horizontal velocity at take-off (TO), the vertical velocity at TO, the speed at TO, the angle of projection at TO, the maximum force exerted in the horizontal (X) and vertical (Y) axes, as well as the force impulse during the compression and lift phases. The landing phase of a rational long jump technique is crucial and may be characterized by two key measurements: the landing distance and the fall-back distance. This study was undertaken with a single exceptional long jumper (who achieved third place in the Universiade in Fukuoka and third place at the Seville World Athletics Championship). It provides valuable information about the dynamic and kinematic factors necessary for achieving excellent performance in this particular athletic discipline. Biomechanical diagnostics focuses on the specific measurable factors of a technique that should form the foundation of the training process. The study findings have significant theoretical and practical implications in the field of applied biomechanics and sports training, particularly in terms of determining the most effective training approaches.

REFERENCES

1. Ballreich G, Bruggemann G. Biomechanik des Weitsprungs. Biomechanik der Leichtathletik. Stuttgart: Ferdinant Enke Verlag; 1986. pp. 28-47
2. Čoh M. Biomechanical diagnostic methods in athletic training. Faculty of Sport, University of Ljubljana; 2008
3. Dempster W. Space Requirements of the Seated Operator. Wright-Patterson Air Force Base, Ohio: WADC Technical Report; 1955. pp. 55-159
4. Gollhofer A, Kyrolainen H. Neuromuscular control of the human leg extensor muscles in jump exercises under various stretch-load conditions. IJSM. 1991;12:34-40
5. Grahman-Smith P, Lees A. A comparison of the information quality between cinemato-graphy and videography long jump technique analysis. Biology of Sport. 1997;14(3):213-225
6. Hay J, Miller J, Canterna R. The techniques of elite male long jumpers. Journal of Biomechanics. 1986;19:855-866
7. Hay J, Nohara H. Techniques used by elite long jumpers in preparation for take-off. Journal of Biomechanics. 1990;23:229-239
8. Komi P. Stretch-shortening cycle: A powerful model to study normal and fatigued muscle. Journal of Biomechanics. 2000;33:1197-1206
9. Lees A, Smith G, Fowler N. A biomechanical analysis of last stride, touchdown, and take off characteristics of the Men's long jump. Journal of Applied Biomechanics. 1994;10(1):61-78
10. McFarland E. Encyclopedia of Sport Science. MacMillen; 1997. p. 185
11. Mendoza L. Individuelle Optimierung der Landeweite beim Weitsprung mit Hilfe der Computersimulation. Leistungssport. 1989;6:35-40

12. Muller H, Hommel H. Biomechanical research project at the VIth World Championships in Athletics, Athens 1997: Preliminary report. New Studies in Athletics. 1997;2–3:43-73

13. Nicol C, Avela J, Komi P. The stretch-shortening cycle. Sports Medicine. 2006;36(11):977-999

14. Nixdorf E, Bruggemann P. Biomechanical analysis of the long jump. In: Scientific Research Project at the Games of the XXIVth Olympiad—Seoul 1988; International Athletic Foundation; 1990. pp. 263-302

15. Tihanyi J. Die physiologische und mechanische Grundprinzipien des Krafttraining. Leistungssport. 1987;17(2):38-44

16. Tom A. The mathematics and physics of pole vault. Theta. 1996;10(2):14-18

17. Zatsiorsky V, Kraemer W. Science and Practice of Strength Training. 2006. Available from: HumanKinetics.com

Chapter 7

A KINEMATIC MODEL IS PROPOSED FOR PROJECT SCHEDULING THAT TAKES INTO ACCOUNT CONSTRAINED RESOURCES AND UNCERTAINTIES

1. INTRODUCTION: AN OVERVIEW

The issue of project scheduling has been extensively examined in academic literature since the 1950s. As stated in reference [1], project scheduling is a crucial procedure in project management. The existing body of research on project scheduling mostly focuses on creating a plan that is both practical in terms of task dependencies and resource availability. This schedule aims to optimize the project's objectives, often the duration, and serves as a reference point for project execution. Baseline schedules are used as a foundation for allocating resources to various operations in order to optimize performance and plan external tasks such as material acquisition, preventative maintenance, and order delivery to consumers, whether they are external or internal [2].

Project scheduling is a challenging task because of limited resources and the interdependencies across operations [3]. Admissible schedules must adhere to constraints such as precedence relations, which dictate that a work cannot begin until certain other tasks have been accomplished, and resource restrictions, which involve limited capacities of precious resources such as labor and equipment [4]. The resource-constrained project scheduling problem (RCPSP) involves scheduling project activities, taking into account various uncertainties, in order to minimize the total duration or makespan of the project. This scheduling must consider precedence and the availability of renewable, non-renewable, and doubly constrained resources [5].

The majority of the variations and expansions of the RCPSP can be condensed and categorized into several modes, generalized time delays, and objective functions, leading to exceedingly intricate optimization issues [1]. The RCPSP, or Resource-Constrained Project Scheduling Problem, is well recognized as a standard problem in the field of project scheduling. Many academics have devised both precise and heuristic scheduling methods to address this problem. Given the intricate nature of the problem, only a limited number of precise algorithms have been documented in the literature, while numerous heuristic solution strategies have also been described [3].

This chapter describes the creation and application of a kinematic model called the Coupled Estimate Technique for project scheduling. The model takes into account limited resources and uncertainties. The Coupled Estimate Technique involves a modelled duration that is influenced by both the planned duration and the variability of resources (aleatory uncertainty). Similarly, the modelled resource is determined by the planned resource and the variability of duration (aleatory uncertainty).

1.1. Adapting the kinematic notions of the robotic manipulator to the Coupled Estimate Technique.

The development of the Coupled Estimate Technique was based on robotic manipulator kinematic concepts. Kinematics is the science of motion that treats motion without regard to the forces/moments that cause the motion [17]. Robotic manipulator or industrial robot consists of several rigid links connected in series (open kinematic chain) by revolute or prismatic joints; one end of the chain is attached to a supporting base, while the other end is free and equipped with a tool (end-effector) to manipulate objects or perform assembly tasks; and the end-effector could be a gripper, a welding torch, an electromagnet or another device [13].

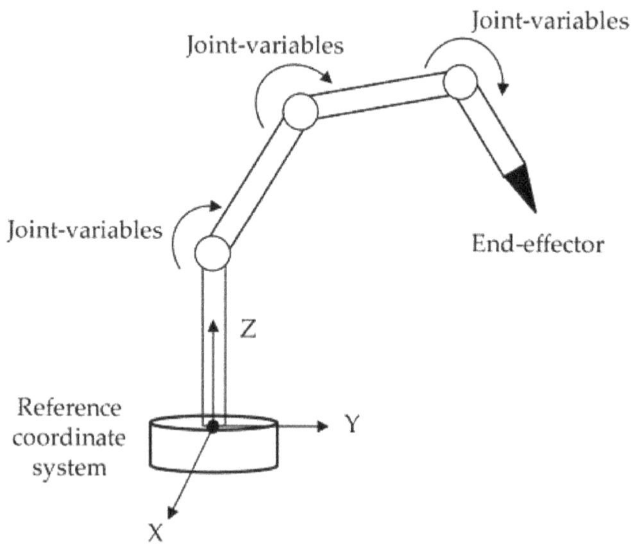

Figure 1. Robotic manipulator or industrial robot.

Robotic manipulator kinematics deals with the analytical description of the spatial displacement of the end-effector with respect to a fixed reference coordinate system, in particular the relations between the joint variables and the position and orientation of the end-effector [15]. There are two (direct and inverse) fundamental problems in robotic manipulator kinematics. Given a set of joint variables, the direct kinematic problem is performed to find the position and orientation of the end-effector relative to the fixed reference coordinate system. Given the position and orientation of the end-effector relative to the fixed reference coordinate system, the inverse kinematic problem is performed to find all possible sets of joint variables that could be used to attain this given position and orientation. Figure 1 illustrates a robotic manipulator or industrial robot.

In analogous way, the Coupled Estimate Technique deals with the analytical study of the geometry of project activities of a precedence diagram with respect to a fixed reference two-dimensional coordinate system (0, T, R) where the abscissa is the duration axis (T) and the ordinate is the resource axis (R). Thus, the precedence diagram may be considered as a kinematic chain (open, closed and/or hybrid), where one end of the chain is attached to a fixed reference two-dimensional coordinate system and represents the begin project activity, while the other end is free and represents the end project activity, as well as the rigid links represent the precedences between project activities and joints are represented by the project activities. Figure 2 presents a precedence diagram analogous to a robotic manipulator in open kinematic chain.

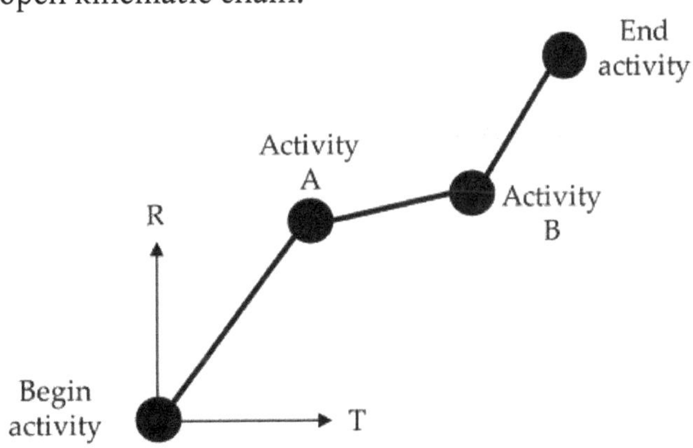

Figure 2. Precedence diagram in open kinematic chain.

1.2. The main effects of the uncertainties in project scheduling

The majority of research efforts related to RCPSP assume complete information about the scheduling problem to be solved and a static and deterministic environment within which the precomputed baseline schedule is executed [6]. However, in the real world, project activities are subject to considerable uncertainties, stemming from various sources, which are gradually resolved during project execution.

The presence of uncertainties in project management is recognized by practitioners and researchers as an important cause underlying the high project failure rate [5]. Project-related uncertainties can lead to numerous schedule disruptions. These uncertainties may stem from the following sources [7]: activities may take more or less time than originally estimated; resources may become unavailable; material may arrive behind schedule; ready times and due dates may have to be changed; new activities may have to be incorporated; activities may have to be dropped due to changes in the project scope; and weather conditions may cause severe delays.

This chapter makes a distinction between uncertainty and risk, assuming that there are degrees of knowledge about the estimations of project duration and resources; the uncertainty may be quantifiable but without probability of occurrence, while the risk is quantifiable and with probability of occurrence [8]. Uncertainty may be aleatory due to the intrinsic variability of projects or epistemic due to the lack of relevant knowledge related to carrying out the project [9]. Thus, quantifying uncertainty is relevant to project scheduling because it sheds light on the knowledge gaps and ambiguities that affect the ability to understand the consequences of uncertainties in project objectives.

1.3. Common approaches for RCPSP under uncertainties

From a modelling viewpoint, there are four approaches for dealing with uncertainty quantification in a scheduling environment where the evolution structure of the precedence network is deterministic: reactive scheduling, stochastic scheduling, scheduling under fuzziness and proactive (robust) scheduling.

1.3.1. Reactive scheduling approach

The reactive (predictive) scheduling models do not try to cope with uncertainty quantification when creating the baseline schedule. Basically, efforts are largely concentrated on repairing, revising or re-optimizing the baseline schedule when an unexpected event occurs [2]. The reactive effort may be based on very simple techniques aimed at a quick schedule restoration; a typical example is the well-known right-shift rule. This rule will move forward in time all the activities that are affected by the delays; often, the precedence diagram must be tailored in order to rearrange the changed activities. The reactive effort also may be based on as full rescheduling of the precedence diagram [4].

1.3.2. Stochastic scheduling approach

Most of the literature on the RCPSP assumes that activity durations are deterministic; however, activity durations are often uncertain. These uncertainties may be due to different sources, such as estimation errors, late delivery of resources, unforeseen weather conditions, unpredictable incidents (machine or worker) and others [6].

The stochastic RCPSP acknowledges that activity durations are not deterministic, i.e. the activity durations are modelled as stochastic variables. Therefore, the stochastic RCPSP aims to schedule project activities with uncertainty quantification of durations in order to minimize the expected project duration subject to precedence and constrained resources [5]. The stochastic project scheduling models view the project scheduling problem as a multistage decision process; the complete schedule (containing all activities) is constructed gradually as time progresses by means of a scheduling policy, exploiting the available information about the uncertainty of activity durations [1].

1.3.3. Fuzzy scheduling approach

Generally, the uncertainty of activity duration in project scheduling was handled by stochastic approaches using a probabilistic-based method. This kind of uncertainty in project duration is associated with randomness. However, for projects never be carried out previously, it is infeasible to determine the probability distribution of activity duration [2].

Therefore, the fuzzy project scheduling approach is used when the probability distributions for the activity durations are unknown due to a lack of historical data and, thus, the activity durations have to be estimated by human experts. In the situations involving the imprecision instead of uncertainty, the project scheduling literature recommends the use of fuzzy numbers for modelling activity durations. Instead of probability distributions, these quantities make use of membership functions, based on possibility theory [3].

1.3.4. Proactive (robust) scheduling approach

Traditionally, the robust schedule may absorb some level of unexpected events (machine breakdowns, staffing problems, unexpected arrival of new orders, early or late arrival of raw material and uncertainties in the duration of processing times) [2]. Thus, in order to minimize the impacts of uncertainties and the need of new scheduling or rescheduling, proactive scheduling approach aims at the generation of a robust baseline schedule that incorporates a certain degree of anticipation of potential variability or of potential disruptions according to the objective values (makespan) [3].

The proactive (robust) project scheduling model has prospered widely in the field of machine scheduling. Redundancy-based techniques related to durations and resources have already found their way to the field of project scheduling. The critical chain project management (CCPM) and Success Driven Project Management (SDPM) methods are becoming increasingly popular among project management practitioners [1].

1.4. Main differences between the Coupled Estimate Technique and the common approaches for RCPSP under uncertainties

This chapter presents a kinematic model named as Coupled Estimate Technique for project scheduling with constrained resources under uncertainties. This technique considers precedence, duration, resources and uncertainties related to project activities in order to analytically model the outcomes of project-related events or conditions (uncertainty) with the potential of favourable but mainly adverse consequences on project objectives (duration and resources). This approach can be used to quantify uncertainties; thus, it can help to solve project scheduling problems related to the following limitations and disadvantages identified in the literature review:

- The literature on project scheduling under risk and uncertainty was clearly conceived in a machine scheduling environment [10], whereas this work presents a project scheduling model from the viewpoint of project management.
- Projects are often subject to considerable uncertainty during their execution, but most of the research on project scheduling deals with only one source of uncertainty most often the duration of activities [11]. With the kinematic model detailed herein, the project activities have a set of attributes represented by the duration and resources as well as the uncertainties related to duration and resources.
- Traditional project scheduling is represented mainly by an activity-on-node network [12]. However, this representation is insufficient and inadequate when the activities of a project have a set of attributes. In the kinematic model, project scheduling is represented by IDEF0 (Integrated Computer-Aided Manufacturing Definition for Functional Modelling).
- Project scheduling methods are focused mainly on the basic RCPSP model [3]. In the kinematic model presented here, project scheduling is considered as a kinematic chain (open (serial), closed (parallel) and/or hybrid) formed by a set of rigid links (precedence of activities) that are connected by joints (project activities) with one fixed extremity (activity that represents the beginning of the project) and one free extremity (activity that represents the end of the project), which are represented by a homogeneous transformation matrix.
- The literature on project scheduling states that the generation of proactive (robust) multi-resource baseline schedules in combination with efficient and effective reactive schedule repair mechanisms constitutes a viable area of future research [10]. The kinematic model presented in this chapter

provides this combination through the direct and inverse kinematic models. It provides evidence of the influences stemming from uncertainties in project activities, enabling the balancing of durations and resources between project activities and between projects.

2. KINEMATIC FUNDAMENTALS FOR THE ROBOTIC MANIPULATORS AND FOR THE MODEL PROPOSED BY THIS CHAPTER

This topic aims to provide an overview of the key principles associated with kinematic problems for robotic manipulators or industrial robots. It also aims to introduce the fundamental concepts related to the kinematic model proposed in this chapter for project scheduling with limited resources and uncertain conditions.

2.1. Kinematic problems for robotic manipulators

In order to describe and represent the spatial geometry of the links of a robotic manipulator with to a fixed reference coordinate system, Denavit and Hartenberg (D-H) proposed a systematic approach utilizing matrix algebra [17]. This systematic approach reduces the direct kinematic problem to finding an equivalent 4x4 homogenous transformation matrix that describes the spatial relationship between two adjacent rigid links with respect to a fixed reference coordinate system [13].

In the inverse kinematic problem, given the position and orientation of the end-effector, we would wish to find the corresponding joint variables of the robotic manipulator [15]. The inverse kinematic problem is not as simple as the direct kinematic problem, because the kinematic equations are nonlinear and their solution is not easy (or even possible) in a close form. The inverse kinematic problem may be solved by various methods, such as inverse transform of Paul, screw algebra of Kohli and Soni, dual matrices of Denavit, dual quaternion of Yang and Freudenstein, iterative of Uicker, geometric approaches of Lee and Ziegler and others. Figure 3 represents a synthesis to the direct and inverse kinematic problem [16].

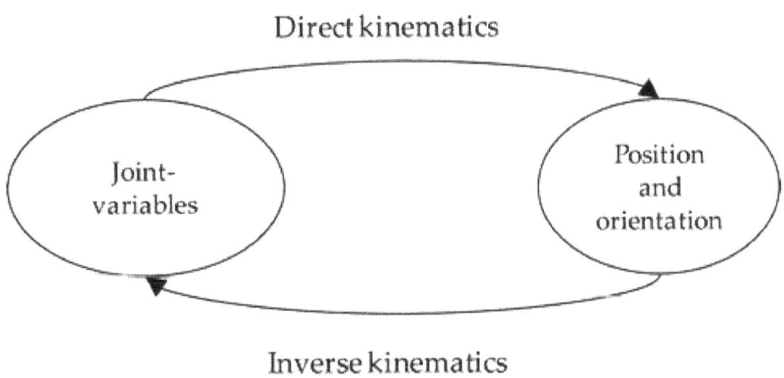

Figure 3. Synthesis to the direct and inverse kinematic problem for robotic manipulators.

2.2. Kinematic models for Coupled Estimate Technique

Basically, the project consists of a set of activities that must be performed in order to complete the project. The kinematic model technique for project scheduling with constrained resources under uncertainties named as Coupled Estimate Technique deals with the movements in the scheduled activities without consideration of the causes of the movements. Each project activity has a set of attributes represented by activity estimates (duration, resource and precedence), activity uncertainties (duration, resources and critical factor) and activity variables (duration and resource).

The outcomes of activities depend of their attributes, and the analytical kinematic model provides a metaschedule to the project. To model a project activity with kinematic model, it is assumed that the activity estimates are the constants, the kinematic equations are the orientation, and the modelled outcomes depend of the uncertainties and variables. Thus, the kinematic model may be direct or inverse type according to Figure 4.

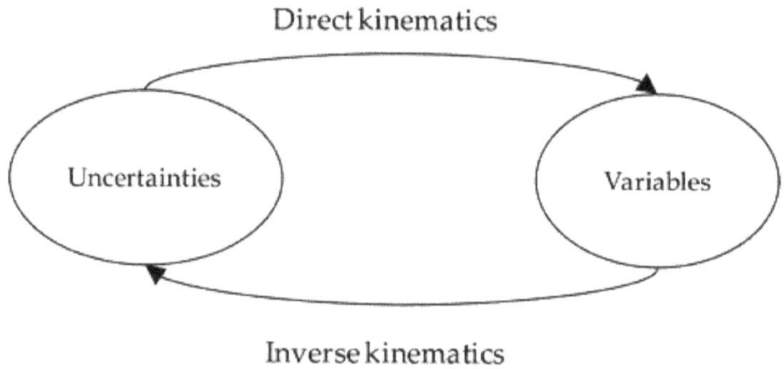

Figure 4. Kinematic model for Coupled Estimate Technique.

Figure 5. (a) Direct kinematic model and (b) inverse kinematic model.

Therefore, given the activity uncertainties determined by the project team, the direct kinematic model or the CET may be used to model the activity variables. Moreover, given the activity variables determined by the project team, using the inverse kinematics or the CET, the activity uncertainties may be modelled. It is important to emphasize that the estimates of the activity attributes must be part of organizational policy used by the team during the project planning. Figure 5 illustrates the kinematic model or the CET with the IDEF0 language.

3. COUPLED ESTIMATE TECHNIQUE CONCEPTS

Coupled Estimative Technique (CET) is a project scheduling artefact to model the project activity with constrained resources under uncertainties. This section presents the main aspects of the CET that provide the mathematical formulation involving the activity estimates (duration, resource and precedence), activity uncertainties (duration, resources and critical factor) and activity variables (duration and resource).

Let each project activity be represented in the two-dimensional Cartesian coordinate system (0, T, R) where the abscissa is the duration axis (T) and the ordinate is the resource axis (R). The value set of the duration and resource for project activity lies between the estimated ordered pair (estimated duration (te) and estimated resource (re)) and the modelled ordered pair (modelled duration (tm) and modelled resource (rm)).

Estimated ordered pair is represented in the estimated coordinate system (0, Te, Re), and the modelled ordered pair is represented in the modelled coordinate system (0, Tm, Rm). The rotation alpha expresses the uncertainty in the estimation of the activity duration, and the rotation beta expresses the uncertainty in the estimation of activity resource. Figure 6 shows the graphical representation of one project activity.

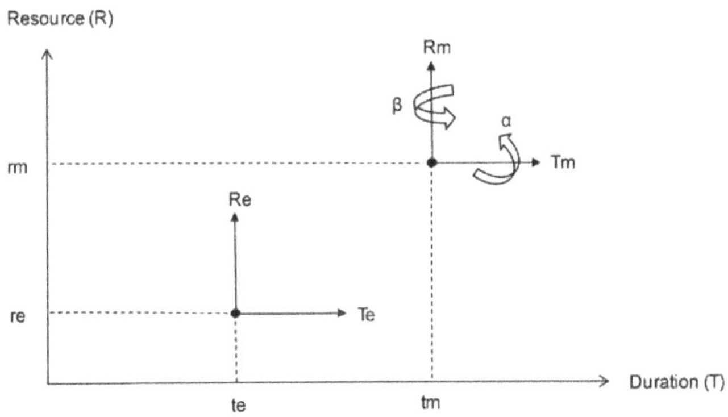

Figure 6. Graphical representation of one project activity.

The mathematical formulation of the CET used to model the project activities is obtained by algebraic operations with homogeneous transformation matrices (4 × 4). They map a project activity from estimated coordinate system (estimated ordered pair) to modelled coordinate system (modelled ordered pair) using the homogeneous coordinate notation. The homogeneous transformation matrix (H) consists of four submatrices [13], according to Eq. (1):

$$H = \begin{vmatrix} \text{Rotation} & \text{Position} \\ \text{Perspective} & \text{Scaling} \end{vmatrix} = \begin{vmatrix} R_{3x3} & P_{3x1} \\ f_{1x3} & 1x1 \end{vmatrix} \quad (1)$$

The rotation submatrix (R3 × 3) transforms one project activity expressed in the estimated coordinate system to the modelled coordinate system. For project scheduling, the rotation submatrices are used to produce the effect of the uncertainty in the estimation of the activity duration by rotation alpha at the duration axis and the effect of the uncertainty in the estimation of activity resource by rotation beta at the resource axis. According to [14, 15], the basic rotation matrices are shown in Eq. (2), rotation alpha at the duration axis and rotation beta at the resource axis:

$$R_{T,\alpha} = \begin{bmatrix} 1 & 0 & 0 \\ 0 & \cos\alpha & -\sin\alpha \\ 0 & \sin\alpha & \cos\alpha \end{bmatrix}; R_{R,\beta} = \begin{bmatrix} \cos\beta & 0 & \sin\beta \\ 0 & 1 & 0 \\ -\sin\beta & 0 & \cos\beta \end{bmatrix} \quad (2)$$

As the duration and resource of the project activity vary between estimated and modelled values, the position submatrix (P3 × 1) represents the coordinates of the estimated ordered pair. Therefore, the first element of the position submatrix is the duration translation or the estimated duration (te), the second element of the position submatrix is the resource translation or the estimated resource (re), and as the project activity is represented in a geometric plane, the third element is null. The position submatrix (P3 × 1) is shown in Eq. (3):

$$P = \begin{bmatrix} te \\ re \\ 0 \end{bmatrix} \quad (3)$$

The perspective transformation submatrix (1 × 3) is useful for computer vision and the calibration of camera models [16]. For the mathematical formulation of the CET, the elements of this submatrix are set to zero to indicate null perspective transformation. The fourth diagonal element is the global scaling factor (1 × 1), which is set to 1 in this work. Values different from 1 are commonly used in computer graphics as a universal scale factor taking on any positive values [17]. The project activity is mathematically represented through homogeneous transformation matrices (4 × 4) which are arranged in four steps to obtain the continuous duration and cost functions as described in Table 1.

Table 1. Steps to Develop the Homogeneous Transformation Matrix of the CET.

Step	Description
1	To represent the effect of the uncertainty in the estimation of the activity duration in estimated resource of the project activity.
2	To represent the effect of the uncertainty in the estimation of the activity resource in estimated duration of the project activity.

Step	Description
3	To represent the joint of the uncertainties (duration and resource) in the estimations (duration and resource) of the project activity.
4	To argue about the range of the uncertainties as well as the critical factor of the project activity.

The first step is performed through the product between the homogeneous matrices ($H_{T,\alpha}$) and ($H_{te; re}$), where ($H_{T,\alpha}$) means a rotation alpha on the duration axis and ($H_{te; re}$) means a translation in the duration and resource axes. The algebraic operation of Eq. (4) corresponds to the effect of the uncertainty in the estimation of the activity duration in estimated resource of the project activity:

$$\begin{bmatrix} 1 & 0 & 0 & 0 \\ 0 & \cos\alpha & -\sin\alpha & 0 \\ 0 & \sin\alpha & \cos\alpha & 0 \\ 0 & 0 & 0 & 1 \end{bmatrix} \cdot \begin{bmatrix} 1 & 0 & 0 & te \\ 0 & 1 & 0 & ce \\ 0 & 0 & 1 & 0 \\ 0 & 0 & 0 & 1 \end{bmatrix} = \begin{bmatrix} 1 & 0 & 0 & te \\ 0 & \cos\alpha & -\sin\alpha & ce.\cos\alpha \\ 0 & \sin\alpha & \cos\alpha & ce.\sin\alpha \\ 0 & 0 & 0 & 1 \end{bmatrix} \quad (4)$$

The first element of the position submatrix (P3 × 1) in Eq. (4) represents the activity duration; the second represents the estimated resource varying according to the uncertainty in the estimation of the activity duration. As the position submatrix (P3 × 1) is a spatial geometric representation (abscissa, ordinate and cote), the third element may be disregarded because the project activity is a plane geometric representation (abscissa and ordinate).

The second step listed in Table 1 performs the product between the homogeneous matrices ($H_{T,\beta}$) and ($H_{te; re}$). The former is the homogeneous matrix composed by a rotation beta on the resource axis, and ($H_{te; re}$) means a translation in the duration and resource axes. The algebraic operation of Eq. (5) corresponds to the effect of the uncertainty in the estimation of the activity resource in estimated duration of the project activity:

$$\begin{bmatrix} \cos\beta & 0 & \sin\beta & 0 \\ 0 & 1 & 0 & 0 \\ -\sin\beta & 0 & \cos\beta & 0 \\ 0 & 0 & 0 & 1 \end{bmatrix} \cdot \begin{bmatrix} 1 & 0 & 0 & te \\ 0 & 1 & 0 & re \\ 0 & 0 & 1 & 0 \\ 0 & 0 & 0 & 1 \end{bmatrix} = \begin{bmatrix} \cos\beta & 0 & \sin\beta & te.\cos\beta \\ 0 & 1 & 0 & re \\ -\sin\beta & 0 & \cos\beta & -te.\sin\beta \\ 0 & 0 & 0 & 1 \end{bmatrix} \quad (5)$$

The first element of the submatrix (P3 × 1) in Eq. (5) represents the activity duration varying according to the uncertainty in the estimation of the activity resource; the second, the activity resource. The third element of the submatrix (P3 × 1) in Eq. (5) may be discarded because the project activity is mapped at the geometric plane. The negative signal arises because the projection of the duration translation is at the negative semi axis of the cote coordinate.

The third step prescribed in Table 1 performs the sum between Eqs. (4) and (5). The summation shown in Eq. (6) represents the Coupled Estimate Technique (CET) overall homogeneous matrix that provides the joint effect of the uncertainties (duration and resource) in the estimations (duration and resource) of the project activity:

$$\begin{bmatrix} 1+\cos\beta & 0 & \sin\beta & te\,.\,(1+\cos\beta) \\ 0 & 1+\cos\alpha & -\sin\alpha & ce\,.\,(1+\cos\alpha) \\ -\sin\beta & \sin\alpha & \cos\beta+\cos\beta & ce\,.\,\sin\alpha - te\,.\,\sin\beta \\ 0 & 0 & 0 & 2 \end{bmatrix} \quad (6)$$

The first element of the position submatrix (P3 × 1) in Eq. (6) represents the modelled duration in function of the estimated duration and uncertainty in the estimation of the activity resource. The second element represents the modelled resource in function of the estimated resource and uncertainty in the estimation of the activity duration. As the project activity is mapped at the geometric plane, the third element may be disregarded. However, it might be used to represent other project goals, e.g. the quantification of the activity performance or quality.

And finally, the fourth step in Table 1 performs some orientations about the range of the uncertainties (duration and resource) and the critical factor of the project activity:

- The uncertainties must range between 0° (highest degree of uncertainty) and 89° (lowest degree of uncertainty). When alpha and beta equal to 0°, the modelled duration equals to the double of estimated duration (2te), and the modelled resource equals to the double of estimated resource (2re).
- When alpha and beta equal to 90°, there are no uncertainties or certainties. Therefore, the modelled duration equals to the estimated duration (te), and the modelled resource equals to the estimated resource (re).
- The certainties must range between 91° (lowest degree of certainty) and 180° (highest degree of certainty). When alpha and beta equal to 180°, the duration and resource modelled are null; this means a dummy activity, i.e. a project activity without duration and resource.
- The critical factor (theta) of the project activity must be unitary for cases where the modelled value is less than or equal to double of the estimated value, and the critical factor (theta) must be greater than one for cases where the modelled value is greater than double what was estimated value.

Table 2. Mathematical formulation of the CET.

Modelled variables	Equations
Duration	$tm = \theta.\ te.\ (1+\cos\beta)$
Resource	$rm = \theta.\ re.\ (1+\cos\alpha)$

Table 2 shows the mathematical formulations for one project activity of the Coupled Estimate Technique.

4. IMPLEMENTATION OF THE DIRECT KINEMATIC MODEL WITH CET

To demonstrate how the direct kinematic model is implemented, this section presents a didactic example with scheduling of duration in days, one type of resource (financial in dollar) and the critical factor unitary for all activities present in the critical path, according to the four sequential tasks. The main objective is to obtain the modelled schedule or project duration modelled (pdm) and modelled budget or project resource modelled (prm) with the effects of uncertainties of the estimates in the planned schedule or project duration estimated (pde) and planned budget or project resource estimated (pre) of the project. Assuming a critical path with the activities in the project precedence diagram which have the attributes:

- A0: Project beginning
- A1: Duration estimation equal to 30 days with uncertainty equal to 45°; resource estimation equal to \$100 with uncertainty equal to 45°
- A2: Duration estimation equal to 20 days with uncertainty equal to 45°; resource estimation equal to \$50 with uncertainty equal to 45°
- A3: End of project

As task 1 for implementation of the direct kinematic model with CET, the planned schedule and budget without the uncertainties must be determined; thus, the planned schedule or project duration estimated (pde) is equal to 50 days, and the planned budget or project resource estimated (pre) is equal to \$150.

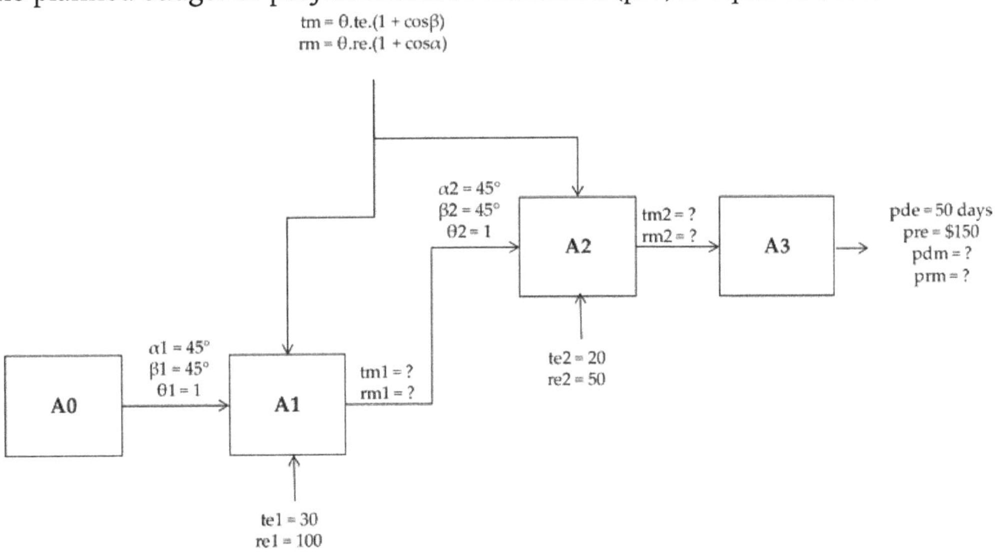

Figure 7. Project precedence diagram with IDEF0 for direct kinematic model with CET.

In task 2, the project precedence diagram is modelled with IDEF0 from the attributes of the project activities. Figure 7 presents the IDEF0 with:

- Activities estimates (te1, re1; te2, re2)
- Activity uncertainties (α1, β1, θ1; α2, β2, θ2)
- Activity variables to be modelled (tm1, rm1; tm2, rm2)
- Project duration estimated (pde) and project resource estimated (pre)
- Project duration modelled (pdm) and project resource modelled (prm) to be modelled

During the third task, the modelled variables are calculated according to the kinematic equations of the CET presented in Table 2 and Figure 4. From the estimates and uncertainties of the activities, the modelled variables of project activities are presented in Table 3.

Table 3. Modelled variables of the project activities.

Activities	Estimates	Uncertainties	Modelled variables
A1	te1 = 30 days	α1 = 45°	51 days
	re1 = $100	β1 = 45°	$170
A2	te2 = 20 days	α2 = 45°	34 days
	re2 = $50	β2 = 45°	$85

And finally, the fourth task describes that the project duration modelled (pdm) is equal to the sum of the modelled durations, and project resource modelled (prm) is equal to the sum of the modelled resources. Table 4 presents the project scheduling modelled according to the direct kinematic model with CET.

Table 4. Project scheduling modelled according to the direct kinematic model with CET.

Project	Estimates	Modelled	Effects
Duration	pde = 50 days	pdm = 85 days	35 days besides of planned
Resource	pre = $150	prm = $255	$85 besides of planned

Analyzing the information in Table 4, due to the uncertainties in the estimation of the project duration, the schedule should be increased in 35 days that represent 70% besides of the project duration estimated. In the same manner, due to the uncertainties in the estimation of the project resource, the budget should be increased in $85 that represents 70% besides of the project resource estimated.

5. IMPLEMENTATION OF THE INVERSE KINEMATIC MODEL WITH CET

To demonstrate how the inverse kinematic model is implemented, this section presents a didactic example with scheduling of duration in days, one type of resource (financial in dollar) and the critical factor unitary for all activities present in the critical path, according to the four sequential tasks. The main objective is to obtain the modelled uncertainties in the estimation of the project durations and resources from estimated and performed values of the project activities. Assuming a critical path with the activities in the project precedence diagram which have the attributes:

- A0: Project beginning
- A1: Duration estimation equal to 30 days and duration performed equal to 51 days; resource estimation equal to \$100 and resource performed equal to \$170
- A2: Duration estimation equal to 20 days and duration performed equal to 34 days; resource estimation equal to \$50 and resource performed equal to \$85
- A3: End of project

As task 1 for implementation of the inverse kinematic model with CET, the planned schedule and budget as well as the performed schedule and budget with the effects of the uncertainties must be determined; thus, the planned schedule or project duration estimated (pde) is equal to 50 days, the planned budget or project resource estimated (pre) is equal to \$150, the performed schedule or project duration performed (pdp) is equal to 85 days and the performed budget or project resource performed (prp) is equal to \$255.

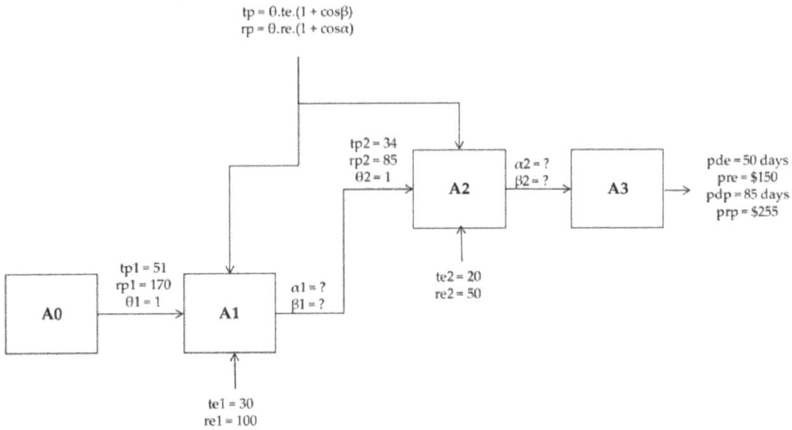

Figure 8. Project precedence diagram with IDEF0 for inverse kinematic model with CET.

In task 2, the project precedence diagram is modelled with IDEF0 from the attributes of the project activities. Figure 8 presents the IDEF0 with:

- Activity estimates (te1, re1; te2, re2)
- Activity uncertainties (α1, β1, θ1; α2, β2, θ2) to be modelled

- Activity variables, which for inverse kinematic model, the performed duration and resource (tp1, rp1; tp2, rp2)
- Project duration estimated (pde) and project resource estimated (pre)
- Project duration performed (pdp) and project resource performed (pre)

For the third task, the kinematic equations of the CET presented in Table 2 must be tailored where the modelled variables are substituted by the performed variables according to Figure 5. From the estimated and performed values, the modelled uncertainties of project activities are presented in Table 5.

Table 5. Modelled uncertainties of the project activities.

Activities	Estimate	Performed	Modelled uncertainties
A1	te1 = 30 days	51 days	$\alpha 1 = 45°$
	re1 = \$100	\$170	$\beta 1 = 45°$
A2	te2 = 20 days	34 days	$\alpha 2 = 45°$
	re2 = \$50	\$85	$\beta 2 = 45°$

In task 4, the uncertainty of the project must be analyzed. There are some ways to determine the project uncertainty; frequently, this choice is realized by the project team taking account the nature and challenges of the project. The following are some suggestions:

- The project uncertainty may be determined by the average between the uncertainties of the activities, where for the case studied in this section, the project uncertainty for the duration and resource are equal to 45°.
- The project uncertainty may be determined by the greater or lower uncertainty of the activities.
- The project uncertainty may be determined by the sum with the weight of uncertainties of the activities.

6. CONCISE SUMMARY

This chapter exemplified a fascinating instance of cross-pollination between various fields, where the concepts of manipulator kinematics (both direct and inverse) were customized and applied to offer a groundbreaking solution for project scheduling that takes into account limited resources and uncertainties. The homogeneous transformation matrices (4 × 4) were employed to represent the schedule (time) and budget (resource) of the projects, while considering the uncertainties in the estimates.

From a project management standpoint, the direct kinematic model can be employed for project scheduling to represent the schedule and budget when the duration and resource estimates are not fully known. Consequently, there are varying levels of uncertainty associated with duration and resource estimation.

The inverse kinematic model can be utilized in realized projects to accurately model and quantify the uncertainty associated with estimating the duration and resources required. The estimated uncertainties of the implemented project might be utilized for scheduling comparable projects.

The modeled results offer valuable insights that can optimize the project management processes, specifically in terms of project scheduling and budgeting. Additionally, they offer guidance on project risk management procedures, aiding in the identification, analysis, and mitigation of uncertainties that are not yet certain but have the potential to impact project objectives (such as length and resources) in either positive or, more commonly, negative ways.

Nevertheless, the model serves as a hypothesis of a phenomenon and, consequently, an estimation of the behavior of that phenomenon. Therefore, it is essential for a project team to thoroughly evaluate the projected results of the kinematic model for project scheduling, taking into account limited resources and uncertainties.

REFERENCES

1. Besikci U, Bilge U, Ulusoy G. Multi-mode resource constrained multi-project scheduling and resource portfolio problem. European Journal of Operational Research. 2015;240:22-31
2. Maravas A, Pantouvakis J. Project cash flow analysis in the presence of uncertainty in activity duration and cost. International Journal of Project Management. 2012;30:374-384
3. Herroelen W, Reyck B, Demeulemeester E. Resource-constrained project scheduling: A survey of recent developments. Computers & Operations Research. 1998;25:279-302
4. Peteghem VV, Vanhoucke M. An experimental investigation of metaheuristics for the multi-mode resource-constrained project scheduling problem on new dataset instances. European Journal of Operational Research. 2014;235:62-72
5. Yang Q, Lu T, Yao T, Zhang B. The impact of uncertainty and ambiguity related to iteration and overlapping on schedule of product development projects. International Journal of Project Management. 2014;32:827-837
6. Balouka N, Cohen I, Shtub A. Extending the multimode resource-constrained project scheduling problem by including value considerations. IEEE Transactions on Engineering Management. 2016;63(1):4-15
7. Xu J, Zheng H, Zeng Z, Wu S, Shen M. Discrete time–cost–environment trade-off problem for large-scale construction systems with multiple modes under fuzzy uncertainty and its application to Jinping-II Hydroelectric Project. International Journal of Project Management. 2012;30:950-966

8. Scholten L, Schuwirth N, Reichert P, Lienert J. Tackling uncertainty in multi-criteria decision analysis—An application to water supply infrastructure planning. European Journal of Operational Research. 2015;242:243-260
9. Ward S, Chapman C. Transforming project risk management into project uncertainty management. International Journal of Project Management. 2003;21:97-105
10. Hans EW, Herroelen W, Leus R, Wullink GA. Hierarchical approach to multi-project planning under uncertainly. The International Journal of Management Science. 2007;35:563-577
11. Klerides E, Hadjiconstantinou E. A decomposition-based stochastic programming approach for the project scheduling problem under time/ cost trade-off settings and uncertain durations. Computers and Operations Research. 2010;37:2131-2140
12. Vonder SV, Demeulemeester E, Herroelen W, Leus R. The trade-off between stability and makespan in resource-constrained project scheduling. International Journal of Production Research. 2006;44:215-236
13. Fu KS, Gonzalez RC, Lee CSG. Robotics: Control, Sensing, Vision and Intelligence. New York: McGraw-Hill Book Company; 1987
14. Lipschutz S. Linear Algebra. Schaum Outline, McGraw-Hill. 4th ed2011
15. Richard P. Robot Manipulators—Mathematics, Programming and Control. Cambridge, MA: MIT Press; 1982
16. Goldman R. Design of an Interactive Manipulator Programming Environment. Ann Arbor, MI: UMI Research Press; 1985
17. Craig JJ. Introduction to Robotics—Mechanical and Control. 3rd ed., Hardcover ed2004

Chapter 8 WMR KINEMATIC CONTROL UTILIZING UNDERACTUATED MECHANISMS FOR GOAL ORIENTATION AND AVOIDANCE

1. INTRODUCTION: AN OVERVIEW

Today, a wide variety of advanced robotic platforms capable of performing difficult tasks are built using underactuated mechanisms that were developed by ancient civilizations. Underactuated mechanisms are more efficient than redundantly kinematic systems because they effectively utilize the law of conservation of energy. Underactuated systems, in contrast to redundant systems, have a smaller number of actuators and fewer independent control variables. As a result, they are able to more effectively utilize the forces of inertia and gravity.

The redundant systems possess a greater quantity of control variables in comparison to the variables present in the working space. Therefore, each controlled actuator must exert forces to fight the effects of gravity and inertia in order to establish its own kinematic behaviors. One important part of robotics engineering is the development of physical models and control systems that enable a robot to navigate autonomously. Like any living organism, the capacity for deliberate navigation is essential for an intelligent robot. More than two millennia ago, the Chinese developed the south-seeking chariot (SSC), a device utilized to ensure a constant direction over extensive journeys spanning hundreds of kilometers. Therefore, they essentially developed one of the earliest compass devices that possessed absolute directionality and did not rely on any external components for its operation, relying solely on its internal processes. The invention's compass was utilized to align with a specific orientation at the commencement of a journey, and it was created almost 800 years prior to the development of the magnetic compass. The SSC is essentially a differential gearing system with a profile that points outwards, as shown in Figure 1. The gearing system utilizes gear transmission relations to counteract the chariot's turns, ensuring that the statue arm consistently points in the same direction. Figure 1a illustrates the design of the SSC, while Figure 1b showcases the SSC prototype using a gearing system composed of straight wooden gears. Nevertheless, this system has extensively been examined in contemporary times by multiple authors [1–7].

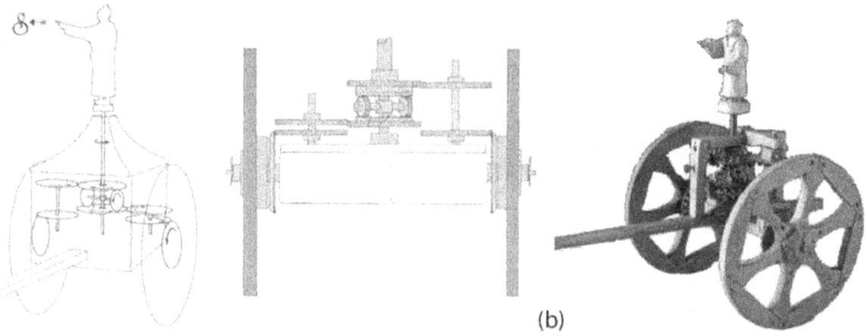

Figure 1. South-pointing chariot. (a) Modern design and (b) modern prototype.

Steering systems are essential for any navigational and path-tracking activity. Mobile robots utilize a variety of exteroceptive sensing technologies, such as ultrasonic sonar, infrared range detectors, cameras, and GPS, to deduce navigation references. However, a frequent drawback of using these devices is the need to acquire external measurements relative to the robot's stationary Cartesian coordinate system and utilize them to estimate orientation using geometry models that accumulate errors over time.

Furthermore, certain proprioceptive sensing systems provide internal readings that are relative to a global inertial system, such as a magnetic compass or GPS. Global proprioceptive sensors do not accumulate measurement errors over time, unlike local exteroceptive sensors. For example, a magnetic compass used as a worldwide orienting system measures the angle with respect to the Earth's magnetic axis. Subsequently, this angle measurement is utilized through mathematical operations to deduce a global target angle that is likely defined in a separate external inertial system. Nevertheless, magnetic compasses are susceptible to interference from adjacent magnetic fields and can experience inaccuracies when rotated to a position perpendicular to the earth's magnetic axis. Consequently, it is necessary to integrate additional types of sensing devices in order to compensate for any missing observations. Moreover, the global positioning system (GPS) is a different kind of global orientation measuring technology that necessitates a minimum of two consecutive position observations over time to determine the immediate global angle of a robot with respect to the earth's geographic north pole. One significant drawback of GPS is its inability to accurately measure locations within buildings, forests, or on overcast days due to the obstruction of electromagnetic signals. This results in the loss of data during certain periods of time.

Various studies have documented the use of navigation systems that integrate multiple orientation measuring devices, including GPS and magnetic compasses, to achieve a high level of accuracy. This method generates sensing readings that are linked to recognized references, such as the magnetic or north pole. Therefore, it is necessary to derive valuable global-specific references using various methods, which may involve solving geometric triangulations or algebraic models in either an inverse or direct manner [8–12].

The reliable navigation of a wheeled mobile robot, whether on a local or global scale, is based on the fundamental principles of controllability and maneuverability. Both robots' capabilities rely on solid driving control models and steering kinematic designs [13, 14]. These models serve as the foundation for planning and motion control in navigation [15, 16], relying on various numerical mathematical solutions in robotics [17]. There exist intricate studies on the navigation of robots that aim to find absolute orientation references [19, 20]. Other navigation systems prioritize collision avoidance and rely on kinematic techniques [18, 21].

This study presents a proposal for the mechanical design, physical model, and control system of a tricycle-wheeled robot that incorporates underactuated mechanical functions. The design of the self-steering robot presented in this study draws inspiration from the south-seeking chariot, specifically to utilize the underactuated mechanical compass that provides the robot with the absolute direction to its destination. Hence, a specialized and intricate gear mechanism system was devised to autonomously navigate between a compass that avoids obstacles and a compass that guides towards the destination (Figure 2). The global goal-reference compass consistently provides the robot with direction information without the need for sensors, even when the robot performs various evasive maneuvers.

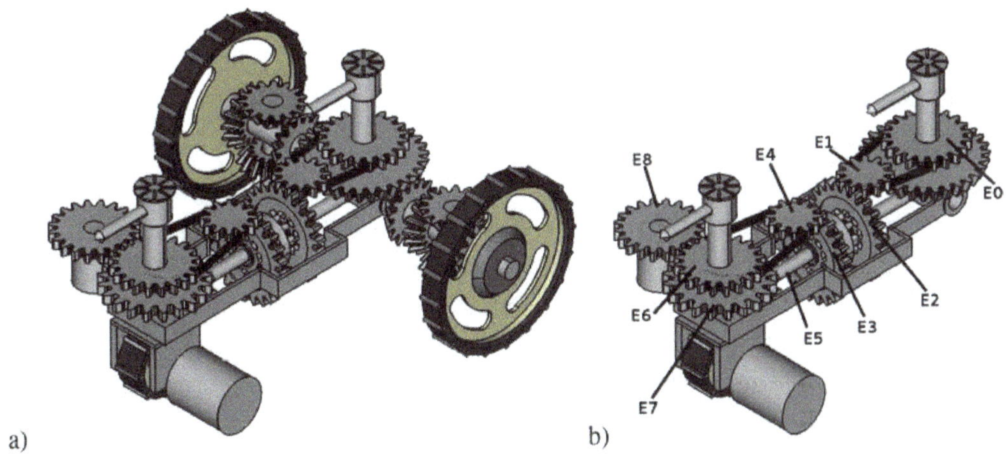

Figure 2. Proposed tricycle robot system. (a) General system view and (b) absolute/local compass mechanism.

A kinematic model is derived for the local/global compass and commutation processes. Furthermore, an examination and disclosure are conducted on a control law that governs the motion of a three-wheeled structure consisting of one drive-steer frontal wheel and a pair of lateral passive wheels. The proposed control law utilizes estimation techniques to determine the robot's posture and incorporates a dynamic switching mechanism between goal-oriented navigation and obstacle avoidance. A single underactuated mechanism directly guides the robot to the destination. Another mechanism with limited control inputs guides the robot along collision-free paths.

This publication presents simulation findings to verify a new technique that blends an old underactuated automaton type with a modern focus on wheeled robots. As of today, the authors are unaware of any similar methodologies mentioned in the scholarly literature. This work does not aim to provide a comparison analysis or assess the efficiency of other state-of-the-art robotic trends. The researchers assert that the initial findings mentioned in this document will develop into a very effective technical method in the next years. The application of this approach will introduce a new method that enables direct and accurate measurement of absolute angles over time. This allows the robot to navigate globally in a more natural way, reducing the computational load required for other algorithmic tasks. Additionally, it complements other sensing devices to enhance control and perception.

The sections of this chapter have been arranged in the following manner. Chapter 2 presents the kinematic models for the absolute/local mechanics. Chapter 3 covers the analysis of the kinematic properties and temporal delays of the commutation system. Chapter 4 introduces the control model and provides an overview of the simulation results. Chapter 5 provides definitive conclusions regarding this effort.

2. ABSOLUTE/LOCAL COMPASS MECHANISM

2.1. Global underactuated compass

The proposed system establishes the robot's chosen Cartesian aim as the definitive reference orientation. The absolute compass balances its bearing by utilizing the differential angle generated by the two lateral wheels (Figure 3a). The mechanical compass autonomously guides the robot towards the global reference point without the need for any additional actuator (Figure 3b). When the robot deviates from the global compass angle, the synchronization chain gradually adjusts the robot's steering wheel until it aligns with the goal's absolute angle. Figure 3c depicts the components of the self-steering mechanism, including: (1) passive lateral wheels, (2) frontal driving wheel with passive steering, (3) wheel differential gearing mechanisms, (4) transmission gears connecting the absolute compass, (5) global goal-angle compass, (6) driving actuator/motor, (7) self-steer synchronization chain, and (8) lateral wheel shaft.

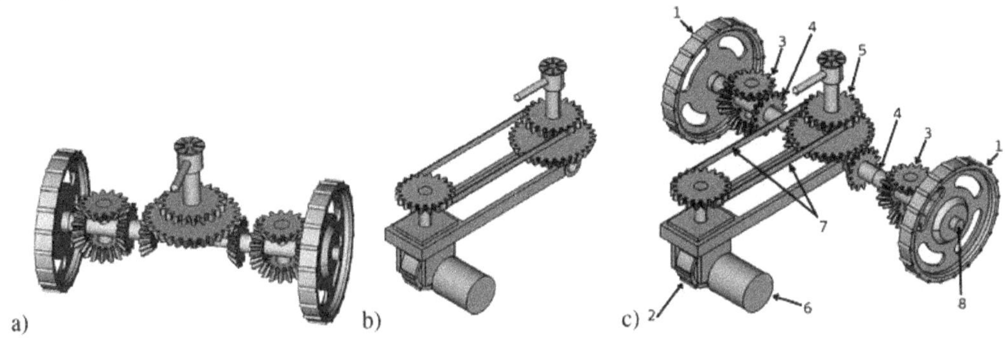

Figure 3. Absolute compass system. (a) Wheel differential mechanisms, (b) self-steer mechanism, and (c) goal-reference self-steer robot mechanism.

In Figure 4, a more detailed depiction of the differential mechanism is illustrated. The absolute compass (5) is composed of three differential systems (3) and (5), which transmit rotary motion from the lateral wheels (1) up to the steering wheel (2) (see Figure 4b). The compass (5) is compensated by the gearing relations (Figure 4b), and the differential rotary motion is transmitted to the gear (5), namely, e_{abs}. It is worth noting that the differential motion (3) of the lateral wheels' shaft (8) is fixed to the main axis (4). The lateral wheels' angular speeds, dubbed w_1 and w_2, are asynchronous w.r.t. the main shaft (4), similarly differential gears e_1 and e_2. The differential shafts of gears dif_1 and dif_2 (Figure 4b) are perpendicularly joined to the main axis (Figure 4a), and the rotary speed is basically the same for both differentials (4). Nevertheless, the angular velocity for gears ($dif1_1$, $dif1_2$, $dif2_1$, and $dif2_2$) might be different for each differential mechanism (3); it would depend on each wheel's (1) instantaneous velocity.

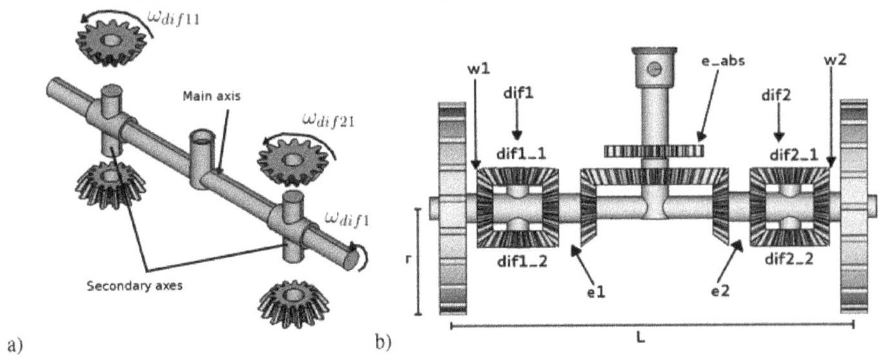

Figure 4. Differential gearing system. a) wheel main shafts, b) absolute direction differential.

Finally, the compass, differential mechanism (5) that is composed of the gears e_1, e_2, and e_{abs} is the mechanism providing the absolute orientation reference toward the robot's goal.

Without loss of generality, it follows that the deduction of the differential angular velocity model transmitted between e_1 and e_2 is provided next. Thus, the angular velocity for ω_{e1} is an averaged differential value:

$$\omega_{e1} = -\frac{\omega_1 - \omega_2}{2} \tag{1}$$

where both lateral angular speeds have clockwise (left-sided wheel) and counterclockwise (right-sided wheel) signs, respectively:

$$\omega_{e1} = -\omega_{e2} \tag{2}$$

The differential angular speed is equivalent to the main shaft rotary speed, such that

$$\omega_{dif1} = \omega_{dif2} = \frac{\omega_1 - \omega_2}{2} \tag{3}$$

and, in general, it is assumed that the gears' (*dif*1 and *dif*2) angular velocities are averaged values:

$$\omega_{dif11} = \omega_{dif12} = \frac{\omega_1 + \omega_{e1}}{2} \tag{4}$$

as well as

$$\omega_{dif21} = \omega_{dif22} = \frac{\omega_2 + \omega_{e2}}{2}. \tag{5}$$

In such a manner, the angular velocity transmitted to the compass *e_abs* poses the following relation:

$$\omega_{e_{abs}} = \frac{r_{e_{abs}}}{r_{e_1}}\omega_{e1}, \tag{6}$$

In the proposed design, it is assumed that the gear *e_abs* doubles its radius w.r.t. gears e_1 and e_2. Therefore, by rewriting Eq. (6)

$$\omega_{e_{abs}} = \frac{1}{2}\omega_{e1}, \tag{7}$$

and by substituting in the previous expression, the differential rotary speed model ω_{e1} since ultimately depends on velocities ω_1 and ω_2:

$$\omega_{e_{abs}} = -\frac{\omega_1 - \omega_2}{4}. \tag{8}$$

Therefore, from the previous expression, let us assume that the robot moves along a straight trajectory line, then $\omega_1 = \omega_2$, and no compass lateral motion is yielded; hence, $\omega_{eabs} = 0$. Following Figure 2b notation for the gears, the steering wheel with gear E_7 is synchronized with ω_{eabs} by the mechanism system illustrated in Figure 5a.

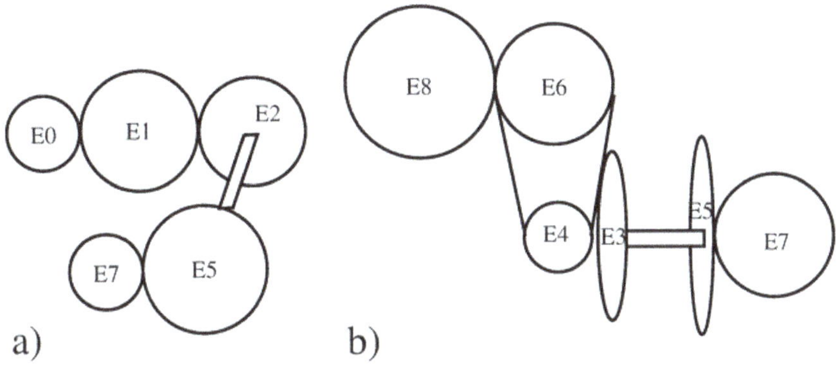

Figure 5. Steering mechanisms. (a) Goal-direction system and (b) evasive compass gearing.

A first gear connection $E0 - E1$ has the following speed relation:

$$\dot{\varphi}_1 = -\frac{r_0}{r_1}\dot{\varphi}_0; \tag{9}$$

Similarly, the gear connection $E_1 - E_2$ is algebraically simplified:

$$\dot{\varphi}_2 = -\frac{r_1}{r_2}\dot{\varphi}_1 = \frac{r_0}{r_2}\dot{\varphi}_0; \tag{10}$$

The parallel connection E_5 with E_2 has the same relation:

$$\dot{\varphi}_5 = \dot{\varphi}_2 = \frac{r_0}{r_2}\dot{\varphi}_0; \tag{11}$$

Finally, the angular velocity for E_7 w.r.t. E_0

$$\dot{\varphi}_7 = -\frac{r_5}{r_7}\dot{\varphi}_5 = -\frac{r_0 r_5}{r_2 r_7}\dot{\varphi}_0. \tag{12}$$

Therefore, equation (12) indicates that E7 exhibits a counter-clockwise rotational motion in reference to the absolute compass direction, with an angular speed that is half of its original value.

2.2. Local evasive compass

Another compass mechanism with local reference frame steers the robot for obstacle avoidance leading the robot along a safe instantaneous angle (Figure 5b). When the robot detects near obstacles, the absolute compass is suspended, and a commutator device switches to the evasive local compass mechanism, activating the gear E_8. Figure 5b describes the gearing mechanism that steers the robot for evasive navigation.

In the kinematic model for evasive steering through E_7, w.r.t. E_8 is deduced next. The relation between E_8 and E_6 is

$$\dot{\varphi}_6 = -\frac{r_8}{r_6}\dot{\varphi}_{8'} \tag{13}$$

and the angular velocity for E_4 is

$$\dot{\varphi}_4 = \frac{r_6}{r_4}\dot{\varphi}_6 = -\frac{r_8}{r_4}\dot{\varphi}_{8'} \tag{14}$$

For the perpendicular connection with E_3, the angular speed model is

$$\dot{\varphi}_3 = -\frac{r_4}{r_3}\dot{\varphi}_4 = \frac{r_8}{r_3}\dot{\varphi}_{8'} \tag{15}$$

For E_3 and E_5 connected in parallel

$$\dot{\varphi}_5 = \dot{\varphi}_3 = \frac{r_8}{r_3}\dot{\varphi}_{8'} \tag{16}$$

Finally, the model for E_7 is

$$\dot{\varphi}_7 = -\frac{r_5}{r_7}\dot{\varphi}_5 = -\frac{r_8 r_5}{r_2 r_7}\dot{\varphi}_8. \tag{17}$$

Let us highlight that for the evasive local compass E_8, the escaping orientation is instantaneously set up by an actuator only when any nearby obstacle is detected. And, such local orientation transmits motion to the steering gear (E_7) through the mechanism of Figure 5b. Thus, the local/global compass mechanisms are physical controllers that substitute models and computer algorithms.

3. ABSOLUTE/LOCAL COMMUTING MECHANISM

The commutation mechanism interactively couples and uncouples either the absolute compass or the local compass (Figure 6a). The commuting device switches into the local compass immediately where any nearby obstacle is detected. Alternatively, it switches to the global compass retaking orientation toward the goal as soon as obstacles are no longer detected. Figure 7a depicts the gearing transmission that commutates the different steering mechanisms, the local compass (the front), and the global compass (the back).

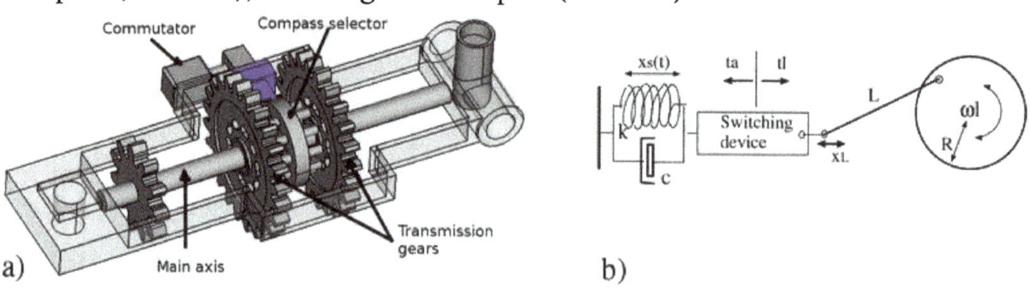

Figure 6. Commutator device. (a) Absolute/evasive motion transmission system and (b) basic commuter mechanism.

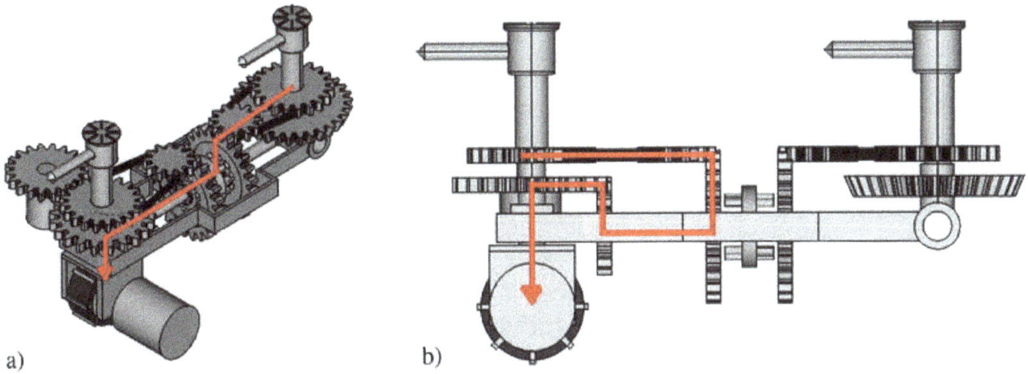

Figure 7. Motion transmission flow (red arrowed lines). (a) Goal-reference and (b) obstacle evasive.

According to Figure 6b, the commutation mechanism yields linear motion (the commutator device of Figure 6a); it activates the local compass by rotating a servomotor and shrinking a sliding crank link *L*. Inversely, the absolute compass is activated by turning off the servomotor, and then a spring-mass-damper system stretches the sliding crank link *L*.

The motion transmission system (Figure 6a) is inspired by the model of a modern vehicle's speed transmission box. For the present case, a common shaft works for two asynchronous speed gearing systems. In addition, Figure 7a and 7b illustrates the motion transmission flow by the red color arrowed lines. When either the local compass mechanism or the global compass mechanism is commutated, the flow transmission motion only concerns with its own gearing mechanism featuring a physical continuous controller.

The angular speed model ω_l [s^{-1}] for the commutation rotary servomotor (Figure 6b) to switch into the local compass was obtained through fitting empirical observations with the next theoretical model:

$$\omega_l = \kappa_1 e^{\kappa_2 t}; \tag{18}$$

where κ_1 [rad/s] is an amplitude factor and κ_2 [1/s] is a slope rate factor. Subsequently, to estimate the commutation time t_l [s] required to mechanically coupling into the local compass mechanisms, the next model is deduced:

$$t_l = \frac{\ln\left(\frac{\omega_l}{\kappa_2}\right)}{\kappa_1}. \tag{19}$$

Moreover, when no obstacles are detected, the commuter device switches into its initial state by coupling the absolute compass. In this case, an underactuated system commutates the state by a spring-mass-damper mechanical system with critically dumped configuration, modeled by second-order linear differential equation. Where *m* [kg] is the mass pulling the spring, *c* [kg/s] is the damping effect constant, k [kg/s [2]] is the spring elasticity constant, and x, $\dot{x}$, and $\ddot{x}$ are the spring distance, velocity, and acceleration, respectively:

$$m\ddot{x} + c\dot{x} + kx = 0 \tag{20}$$

Thus, it may be solved as a first-order linear equation by temporally omitting the second-order term $m\ddot{x}$ such that

$$c\frac{dx}{dt} = -kx, \tag{21}$$

reorganizing and completing the integrals

$$\int_x \frac{dx}{x} = \frac{-k}{c}\int_t dt, \tag{22}$$

Thus, by solving the improper integrals and multiplying both sides of the equation by the Euler number e,

$$e\left[\ln(x) = -\frac{k}{c}t + C_1\right] \tag{23}$$

and a solution is obtained; to simplify let us define $\lambda := -k/c$,

$$x(t) = e^{\lambda t + c_1} \tag{24}$$

Assuming the integration constant $C_1 = 0$ and developing the higher order solutions

$$x(t) = e^{\lambda t};\ \dot{x}(t) = \lambda e^{\lambda t};\ \ddot{x}(t) = \lambda^2 e^{\lambda t}. \tag{25}$$

In addition, by substituting such functions in the next expression

$$m\lambda^2 e^{\lambda t} + c\lambda e^{\lambda t} + ke^{\lambda t} = 0, \tag{26}$$

In order to decrease the commutation time of the sliding linear mechanism, a critically damped system is assumed and modeled by $(c^2 - 4mk) = 0$, with terms $\lambda_1 = \lambda_2$:

$$\lambda_{1,2} = \frac{-c}{2m}, \tag{27}$$

It is desired to speed up as much as possible the commutation time. Then, the mechanical device linear displacement is modeled as a function of time by

$$x(t) = (a_1 + a_2)e^{-ct/2m}. \tag{28}$$

And, solving for the commutation time for switching to the absolute compass t_a,

$$t_a = -\frac{2m}{c}\ln\left(\frac{x}{a_1 + a_2}\right) \tag{29}$$

Figure 8 depicts a sequence of transition times taken by the commuting mechanism. Times t_l are the times behavior of the slider crank servomotor when shrinking the link. Times t_a are the times behavior of the spring-mass-damper device when stretching the link. Such time magnitudes are reached due to mechanical motions. However, such commutation times are physically fast enough, and a robot's trajectory is not affected in the precise points when commutations occur.

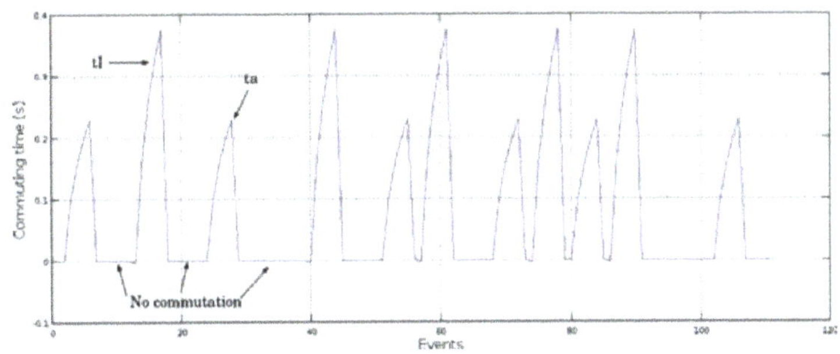

Figure 8. Commuting times: t_a *(absolute mechanism) and* t_l *(local mechanism).*

4. KINEMATIC CONTROL LAW

In this section, a kinematic control law is deduced and analyzed. The proposed controller simultaneously controls driving and self-steering velocities and keeps track of the robot's posture. The robotic platform is a tricycle-type kinematic structure with two lateral passive wheels at the back (Figure 9a) and a central active-drive and passive self-steering wheel at the front (Figure 9b). Each wheel kinematic is described by three parameters, α, ℓ, and r, where α [rad] is the angle of a wheel w.r.t. the x-axis of the robot's fixed inertial system, ℓ [m] is the distance between the robot's centroid and each wheel's contact point with the ground, and r [m] is an ideal wheel's radius. In addition, the robot's kinematic structure considers two controlled kinematic variables β_t and ϕ_t, where β_t [rad] is the instantaneous steering angle of the front-sided wheel and ϕ [rad] is the instantaneous driving wheel's angle.

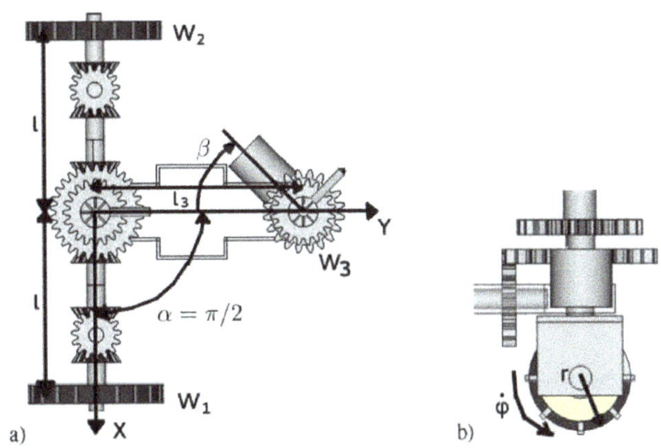

Figure 9. Robot's kinematic structure. (a) Top-view kinematic structure and (b) side-view drive-steer wheel.

The kinematic parameters and variables describing the robot platform are summarized in Table 1 according to each type of wheel (fixed passive, steerable, and drivable active).

Table 1. Tricycle-type robot's wheel kinematic parameters and variables.

Wheel type	α	β_t	ℓ
W_1 (fixed)	0	0	ℓ
W_2 (fixed)	π	0	ℓ
W_3 (drive/steer)	$\pi/2$	—	ℓ_3

The robot's wheel kinematic parameter is modeled by the following constraint equation, which is stated from the wheel plane,

$$J_1(\beta_t)R(\theta_t)\dot{\xi}_t + J_2\varphi_t = 0, \tag{30}$$

Likewise, the kinematic constraint is modeled in the wheel's orthogonal plane by

$$C_1(\beta_T)R(\theta_t)\dot{\xi}_t = 0. \tag{31}$$

where $\mathbf{J}_1, \mathbf{C}_1 \in R^{N\times 3}$ basically describes the wheel kinematic constraints. Matrix $\mathbf{J}_2$ poses the ideal wheel's radius, and $\mathbf{R}(\theta)$ is the Euler rotation matrix. Vector ξ_t describes the state space system, and its components are the robot's posture $\xi = (x, y, \theta, \beta)^T$, as well as the front-sided wheel driving speed ϕ_t, and its product with $\mathbf{J}_2$ establishes a diagonal matrix of the wheels' tangential velocities.

Moreover, for the particular case of fixed wheels and central orientable wheels, the following kinematic constraints apply for individual wheels.

In the wheel's plane

$$\left[-\sin(\alpha+\beta_t) \quad \cos(\alpha+\beta_t) \quad \ell\cos(\beta_t)\right] \cdot \mathbf{R}(\theta_t) \cdot \dot{\xi}_t + r\dot{\varphi}_t = 0. \tag{32}$$

In the wheel's orthogonal plane

$$\left[\cos(\alpha+\beta_t) \quad \sin(\alpha+\beta_t) \quad \ell\sin(\beta_t)\right] \cdot \mathbf{R}(\theta_t)\dot{\xi}_t = 0. \tag{33}$$

where matrix $\mathbf{J}_1$ is the fixed passive back-sided wheels matrix of kinematic constraints, which is obtained by substituting the parameters of Table 1 in expression (32). Such that, $\mathbf{J}_f$ for the fixed wheels is

$$J_f = \begin{pmatrix} -\sin(0) & \cos(0) & \ell\cos(0) \\ -\sin(\pi) & \cos(\pi) & \ell\cos(0) \end{pmatrix}, \tag{34}$$

and by simplifying

$$J_f = \begin{pmatrix} 0 & 1 & \ell \\ 0 & -1 & \ell \end{pmatrix}. \tag{35}$$

Taking into account the kinematic constraints that are similar to the central orientable wheel, the vector model $\mathbf{J}_o$ is obtained:

$$J_o = \left(-\sin\left(\frac{\pi}{2}+\beta_t\right) \quad \cos\left(\frac{\pi}{2}+\beta_t\right) \quad \ell_3\cos\left(\beta_t\right) \right) \tag{36}$$

For the front-sided central orientable wheel, since the angle β_t is nonstationary, and in order to easy its algebraic solution the following trigonometric identities are substituted in (36) and obtain a simplified form for the vector $\mathbf{J}_o$. Thus,

$$\sin(a+b) = \sin(a)\cos(b) + \cos(a)\sin(b),$$

Likewise,

$$\cos(a+b) = \cos(a)\cos(b) - \sin(a)\sin(b);$$

and the following row vector is produced:

$$J_o = \left(-\cos\left(\beta_t\right) \quad -\sin\left(\beta_t\right) \quad \ell_3\cos\left(\beta_t\right) \right). \tag{37}$$

Therefore, for a tricycle-type robotic structure as in the present context, and with parameters of Table 1, the matrix $\mathbf{J}_1$ is consequently defined as

$$J_1 = \begin{pmatrix} J_f \\ J_o \end{pmatrix} = \begin{pmatrix} 0 & 1 & \ell \\ 0 & -1 & \ell \\ -\cos\left(\beta_t\right) & -\sin\left(\beta_t\right) & \ell_3\cos\left(\beta_t\right) \end{pmatrix}, \tag{38}$$

where $\mathbf{J}_f$ represents the fixed wheel kinematic constraints and $\mathbf{J}_o$ represents the central orientable wheel constraints. The constraints denoted in the general form (31) implicate that vector $R(\theta_{t)} \cdot \xi_t$ belongs to the null space C^*_1 denoted by

$$C_1^*(\beta_t) = \begin{pmatrix} C_f \\ C_\beta \end{pmatrix} \tag{39}$$

being $\mathbf{C}_f$ and $\mathbf{C}_\beta$ as vectors of the fixed and the central orientable wheel constraint, respectively, with notation

$$R(\theta_t) \cdot \dot{\xi}_t \in \mathcal{N}\left[C_1^*(\beta_t)\right]. \tag{40}$$

Hence, the robot's posture first-order derivative ξ (t) or state vector is constrained into a distribution defined as

$$\mathcal{N}\left[C_1^*(\beta)\right] = \mathrm{span}\{\mathrm{col}\boldsymbol{\Sigma}(\beta)\}, \tag{41}$$

and the null spacc of any matrix **A**, written as a N(A), is the set of all solutions of $\mathbf{A}\cdot\mathbf{x}=0$ and stated in set notation by

$$\mathcal{N}(\mathbf{A}) = \{x \in \mathbb{R}^x | \mathbf{A} \cdot x = 0\}. \tag{42}$$

Thus, for the particular case of Eq. (41), the null space is

$$\mathcal{N}\left[C_1^*(\beta_t)\right] = \left\{x \in \mathbb{R}^3 \mid \left[C_1^*(\beta_t)\right]x = 0\right\}. \tag{43}$$

A manner to solve for the null space matrix condition is to reduce it to the echelon form Ψ:

$$C_1 = \begin{pmatrix} 1 & 0 & 0 \\ -1 & 0 & 0 \\ -\sin(\beta_t) & \cos(\beta_t) & \ell_3 \sin(\beta_t) \end{pmatrix}, \tag{44}$$

Thus, a vector **x** that satisfies Eq. (43) must be defined then:

$$\begin{pmatrix} 1 & 0 & 0 \\ -1 & 0 & 0 \\ -\sin(\beta_t) & \cos(\beta_t) & \ell_3 \sin(\beta_t) \end{pmatrix} \cdot \begin{pmatrix} x_1 \\ x_2 \\ x_3 \end{pmatrix} = \begin{pmatrix} 0 \\ 0 \\ 0 \end{pmatrix}. \tag{45}$$

The previous expression (45) is a set of linear systems and is rewritten as an augmented matrix $[\mathbf{C}_1|\mathbf{0}]$:

$$[C_1|0] = \begin{pmatrix} 1 & 0 & 0 & | & 0 \\ -1 & 0 & 0 & | & 0 \\ -\sin(\beta) & \cos(\beta) & \ell_3 \sin\beta & | & 0 \end{pmatrix}. \tag{46}$$

The augmented matrix (46) is algebraically solved by reducing it to echelon form, starting by making zeros the first column, but one the first element:

$$\Psi = \begin{pmatrix} 1 & 0 & 0 \\ 0 & 0 & 0 \\ -\sin(\beta_t) & \cos(\beta_t) & \ell_3 \sin(\beta_t) \end{pmatrix}, \tag{47}$$

Then, sum up $\Psi_{1,i}\Psi_{3,1}$ where $\Psi_{3,1} = \sin(\beta_t)$:

$$\Psi = \begin{pmatrix} 1 & 0 & 0 \\ 0 & 0 & 0 \\ 0 & \cos(\beta_t) & \ell_3 \sin(\beta_t) \end{pmatrix}. \tag{48}$$

Since the pivot element $\Psi_{2,2} = 0$, rows 2 and 3 are exchanged:

$$\Psi = \begin{pmatrix} 1 & 0 & 0 \\ 0 & \cos(\beta_t) & \ell_3 \sin(\beta_t) \\ 0 & 0 & 0 \end{pmatrix}. \tag{49}$$

Now, $\Psi_{i,2} = \mathbf{0}$ except its pivot element $\Psi_{2,2}$, thus let us divide $\Psi_{2,i}/\cos(\beta_t)$ to set to 1 such pivoting element:

$$\Psi = \begin{pmatrix} 1 & 0 & 0 \\ 0 & 1 & \ell_3 \tan(\beta_t) \\ 0 & 0 & 0 \end{pmatrix}. \tag{50}$$

Thus, by having Ψ in the reduced form, now the system is solved as the system (45):

$$\begin{pmatrix} 1 & 0 & 0 \\ 0 & 1 & \ell_3 \tan(\beta_t) \\ 0 & 0 & 0 \end{pmatrix} \cdot \begin{pmatrix} x_1 \\ x_2 \\ x_3 \end{pmatrix} = \begin{pmatrix} 0 \\ 0 \\ 0 \end{pmatrix}. \tag{51}$$

Hence, rewriting such a solution in the matrix form

$$\mathrm{x} = \begin{pmatrix} 0 \\ -\ell_3 \tan(\beta) \\ 1 \end{pmatrix} x_3 \tag{52}$$

Being the vector **x** of this null space of matrix $\mathbf{C}_1$, such that

$$\mathcal{N}\left[\mathrm{C}_1^*(\beta)\right] = \mathrm{span}\begin{pmatrix} 0 \\ -\ell_3 \tan(\beta) \\ 1 \end{pmatrix} x_3; \tag{53}$$

For our application purpose, $x_3 = \cos\beta$, then the null space solution N[C∗1(β)] is

$$\mathcal{N}\left[\mathrm{C}_1^*(\beta)\right] = \mathrm{span}\begin{pmatrix} 0 \\ -\ell_3 \sin\beta \\ \cos\beta \end{pmatrix}. \tag{54}$$

Therefore, Eq. (41) is rewritten because for each instant time t, a temporal variable v_t exist .

$$\dot{\xi} = \mathbf{R}^{\mathrm{T}}(\theta_t)\mathbf{\Sigma}(\beta_t)v_t, \tag{55}$$

and

$$\dot{\beta}_t = w_t. \tag{56}$$

where v_t is the instantaneous robot's absolute velocity and w_t is instantaneous robot's yaw rate. As a matter of fact $\Sigma(\beta_t)$ is the null space of the kinematic constraint matrix in the wheel's orthogonal plane (Eq. (54)), defined by

$$\mathbf{\Sigma}(\beta_t) = \begin{pmatrix} 0 \\ -\ell_3 \sin(\beta_t) \\ \cos(\beta) \end{pmatrix}. \tag{57}$$

This state control law is expressed in a more compact form:

$$\dot{\mathbf{z}}_t = \mathbf{B}(z)\mathbf{u}_t; \tag{58}$$

where

$$\mathbf{z} = \begin{pmatrix} \xi_t \\ \beta_t \end{pmatrix}; \; \mathbf{B}(z_t) = \begin{pmatrix} \mathbf{R}^{\mathrm{T}}(\theta_t)\boldsymbol{\Sigma}(\beta_t) & 0 \\ 0 & 1 \end{pmatrix}; \; \mathbf{u} = \begin{pmatrix} v \\ w \end{pmatrix}. \tag{59}$$

Thus, by substituting expression (57) in the model (58), the kinematic control law is rewritten:

$$\begin{pmatrix} \dot{x} \\ \dot{y} \\ \dot{\theta} \\ \dot{\beta} \end{pmatrix} = \begin{pmatrix} \ell_3 \sin\theta \; \sin\beta & 0 \\ -\ell_3 \cos\theta \; \sin\beta & 0 \\ \cos\beta & 0 \\ 0 & 1 \end{pmatrix} \cdot \begin{pmatrix} v \\ w \end{pmatrix}. \tag{60}$$

The robot's posture is essentially modeled by Eq. (60) and validates by the Cartesian trajectory depicted in Figure 10a. For this case, v was set at constant value overtime, and w gradually increased according to the range $\{0, ..., \pi/4\}$. Then, the robot turns into its original angle and then continues straight.

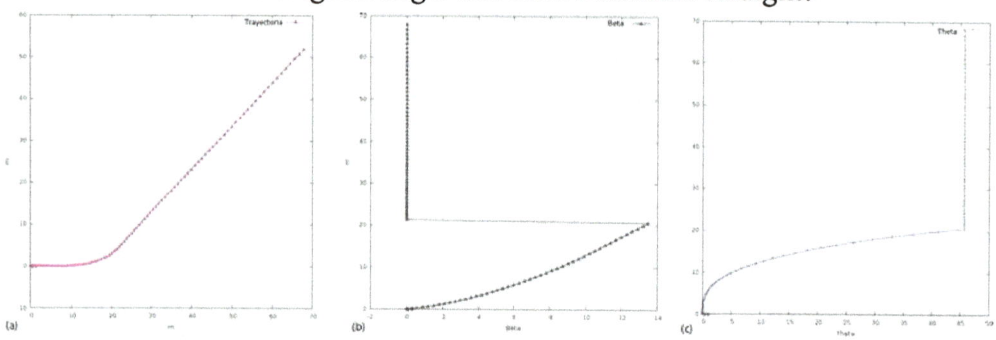

Figure 10. Control law (60) simulation results. (a) Trajectory control with v=cte and w ranging $\beta = \{0, ..., \pi/4\}$, (b) *wheel's steer angle β* behavior, and (c) robot's angle θ behavior.

Figure 10b and 10c shows β and θ behaviors w.r.t. orientation error θ_E, when $\theta_E = 0$ the robot is bearing toward the desired goal. Actually, β depicts how the steering wheel performs from its starting angle up to the final angle. Because of the absence of obstacles, the steering wheel's angle experiences no commutation toward the local mechanism. However, the absolute compass controls the wheel from the beginning and gradually steers until orientation is aligned to the global destination. Its behavior along the navigation task is illustrated in plot 10b.

Moreover, in order to instantaneously estimate the deviation angle θ_E, let us define the global destination θ_G [rad]. And, such desired orientation is perturbed by escaping angle θ_R [rad] due to relative obstacle locations (Figure 11a). Likewise, θ_R is established every time the local compass mechanism is switched to. Obstacles are detected within a territorial radius d_1, or ignored when the robot moves farther away from the scope d_1 (Figure 11b).

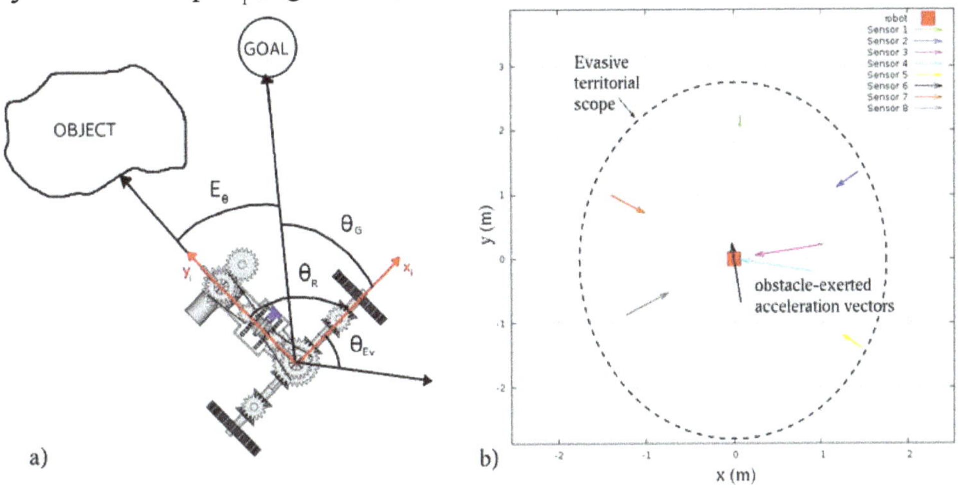

Figure 11. Evasive system. (a) θ_R *and* θ_G *kinematics and (b) evasive territorial scope and obstacle-repulsive acceleration vectors.*

For instance, sensor 1 detects a far obstacle, and its repulsive acceleration influence is low. However, sensor 6 yields a greater repulsive acceleration influence since it is nearer to the robot's location.

Thus, let us postulate:

$$\theta_E = \theta_R - \theta_G \tag{61}$$

From the control law, the state vector **z** is now estimated by directly using the instantaneous error angle θ_E:

$$\dot{\mathbf{z}}\left(\dot{x}, \dot{y}, \theta_E, \beta\right) = \mathbf{B}(z) \cdot \mathbf{u}(v, w), \tag{62}$$

and by obtaining the inverse solution for **u**, now the input system involves error angle:

$$\mathbf{u}(v, w) = \mathbf{B}(z)^{-1} \cdot \dot{\mathbf{z}}\left(\dot{x}, \dot{y}, (\theta_R - \theta_G), \beta\right). \tag{63}$$

The robot's controlled orientation is fed back by θ_E and calculated by numerical successive approximations. This controller is a recursive system that involves both, the direct and the inverse solutions, to reduce the numerical error θ_E.

The instantaneous escaping angle θ_R is obtained within a local coordinate system by directional derivatives, which approach a cosine function, similarly to the work reported in Ref. [20]. A gradient operator of a cosine navigation function approximates a repulsive partial differential equation to evade any observable obstacle. The robot's repulsive acceleration vector function aα=(ax,ay)$^\top$ produces its magnitude effect w.r.t. the obstacle's range measurement δ_α [m]:

$$\mathfrak{a}_\alpha(x,y) = -\nabla_{x,y} k_\alpha (k_2 - r_\alpha) \cdot \cos(\phi_\alpha) \tag{64}$$

where k_2 [m] is the obstacle diametrical territory and k_α [ms^{-2}] is an adjustment constant factor of the physical acceleration amplitude. Thus, the Cartesian geometric model (not measurement) r_α describes the distance between the robot $\mathbf{x}_r = (x_r, y_r)^\top$ and any obstacle $\mathbf{x}_o = (x_o, y_o)^\top$, which is geometrically modeled by

$$r_\alpha = \sqrt{x^2 + y^2} = \sqrt{(x_r - x_o)^2 + (y_r - y_o)^2} \tag{65}$$

The angle limits $\{0 \le \varphi_\alpha \le \pi/2\}$ is a transformation relationship of the obstacle range δ_α, and the condition $\delta_\alpha < d_l$ limits the repulsive acceleration effects:

$$\phi_\alpha = \frac{\delta_\alpha \pi}{2 d_l}; \text{ where } d_l \propto \frac{\pi}{2} \text{ and } d_l = \delta_\alpha \Rightarrow \|\mathfrak{a}\| = 0 \tag{66}$$

Substituting the functional form of δ_α in (64) and temporally considering $k_\alpha = 1$ for analysis purpose

$$\mathfrak{a}_\alpha(x,y) = -\nabla_{x,y}\left(\left(k_2 - \sqrt{(x_r - x_o)^2 + (y_r - y_o)^2}\right) \cdot \cos(\phi_\alpha)\right) \tag{67}$$

and algebraically expanding

$$\mathfrak{a}_\alpha(x,y) = -\nabla_{x,y}\left(-k_2 \cos(\phi_\alpha) + \sqrt{(x_r - x_o)^2 + (y_r - y_o)^2} \cdot \cos(\phi_\alpha)\right) \tag{68}$$

Thus, applying the gradient operator $\nabla_{x,y}$ w.r.t. x and y components and algebraically simplifying

$$\frac{\partial \mathfrak{a}_\alpha}{\partial x} = \frac{(x_r - x_o) \cdot \cos(\phi_\alpha)}{\sqrt{(x_r - x_o)^2 + (y_r - y_o)^2}} \tag{69}$$

and

$$\frac{\partial \mathfrak{a}_\alpha}{\partial y} = \frac{(y_r - y_o) \cdot \cos(\phi_\alpha)}{\sqrt{(x_r - x_o)^2 + (y_r - y_o)^2}} \tag{70}$$

Therefore, the obstacle-repulsive directional vector expressed in terms of Cartesian components XY is

$$\mathfrak{a}_\alpha(x,y) = k_\alpha \frac{\cos(\phi_\alpha)}{\sqrt{(x_r - x_o)^2 + (y_r - y_o)^2}} \cdot \begin{pmatrix} x_r - x_o \\ y_r - y_o \end{pmatrix} \tag{71}$$

Thus, the model for multiple obstacles α produces a controlled escaping direction in robot's local coordinate framework:

$$\mathbf{a}_T = k_\alpha \sum_\alpha \frac{\cos(\phi_\alpha)}{\sqrt{(x_r - x_o)^2 + (y_r - y_o)^2}} \cdot \begin{pmatrix} x_r - x_o \\ y_r - y_o \end{pmatrix} \tag{72}$$

It follows that the evasive acceleration magnitude is defined by

$$\|\mathbf{a}_T\| = \sqrt[2]{\left(\frac{\partial \mathbf{a}_\alpha}{\partial x}\right)^2 + \left(\frac{\partial \mathbf{a}_\alpha}{\partial y}\right)^2}, \tag{73}$$

and the instantaneous escaping angle θ_R

$$\theta_R = \arctan\left(\frac{\left(\frac{\partial a}{\partial y}\right)}{\left(\frac{\partial a}{\partial x}\right)}\right). \tag{74}$$

Finally, the proposed kinematic controller has the following scheme described in Algorithm 1.

Algorithm 1. Local/global robot's underactuated system controller

---

$\omega_l = 0.0$
while $(\|\xi_t\| - \|(x_G, y_G)\|) > 0$ **do**
    switching to either local or global mechanism
    if $\delta_\alpha > 0$
        Commuter's motor speed $\omega_l = k_1 e^{k_2 t}$
        Servo's angle $\phi = \frac{\omega_l \kappa_1}{\ln\left(\frac{\omega_l}{\kappa_2}\right)}$
        Avoidance acceleration $\mathbf{a}_T = k_\alpha \sum_\alpha \frac{\cos(\phi_\alpha)}{r_\alpha} (\mathbf{x}_r - \mathbf{x}_\alpha)$
        Avoidance angle $\theta_R = \arctan\left(\frac{\partial a/\partial y}{\partial a/\partial x}\right)$
        $\|\mathbf{a}_T\| = \sqrt[2]{(\partial a/\partial x)^2 + (\partial a/\partial y)^2}$
    else
        Turns servo off $\omega_l = 0.0$
        Release slider crank $x_a = r\cos(\phi)\sqrt[2]{(a_1 + a_2)^2 e^{-ct/m} - (r\sin(\phi))^2} + \Delta s$
        $\theta_G = -\frac{\varphi_{w1} - \varphi_{w2}}{4}$
    $\theta_E = \theta_R - \theta_G$
    $\mathbf{u} = \left(\frac{\|\mathbf{a}_t\|}{\Delta t}, \frac{d\theta_E}{dt}\right)^T$
    $\dot{\mathbf{z}}_t = \mathbf{B}(\theta_E, \beta) \cdot \mathbf{u}_t$
    $\mathbf{u}_{t+1} = \mathbf{B}^{-1} \cdot \dot{\mathbf{z}}$

Figure 12 confirms the accuracy of our kinematic model approach by illustrating how the robot's path reaches the final Cartesian target (-7, -8) in global coordinates. Furthermore, Figure 12a illustrates the Cartesian trajectory that navigates across many obstacles, while also demonstrating the local and global steering modes. Similarly, Figure 12b demonstrates the relationship between component X and the robot's angle during the same navigational task, serving as a means of validation.

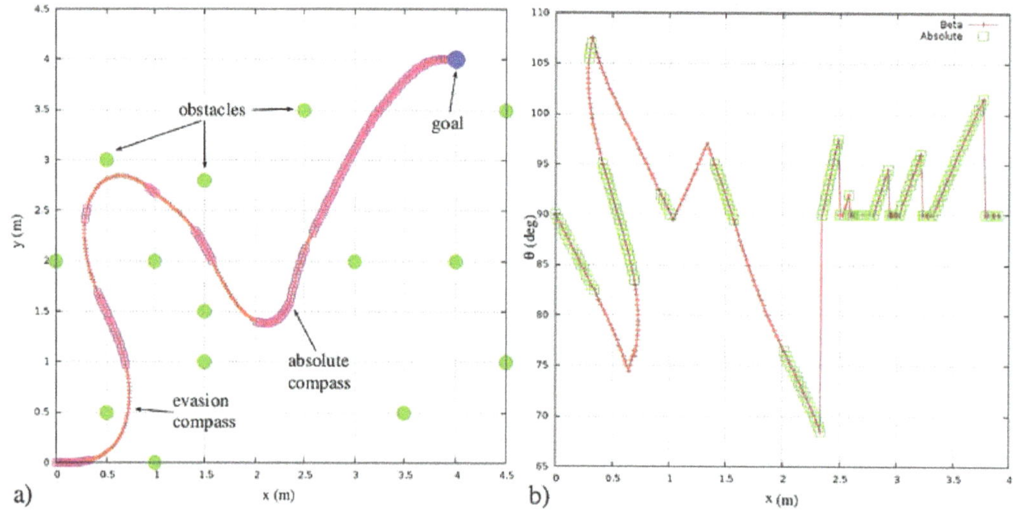

Figure 12. Controlled robot's course v = 1.0 rad/s and d_l = 0.5 m. (a) Cartesian space and (b) component X versus robot's angle.

5. CONCISE SUMMARY

This chapter aimed to introduce the study of a different kinematic steering controller through the utilization of underactuated mechanical compasses. It sought to illustrate the practicality, controllability approach, and inherent efficiency of this alternative controller. Although the installation of a compass mechanism is hard, it is advocated because it enables the direct establishment of the global aim as an absolute reference. Furthermore, the compass mechanism autonomously guides the robot towards the target while also evading obstacles. The functional features of this robotic platform provide significant advantages over traditional orientation systems. Unlike sensor devices, the global orientation does not rely on feedback or require complex computations to estimate the global destination orientation instantly. In contrast to kinematic redundant structures, the suggested solution restricts the consumption of electric energy to a single driving actuator, rather than the steering actuators. Steering controllers no longer allocate algorithmic computational resources, instead reserving them for other robotic algorithmic duties or for enhancing additional hardware devices. Software algorithms outperform mechanical controllers in terms of speed, despite the fact that the mechanisms themselves are already quick enough for the robot's mobility rates. A crucial matter to consider is the commutation linear actuator, which can be either passive or active. It operates by combining a slider crank mechanism with a spring-mass-damper system. The commutator device demonstrated sufficient speed, therefore negating the necessity for advanced active linear actuators. It was determined that the transition times between local and global compasses did not have a detrimental impact on the performance of the system or the stability of the trajectory. The robot's local and global navigation systems enabled it to safely reach its objective without encountering any collisions.

The controlled trajectories obtained consisted of interconnected segments with no interruptions observed. From an implementation standpoint, we have determined that the proposed approach is not inherently more effective or less effective than a standard strategy that utilizes redundant actuators and multiple sensors. In reality, a conventional method is simpler to physically execute. However, an excessive reliance on traditional methods is a disadvantage when the number of driving wheels increases, not to mention the addition of more steering actuators. Therefore, a system becomes more discrete and loses stability in its regulated motions and trajectories as it becomes more redundant.

This is a fundamental distinction from the inherently continuous motions generated by underactuated systems. A concise linear kinematic control algorithm for switching between the use of different compasses was created, providing both direct and inverse solutions. The system effectively integrated the control of the underactuated mechanism with a concise algorithm for detecting impediments and estimating the immediate escape direction. It has been discovered that modifying the kinematic evasion parameters can alter the continuity of the robot's trajectory, hence affecting the reaction of the mechanism's commutation. The validation of inter-mechanism interactions and functions was achieved using simulation results. In the context of the suggested kinematic structure, the local/global compass demonstrated comparable efficiency to other redundant systems of a similar nature. The future research will prioritize the restoration of proper functioning after instances of slips, sliding, or collision dynamics that cause the compass orientation to become uncalibrated. Furthermore, the incorporation of both orientation analysis and positioning sensors is essential to enhance and optimize this underactuated technique.

REFERENCES

1. Li SH. The South-Pointing Carriage and the Mariner's Compass. Taipei: Yee Wen Pub. Co.; 1959
2. Lu ZM. An analysis of the ancient Chinese South-Pointing Chariot. Journal of Sichuan University. 1979;2:95-101

3. Muneharu M, Satoshi K. Study of the mechanics of the South-Pointing Chariot (the South Pointing Chariot with the bevel gear type differential gear train). Transactions of the Japan Society of Mechanical Engineers. 1990;56(C):462-466
4. Hong-Sen Y. Chap. 7. South-pointing Chariots, Reconstruction Designs of Lost Ancient Chinese Machinery, vol. 3. Dordrecht: Springer; 2007
5. Hong-Sen Y, Chun-Wei C. A systematic approach for the structural synthesis of differential-type South Pointing chariots. JSME International Journal Series C Mechanical Systems, Machine Elements and Manufacturing. 2006;49((3), SI on Advanced Technology of Vibration and Sound):920-929
6. Santander M. The Chinese South-Seeking chariot: A simple mechanical device for visualizing curvature and parallel transport. American Association of Physics Teachers. September 1992;60(9)
7. Junmin W, Xiangyu Y, Wei L. Integration of hardware and software designs for object grasping and transportation by a mobile robot with navigation guidance via a unique bearing-alignment mechanism. IEEE/ASME Transactions on Mechatronics. 2016;21(1):576-583
8. Al-Faiz MZ, Mahameda GE. GPS-based navigated autonomous robot. International Journal of Emerging Trends in Engineering Research. 2015;3(4)
9. Sioma A, Blok S. Finding bearing in robot navigation with the use of the Kalman filter. Solid State Phenomena. 2013;199:241-246
10. Georgiou E., Dai J.S., Luck M., The KCLBOT: A double compass self-localizing maneuverable mobile robot. ASME. International Design Engineering Technical Conference and Computers and Information in Engineering Conference, Vol.3, pp. 427–435, 2011
11. Chen w, Zhang T. An indoor mobile robot navigation technique using odometry and electronic compass. International Journal of Advanced Robotic Systems. May-Jun 2017:1-15
12. Zhenhai H. y Shengguo H., Integrated navigation system based on differential magnetic compass and GPS, International Conference on Information Engineering and Computer Science, 2009
13. Parhi D, Deepak B. Kinematic model of three wheeled mobile robot. Journal of Mechanical Engineering Research. 2011:307-318
14. Campion G, Chung W, 17 C. Springer Handbook of Robotics. In: Handbook of Robotics. Springer-Verlag Berlin Heidelberg; 2008
15. Morin P, Samson C. Chap. 34. Springer Handbook of Robotics. In: Handbook of Robotics. Springer-Verlag Berlin Heidelberg; 2008

16. Minguez J, Lamiraux F, Lamound J. Chap. 35 Springer Handbook of Robotics. In: Handbook of Robotics. Springer-Verlag Berlin Heidelberg; 2008

17. Martinez-Garcia EA. Numerical modelling in robotics. OmniaScience. 2015

18. Alonsini N.I., Low cost obstacle detection system for wheeled mobile robot. UKACC International Conference on Control, pp. 529–533, 2012

19. Martinez-Garcia EA. Robotic DCVG Planning for Searching Flaws on Buried Pipelines. Lap Lambert Academic; 2017

20. Castro Jiménez L, Martínez-García EA. Thermal image sensing model for robotic planning and search. Sensors. 2016;1253:1-27

21. Mujahed M., Fischer D. y Mertsching B., Tangential gap flow (TGF) navigation: A new reactive obstacle avoidance approach for highly cluttered environments, Robotics and Autonomous Systems, Vol.84, pp.15–30, 2016

9 Chapter

A NOVEL APPROACH TO THE IDENTIFICATION OF KINEMATIC PARAMETERS IN LASER TRACKERS

1. INTRODUCTION: AN OVERVIEW

A critical need exists for the development of more precise large-scale measurement systems in sectors with high-quality requirements, such as the naval, aeronautic, or spatial industries, for the verification of large parts and facilities. In these industries, the precise positioning of each part in large assemblies is contingent upon the dimensional accuracy of large parts, which necessitates the use of long-range accurate measurement devices [1–4].

Their applications are extensive, including deformation analysis [2], inspection, calibration of an industrial robot [7], reverse engineering [8], and machine tool volumetric verification.

The laser tracker (LT) provides substantial benefits, including precision, portability, high-speed data acquisition, and flexibility. Nevertheless, the mechanical assembly is a significant source of errors, such as offsets or eccentricities, which result in measurement errors. One of the drawbacks of these measurement systems is that the user is unable to determine whether the LT is measuring accurately. The LT's performance is assessed through the use of existing standards. Nevertheless, these assessments necessitate expensive, specialized apparatus. Furthermore, they necessitate specialized apparatus and a lengthy process.

The geometric parameters are identified during the calibration process to enhance the accuracy of the system. Nevertheless, there are a limited number of studies that examine the calibration of LTs.

The geometric error model's parameters are determined by measuring a series of reflectors situated at fixed locations from various LT locations. This is the foundation of the calibration procedure. One of the benefits of this method is that it does not necessitate specialized equipment, allowing any user of a measurement system to execute the LT calibration. Additionally, the time necessary to calibrate the TL is significantly reduced in comparison to the time required to conduct functional tests as recommended by ASME B89.4.19 [9].

The objective of this chapter is to demonstrate a straightforward and expeditious calibration procedure that involves measuring a mesh of reflectors in order to enhance the LT's accuracy. In order to assess the effectiveness of the procedure, we have created a synthetic point generator and a kinematic error model. Subsequently, a sensitivity analysis was conducted to ascertain the most suitable instrument calibration measurement positions, and an experimental trial was conducted to identify the geometric parameters.

2. LASER TRACKER

A long-range metrological device (LT) is capable of achieving an accuracy of several tens of micrometers. An interferometer is mounted on a two-degrees-of-freedom rotatory gimbal to form the LT. The laser beam is reflected by a spherically mounted retroreflector (SMR), which sends the beam back to the interferometer. Position sensitivity devices detect any displacement of the SMR and adjust the LT position to correspond with the SMR's position. The spherical coordinates of the SMR center according to the LT reference system are determined by the distance measured by the interferometer, (d), and the readings of the encoders located in each of the two rotatory axes, (θ, ϕ), as illustrated in Figure 1. Equations (1)–(3) can be employed to determine the Cartesian coordinates that correspond to the spherical coordinates.

$$x = d \times \cos(\theta) \times \sin(\varphi) \quad (1)$$

$$y = d \times \sin(\theta) \times \sin(\varphi) \quad (2)$$

$$z = d \times \cos(\theta) \quad (3)$$

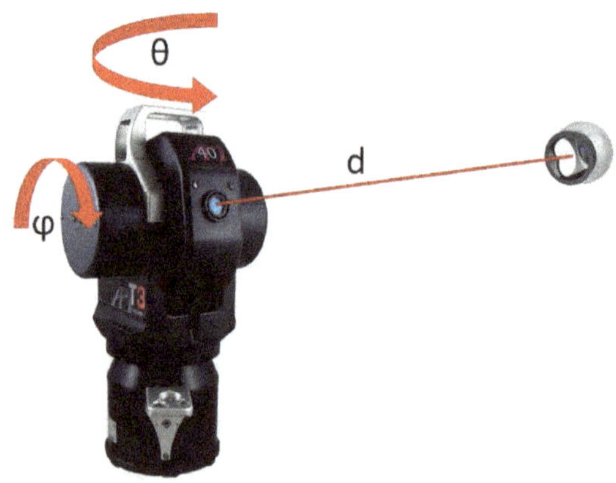

Figure 1. Laser tracker working principle.

3. LASER TRACKER KINEMATIC MODEL

The LT can be considered as an open kinematic chain with three joints: two rotary joints in the gimbal and a prismatic joint corresponding to the laser beam. The kinematic model is a mathematical expression that determines the position of the final joint of the kinematic chain (the SMR centre) with reference to the LT frame. We have used the Denavit-Hartenberg (D-H) [5] formulation to develop the kinematic model. This method defines the coordinate transformation matrices between each two consecutive reference systems j and j - 1 as the product of the rotation and translation matrices from j - 1 joint to j joint. The DH model needs four characteristic parameters (distances d_j, a_j, and angles θ_j, α_j) for these matrices. Figure 2 shows the relationships between the consecutive reference systems. Knowing the kinematic parameters corresponding to every joint, the homogeneous transformation matrix from reference system j to j - 1 is the result of the product of the four rotation and translation matrices shown in Eq. (4).

$$
{}^{j-1}A_j = T_{z,d}R_{z,\theta}T_{x,a}R_{x,\alpha} = \begin{bmatrix} 1 & 0 & 0 & 0 \\ 0 & 1 & 0 & 0 \\ 0 & 0 & 1 & d_j \\ 0 & 0 & 0 & 1 \end{bmatrix} \times \begin{bmatrix} \cos\theta_j & -\operatorname{sen}\theta_j & 0 & 0 \\ \operatorname{sen}\theta_j & \cos\theta_j & 0 & 0 \\ 0 & 0 & 1 & 0 \\ 0 & 0 & 0 & 1 \end{bmatrix} \times \begin{bmatrix} 1 & 0 & 0 & a_j \\ 0 & 1 & 0 & 0 \\ 0 & 0 & 1 & 0 \\ 0 & 0 & 0 & 1 \end{bmatrix}
$$

$$
\times \begin{bmatrix} 1 & 0 & 0 & 0 \\ 0 & \cos\alpha_j & -\operatorname{sen}\alpha_j & 0 \\ 0 & \operatorname{sen}\alpha_j & \cos\alpha_j & 0 \\ 0 & 0 & 0 & 1 \end{bmatrix} = \begin{bmatrix} \cos\theta_j & -\cos\alpha_j \operatorname{sen}\theta_j & \operatorname{sen}\alpha_j \operatorname{sen}\theta_j & a_j\cos\theta_j \\ \operatorname{sen}\theta_j & \cos\alpha_j\cos\theta_j & -\operatorname{sen}\alpha_j\cos\theta_j & a_j\operatorname{sen}\theta_j \\ 0 & \operatorname{sen}\alpha_j & \cos\alpha_j & d_j \\ 0 & 0 & 0 & 1 \end{bmatrix}
$$

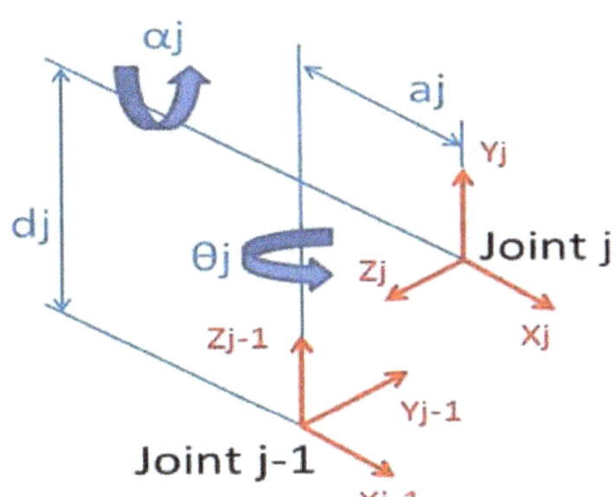

Figure 2. Denavit-Hartenberg model.

Being $T_{m,n}$ and $R_{m,n}$, the homogeneous translation matrices corresponding to translation (T) or rotation (R) n along axis m.

Following the DH model, the LT kinematic model has been determined as shown in Figure 3. The position of the reflector referring to the LT reference system is defined by Eq. (5).

$$ {}^{0}T_{3} = {}^{0}A_{1}{}^{1}A_{2}{}^{2}A_{3} \tag{5}$$

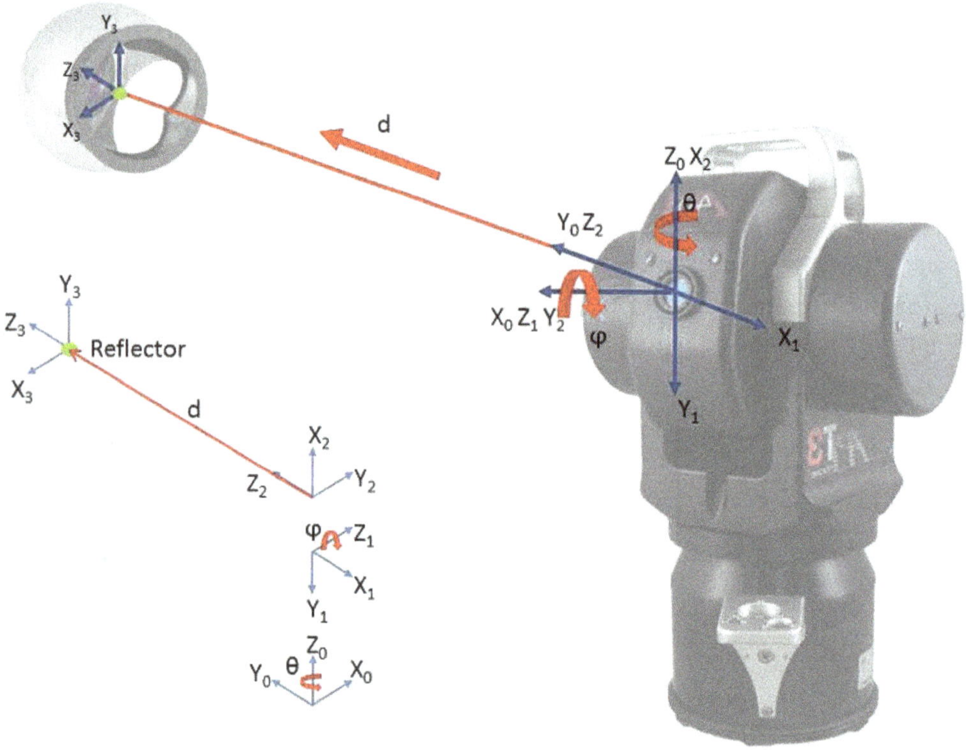

Figure 3. Laser tracker kinematic model.

This chapter is based on the LT API Tracker 3. Table 1 shows the kinematic parameters corresponding to this LT model.

Table 1. Laser tracker kinematic parameters.

i	α_i (°)	a_i (mm)	d_i (mm)	θ_i (°)
1	-90	0	0	θ - 90
2	90	0	0	ϕ - 90
3	0	0	d	-90

4. LASER TRACKER ERROR MODEL

Relative positions between two consecutive joints are defined by the DH model. These are the nominal positions, but they are conditioned by the LT errors. This means that reference frames are not at its expected position as shown in Figure 4. An error model that fits the geometry of the joints is necessary to modify the kinematic model.

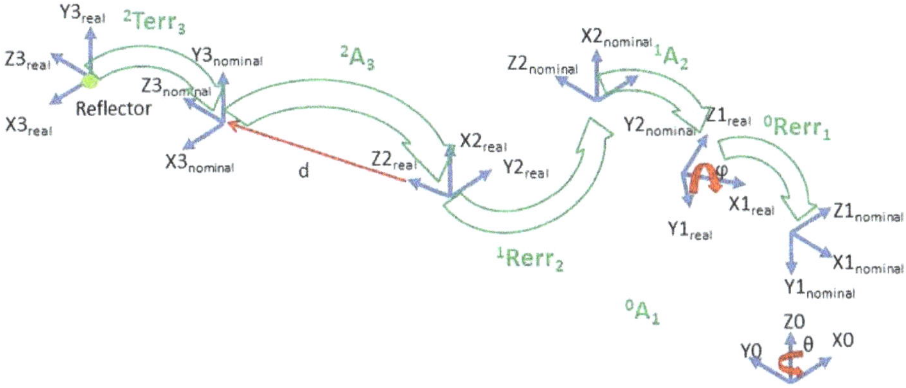

Figure 4. Error model.

In this chapter, the error model shown in Figure 5 has been used. This model is based on six degrees of freedom error for each joint. Mathematically, the error is stated as a transformation matrix between the nominal joint frame and the real frame. The matrices are different for rotary joints (Eq. (6)) and linear joints (Eq. (7)). For each error matrix, a new set of six error parameters is defined (δx, δy, δz, εx, εy, εz). The calibration procedure will calculate the optimum parameter values to minimize the LT error.

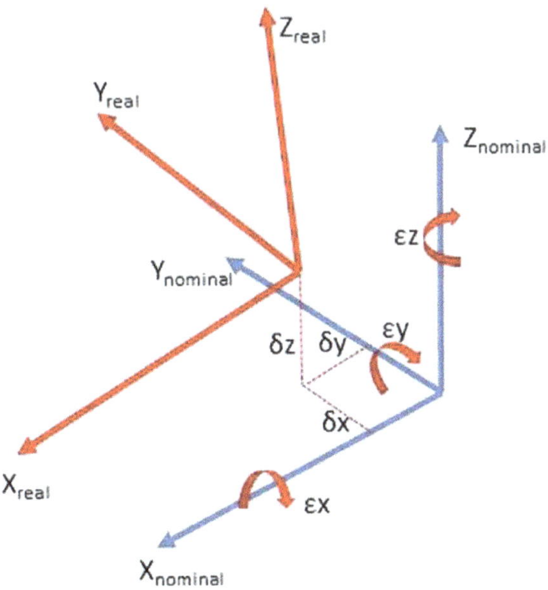

Figure 5. Error parameters.

$$R_{err} = \begin{bmatrix} \cos\varepsilon_Y \cdot \cos\theta_Z & -\cos\varepsilon_Y \cdot \sin\theta_Z & \sin\varepsilon_Y & \delta_X \\ \cos\varepsilon_X \cdot \sin\theta_Z + \sin\varepsilon_X \cdot \sin\varepsilon_Y \cdot \cos\theta_Z & \cos\varepsilon_X \cdot \cos\theta_Z - \sin\varepsilon_X \cdot \sin\varepsilon_Y \cdot \sin\theta_Z & -\sin\varepsilon_X \cdot \cos\varepsilon_Y & \delta_Y \\ \sin\varepsilon_X \cdot \sin\theta_Z - \cos\varepsilon_X \cdot \sin\varepsilon_Y \cdot \cos\theta_Z & \sin\varepsilon_X \cdot \cos\theta_Z + \cos\varepsilon_X \cdot \sin\varepsilon_Y \cdot \sin\theta_Z & \cos\varepsilon_X \cdot \cos\varepsilon_Y & \delta_Z \\ 0 & 0 & 0 & 1 \end{bmatrix} \quad (6)$$

$$T_{err} = \begin{bmatrix} 1 & -\varepsilon_\psi & \varepsilon_\theta & \varepsilon_x \\ \varepsilon_\psi & 1 & -\varepsilon_\phi & \varepsilon_y \\ -\varepsilon_\theta & \varepsilon_\phi & 1 & \varepsilon_z \\ 0 & 0 & 0 & 1 \end{bmatrix} \quad (7)$$

Including the error matrices in Eqs. (5) and (8), which describes the kinematic model with the error model, is obtained.

$$^0T_3 = {}^0A_1^0 Rerr_1{}^1A_2^1 Rerr_2{}^2A_3{}^2Terr_3 \quad (8)$$

By checking the behavior of the error model, it has been proven that errors depend on the joint position (the rotation angle in rotary joints and distance of the interferometer). This means that error parameters must have a formulation depending on the joint position. For the error parameters corresponding to the linear error matrix, we have used a polynomial function (see Eq. (9)) and for the error parameters corresponding to the rotary error matrices, a Fourier shape function is more convenient because of its periodic behavior (see Eq. (10)).

$$\phi = A\phi \cdot sen\left(\frac{2\pi}{T\phi} \cdot \theta_z + \varphi\phi\right) \quad (9)$$

$$\phi = \phi_1 + \phi_2 \cdot d + \phi_3 \cdot d^2 \quad (10)$$

For $\varphi = \delta x, \delta y, \delta z, \varepsilon x, \varepsilon y, \varepsilon z$.

5. MODEL VALIDATION

The kinematic error model must be validated. The validation has been performed first with synthetic values and then with real values.

5.1. Synthetic data validation

A parametric generator of meshes of reflectors with known errors has been programmed. This algorithm generates a mesh of synthetic reflector coordinates with nominal position values. Then a set of error parameters is introduced, and the theoretical measurements of an LT, affected by the error parameters introduced, are calculated.

These measurements are introduced in the calibration procedure to calculate the error parameters with the optimization of the function described in Eq. (11) and correct the measurements. Finally, the initial measurements and the corrected ones are compared with the nominal reflector positions according to Eq. (12) to calculate the calibration accuracy improvement achieved.

$$\phi = \sum_{i=1}^{n}\left((x_i - x_{nomi})^2 + (y_i - y_{nomi})^2 + (z_i - z_{nomi})^2\right) \quad (11)$$

$$err_i = (x_i - x_{nomi})^2 + (y_i - y_{nomi})^2 + (z_i - z_{nomi})^2 \tag{12}$$

The error parameter set is shown in Table 2.

Table 2. Synthetic error parameters.

—	δX (μm)	δY (μm)	δZ (μm)	εX (μrad)	εY (μrad)	εZ (μrad)
θ 0Rerr1	10	10	10	10	10	10
ϕ 1Rerr2	10	10	10	10	10	10
d 2Terr3	10	10	10	10	10	10

Three meshes of synthetic reflectors have been generated:

Flat YZ plane mesh 15 × 14 = 210 reflectors (Figure 6).

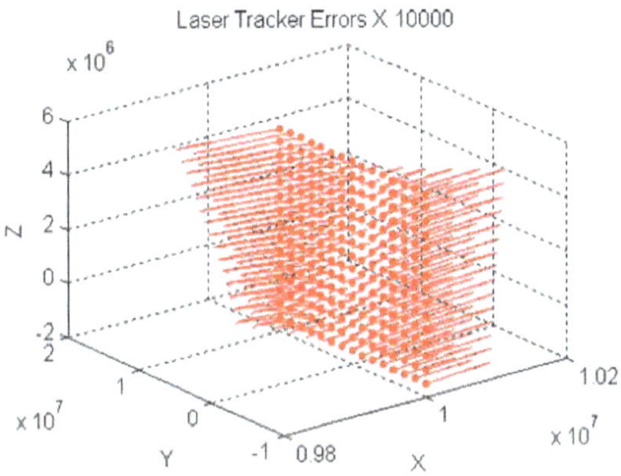

Figure 6. Errors in a plane mesh.

X = 5.000 mm constant.

Y = 10.000 ÷ 10.000 Δ 1.420 mm.

Z = 1.500 ÷ 5.000 Δ 500 mm.

Cubic XYZ mesh 6 × 6 × 6 = 216 reflectors (Figure 7).

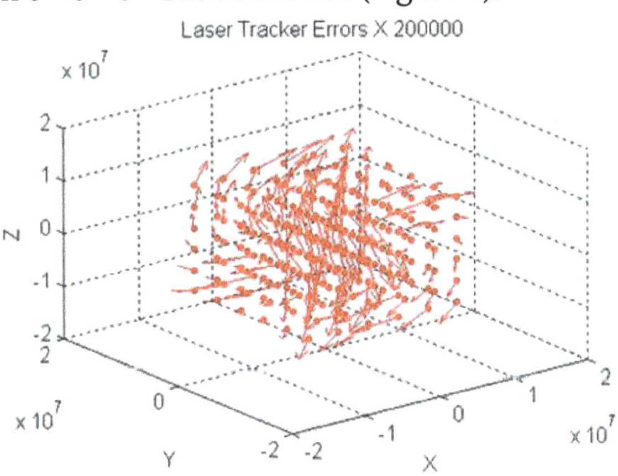

Figure 7. Errors in a cubic mesh.

X – 10,000 ÷ 10,000 Δ 4,000 mm.

Y – 10,000 ÷ 10,000 Δ 4,000 mm.

Z – 10,000 ÷ 10,000 Δ 4,000 mm.

Spherical HVR mesh 12 × 8 × 5 = 480 reflectors (Figure 8).

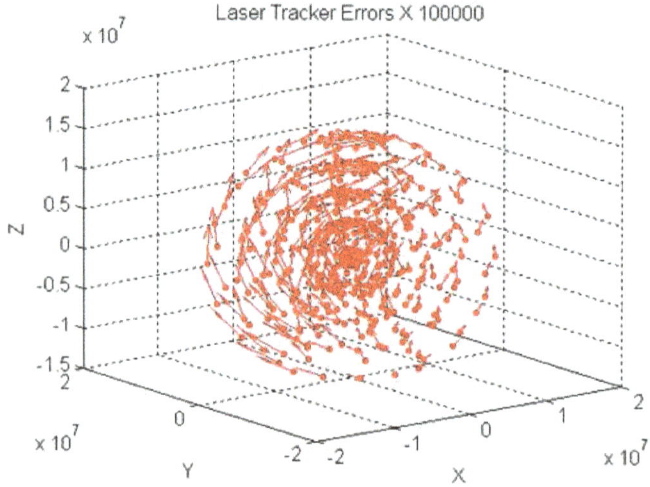

Figure 8. Errors in a spherical mesh.

H 0° ÷ 360° Δ33°.

V 77° ÷ – 60° Δ20°.

R 1,000 ÷ 15,000 Δ 3,500 mm.

The calibration procedure calculates the parameter errors and corrects the generated measurements. As the error parameters are calculated through a mathematical optimization, the result do not gives exactly the nominal parameters but provides a set of parameters that minimizes the LT error. In fact, the calibration reduces the LT error more than 98%. For example, Table 3 shows the calculated error parameters corresponding to the spherical mesh.

Table 3. Calculated error parameters in the spherical mesh.

—	δ_X (μm)	δ_Y (μm)	δ_Z (μm)	ε_X (μrad)	ε_Y (μrad)	ε_Z (μrad)
θ	10.001	9.977	0.000	9.994	9.994	-5.101
ϕ	10.045	0.000	9.969	9.986	-5.137	9.416
d	10.011	9.969	10.018	9.966	10.574	0.000

5.2. Real data validation

The validation with synthetic data shows that the programmed algorithms are working properly but, as the errors have been generated with the error model purposed, it was expected to obtain a good calibration result. It is necessary to check the calibration behavior with real data. To do it, an experiment has been performed. A set of 17 reflectors has been placed over the table of a coordinates measuring machine (CMM). Then the positions of these reflectors have been measured with the CMM and the LT from five different positions as shown in Figure 9.

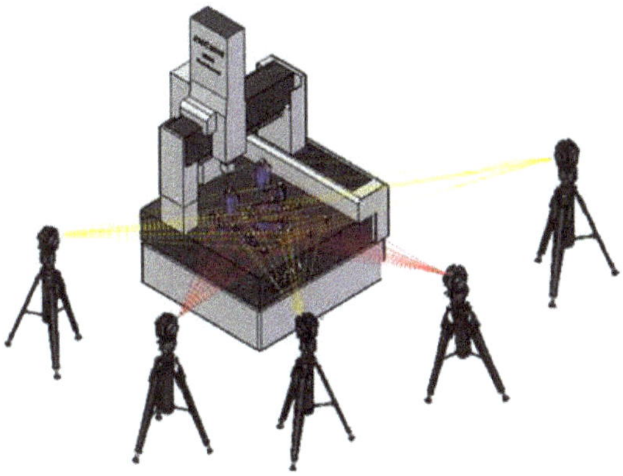

Figure 9. Real data validation experiment.

The estimation of the error parameters have been performed on the basis that the distances between every pair of reflectors must be the same regardless the LT position from which they have been measured. Eq. (13) compares the distances measured by the CMM and the LT.

$$\Phi = \sum_{i=1}^{C_{n,r}} \sum_{K=1}^{LT} \left(d_{mik} - d_{iCMM} \right) \tag{13}$$

Being d_{mik}, the distance measured between the i-esim pair of reflectors from the k-esim LT position and d_{iCMM} the same distance measured by the CMM. $C_{n,r}$ is the number of possible pairs of reflectors. In this case i=$C_{2,17}$ = 136 pairs of reflectors. Figure 10 and Table 4 show the initial and residual errors of the reflectors mesh.

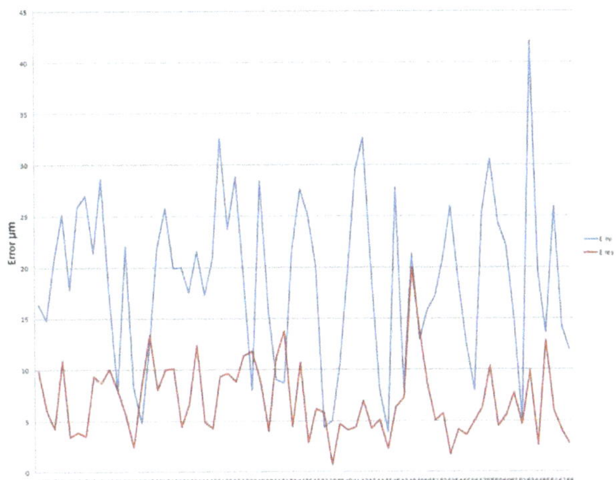

Figure 10. CMM reflectors mesh calibration.

Table 4. Residual errors in the CMM mesh calibration.

	E_ini (μm)	E_res (μm)
Maximum	42.057	19.935
Average	18.785	7.045

In the real calibration object of this work, we measure a mesh of reflectors out of the metrological laboratory, and there will be no nominal data to calculate the error parameters. To simulate the calibration procedure behavior and its requirements under real conditions, we have used the CMM measurements with a new optimization function. This function is equivalent to Eq. (11) but instead of comparing CMM and LT distances, we compare distances from every pair of LT positions as shown in Eq. (14).

$$f = \sum_{i=1}^{m-1} \sum_{j=i+1}^{m} \sum_{k=1}^{n-1} \sum_{l=k+1}^{n} \left| d_{kl}^{i} - d_{kl}^{j} \right| \tag{14}$$

With this optimization criterion, we found out that calibration result increased the LT error. That is due to the fact that the mathematical optimization matches the distances but to a value different from the nominal. This means that it is necessary to introduce a calibrated distances gauge in the reflectors mesh to determine its behavior. After several simulations, a gauge of four reflectors gives the best results, and the objective function is as shown in Eq. (15)

$$f = \sum_{i=1}^{m-1} \sum_{j=i+1}^{m} \sum_{k=1}^{n-1} \sum_{l=k+1}^{n} \left| d_{kl}^{i} - d_{kl}^{j} \right| + \sum_{i=1}^{m} \sum_{k=1}^{3} \sum_{l=k+1}^{4} \left| d_{kl}^{i} - d_{kl}^{CMM} \right| \tag{15}$$

To evaluate the calibration results, we have established two different criteria: (1) a distances criterion that evaluates the differences of distances between every pair of reflectors measured from every pair of positions of LT according to Eq. (16) and (2) a coordinates criterion that evaluates the position error for every reflector measured from every LT position (see Eq. (17)). This second criterion requires first to transform all measurements to the same reference system.

$$err = \frac{\sum_{m=1}^{4} \sum_{n=2}^{5} \sum_{i=1}^{16} \sum_{j=2}^{17} \left(\sqrt{\left(x_i^m - x_j^m\right)^2 + \left(y_i^m - y_j^m\right)^2 + \left(z_i^m - z_j^m\right)^2} - \sqrt{\left(x_i^n - x_j^n\right)^2 + \left(y_i^n - y_j^n\right)^2 + \left(z_i^n - z_j^n\right)^2} \right)}{C_{5,2} C_{17,2}} \tag{16}$$

$$err = \frac{\sum_{m=1}^{4} \sum_{n=2}^{5} \sum_{i=1}^{17} \sqrt{\left(x_{iSRO}^{m} - x_{iSRO}^{n}\right)^2 + \left(y_{iSRO}^{m} - y_{iSRO}^{n}\right)^2 + \left(z_{iSRO}^{m} - z_{iSRO}^{n}\right)^2}}{17 \times C_{5,2}} \tag{17}$$

The results of the calibration following both criteria are shown in Tables 5 and 6.

Table 5. Distances criteria evaluation.

	SMR distance error (µm)						
	LT1	LT2	LT2	LT4	LT5	LT1-5	Improvement
Average	14.31	7.79	9.16	7.42	7.07	9.15	62.67%
Maximum	45.73	25.06	30.76	36.83	28.06	45.73	—

Table 6. Coordinates criteria evaluation.

	SMR coordinates error (µm)						
	LT1	LT2	LT2	LT4	LT5	LT1-5	Improvement
Average	20.26	10.25	11.78	9.80	10.28	12.47	41.79%
Maximum	33.07	17.14	22.20	22.53	23.10	33.07	—

6. SENSITIVITY ANALYSIS

A sensitivity analysis was conducted to determine the most suitable SMR positions for calibration purposes. This analysis examines the impact of each error parameter on the global LT measurement error. The synthetic data generator has generated numerous synthetic measurement meshes that serve as error parameters in the error model. Consequently, 18 apertures are required. All of them were generated using the same nominal coordinates as the spherical mesh generated in the synthetic data validation in chapter 5.1. A single error parameter, with a value of 1µm for linear error parameters and 1µrad for angular error parameters, affects each mesh.

Figure 11 shows the error produced by every error parameter through the spherical mesh.

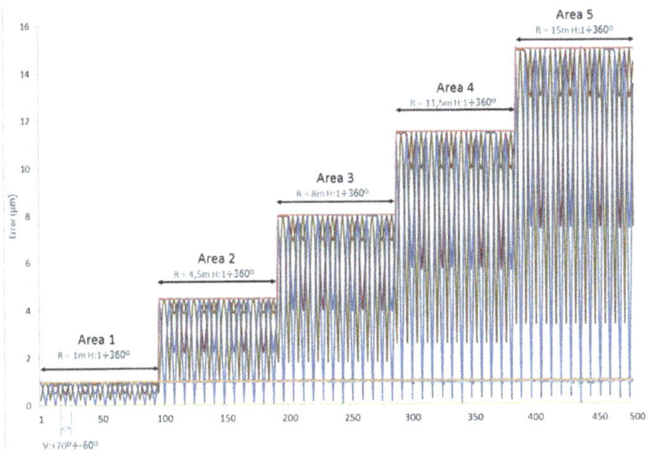

Figure 11. Sensitivity analysis.

As a result, it is deduced that all distance error parameters (δ) produce constant errors. Errors due to parameters εx_1, εy_1 and εz_1 depend on θ, ϕ and *d*, errors due to parameters εx_2 and εy_2 depend on ϕ and d and errors due to parameters εz_2, εx_3 and εy_3 depend only on d. Finally, parameter εz_3 produces no errors. Taking a closer look at the parameters that produce variable errors, we can see in Figure 12 that the maximum and minimum error values correspond to maximum, minimum and zero values of ϕ and also on θ every $\pi/4$.

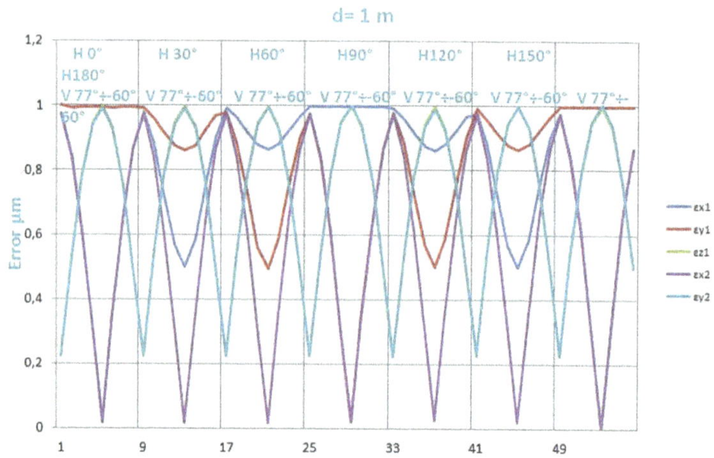

Figure 12. Sensitivity analysis of variable errors.

7. EXPERIMENTAL SETUP

The sensitivity analysis proves that extreme and zero values of tilt angle are the best. It also proves that pan angle ranges must be at least 90°. According to these requirements, 24 reflectors have been placed in a corner of the laboratory. Eight of them spread on the floor (minimum tilt angle), a second set of eight reflectors on the wall at the LT height (zero tilt angle) and the last eight reflectors in the upper part of the floor (maximum tilt angle). LT has been placed at five different positions covering always a pan angle of 90° as shown in Figures 13 and 14.

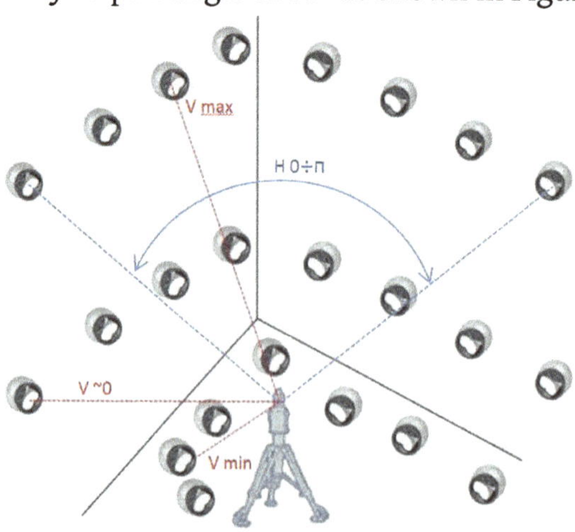

Figure 13. Calibration experimental setup.

Figure 14. View of the experimental calibration process.

8. CALIBRATION RESULTS

The calibration has been performed following the model in Eq. (15) and evaluated according to Eqs. (16) and (17) in the same way as it has been done with the CMM measurements. A gauge of four reflectors has been also included in the mesh of reflectors, and this gauge has been measured in the CMM to know the real distances among its reflectors.

Figures 15 and 16 show the calibration result. In Table 7, the numerical values of the calibration in function of the evaluation method can be appreciated.

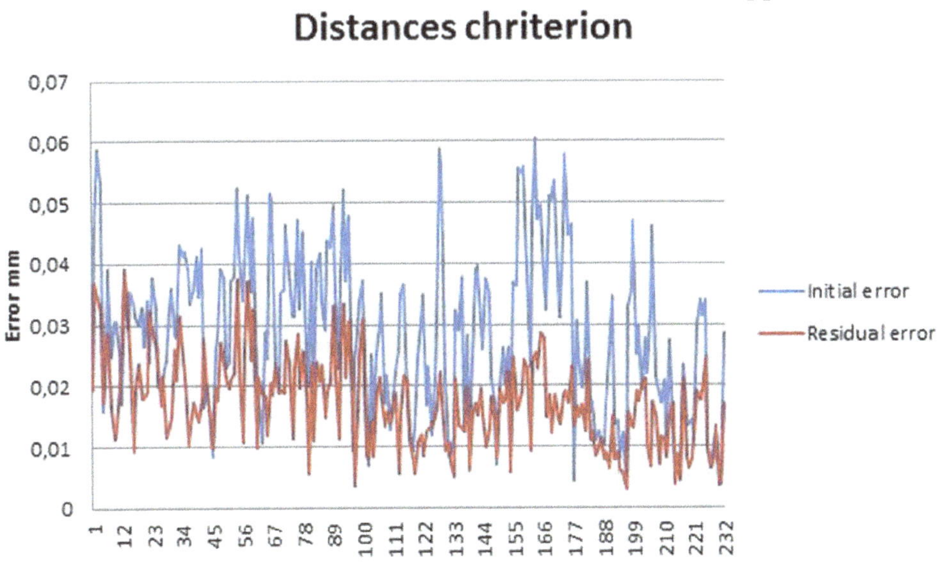

Figure 15. Calibration results distances criteria.

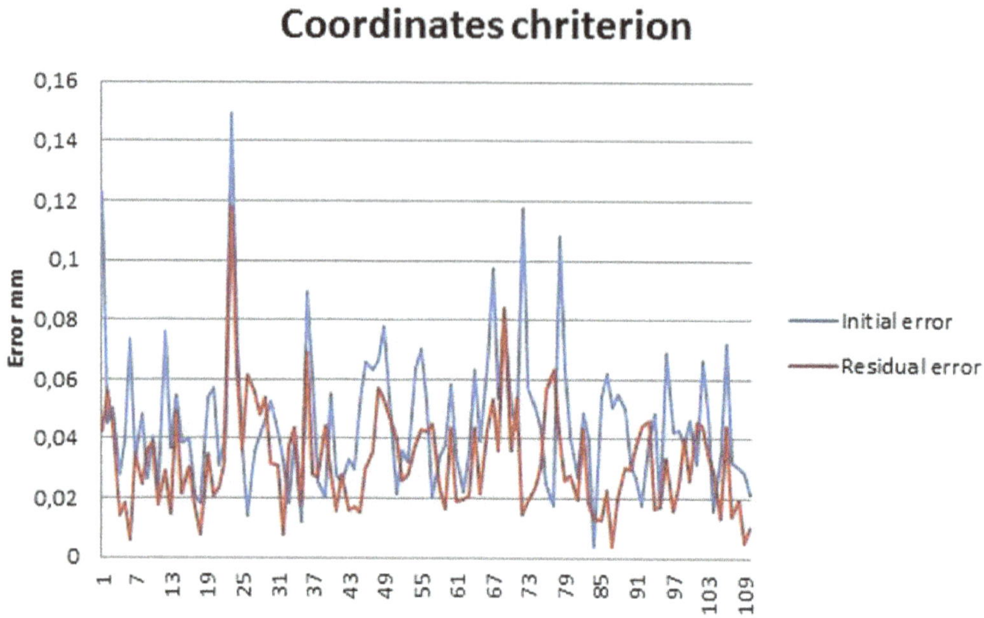

Figure 16. Calibration results coordinated criteria.

9. CALIBRATION VERIFICATION

Calibration results show an LT accuracy improvement according to both criteria used, but as we do not have the SMRs real positions, it is not possible to ensure without any doubt that the calibration procedure increases the LT accuracy.

A new verification is therefore necessary to assess the calibration procedure behavior. In Section 5, a set of SMRs has been measured with the LT from five different positions. These SMRs were placed on a CMM table and measured also with the CMM. These accurate measurements can be used as nominal data to check whether the calibration has improved the LT accuracy or not. The error parameters obtained in the calibration procedure have been applied to the measurements made in the CMM table. The corrected values can be then compared to the CMM nominal measurements and can be seen in Figures 17 and 18 and its values in Table 8.

Table 7. Calibration results.

		Criteria	
		Coordinates	Distances
Initial error (μm)	Maximum	149	60
	Average	46	29
Residual error (μm)	Maximum	118	39
	Average	39	17
Improvement	%	14.47	40.10

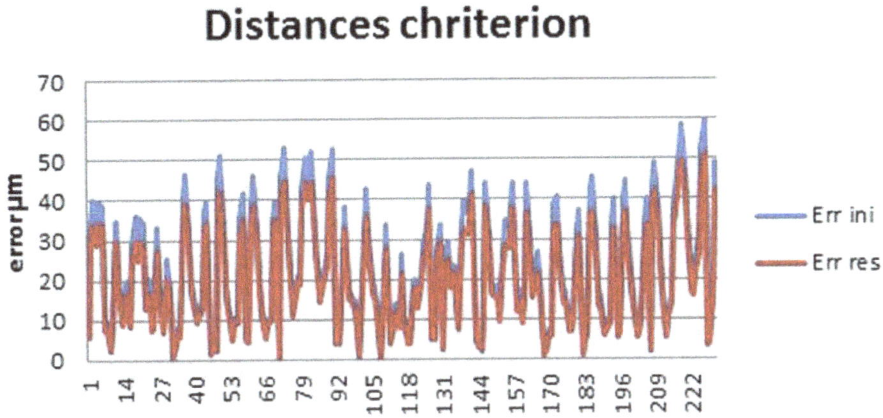

Figure 17. Calibration verification distances criteria.

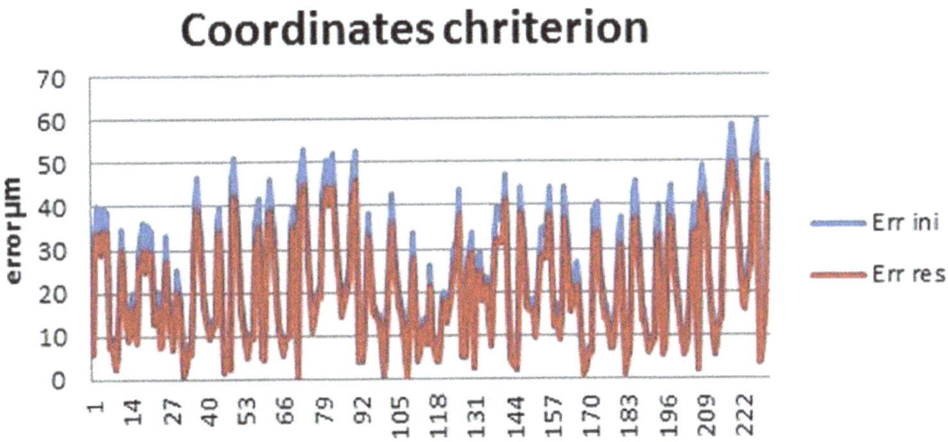

Figure 18. Calibration verification coordinates criteria.

Table 8. Calibration verification results.

		Criteria	
		Coordinates	Coordinates
Initial error (µm)	Maximum	54	84
	Average	21	24
Residual error (µm)	Maximum	41	53
	Average	17	18
Improvement	%	17.98	25.53

10. CONTRIBUTION OF THE SMR INCIDENCE ANGLE IN THE MEASUREMENT UNCERTAINTY

In some of the measurements made, the position of the SMR could not be reached by the LT beam. The SMR maximum viewing angle is within ±30°, and they were placed facing a theoretical point in the middle of the LT selected positions. However, as the SMR positions and orientations are fixed along all the measurements, and they are manually placed, there is the possibility that some of them could not be visible from all the LT positions because the incidence angle was out of the SMR viewing range.

The incidence angle of the laser beam in the SMR has an important influence in the measurement accuracy, and an experiment has been performed to measure the contribution of the SMR incidence angle in the measurement uncertainty.

An SMR with its magnetic holder has been placed on a rotary worktable. The SMR has been centred; so its centre will be in the worktable rotation axis. The SMR was centred using the rotary worktable pencil and its centring accuracy has been measured to be in the range of ±0.1 μm.

The SMR has been initially located with its incidence angles equal to 0° facing to an interferometer laser beam. The interferometer measurement has been then reset to make this position as the zero length measurement.

The SMR has been then rotated on its horizontal and vertical angles within its incident available range of ±30° in 7.5° steps as shown in Figure 19.

Figure 19. SMR incidence angle error measurement.

Data measured by the interferometer are shown in Table 9. An important dependence on the angle variation can be seen, showing the influence of the vertex position error, that is, the distance between the optical centre of the CCR and the SMR sphere.

Table 9. Influence of the laser incidence angle in the SMR.

		θ (deg)								
θ (deg)		-30	-22.5	-15	-7.5	0	7.5	15	22.5	30
	-30	13.4	13	12.4	11.5	10	8.3	6.5	4.3	2.6
	-22.5	8.1	7.9	7.6	6.9	6.2	4.3	3.2	1.6	0
	-15	5	4.8	4.5	3.8	3	2.1	0.8	-0.5	-1.6
	-7.5	2.4	2.2	2	1.5	0.9	-0.2	-1.1	-1.8	-2.8
	0	1.2	1	0.8	0.4	0	-0.7	-1.5	-2.4	-3.4
	7.5	0.8	0.6	0.4	0	-0.5	-1.1	-1.8	-2.6	-3.6
	15	0.3	0	-0.2	-0.5	-1	-1.7	-2.4	-3	-3.9
	22.5	0	-0.2	-0.5	-0.8	-1.3	-2	-2.7	-3.3	-4.1
	30	-0.2	-0.5	-0.8	-1	-1.6	-2.4	-3.1	-3.6	-4.4

11. CONCISE SUMMARY

A novel LT kinematic calibration has been introduced and verified by comparing the calibration results with nominal data obtained using a CMM. The method is unique in that the LT user can perform a final calibration of the LT at the site just prior to measuring with the LT under actual working conditions. This can significantly facilitate the LT measurement procedure by guaranteeing accurate calibration at the time of measurement. A calibrated gauge and a set of reflectors are the sole items required for the calibration process, which must be situated at the measurement site.

The kinematic error paradigm has been established. Additionally, synthetic and nominal data have been employed to validate this model. The optimal configuration for the experimental setup has been determined through the examination of the impact of each error parameter on the global error of the LT.

The factory calibration of the LT was enhanced by the calibration procedure, which was conducted with a previously calibrated LT.

An essential contribution to the measurement errors is demonstrated by the laser incidence angle's influence on the measurement uncertainty.

In comparison to conventional methods, the kinematic calibration model that was devised provides significant benefits. In order to conduct the calibration, numerous measurements are necessary, and the current standards necessitate stringent temperature conditions. The intended method can be employed in two distinct manners. Initially, the distance error calculated for each pair of reflectors measured from various LT locations provides a dimensional value of the LT accuracy, which will assist the user in determining whether or not a device calibration is required. In other words, the final user can conduct the calibration between the programmed calibrations if it is required, without the need for a metrological laboratory. It can also be employed to either complete existing calibration standards or create new ones.

12. PROSPECTIVE ENDEAVORS

It is feasible to locate two viable LT variants from distinct manufacturers. The proposed method is applicable for the laser tracker (LT) with the laser source located in the revolving head. The alternative model is distinguished by the placement of the laser source within the stationary reference frame of the LT. This implies that a rotating mirror must be connected to the stationary axis in order to redirect the laser beam emitted from the source towards the SMR. The calibration approach used in this study can also be applied to the second LT constructive model by adjusting the kinematic model to match the LT geometry and the laser beam path.

In addition to advancing the kinematic model, conducting additional tests is necessary to examine the performance of the calibration method in various scenarios, including varied measurement ranges, temperatures, and the number and arrangement of reflectors.

REFERENCES

1. Wang Z, Mastrogiacomo L, Franceschini F, Maropoulos P. Experimental comparison of dynamic tracking performance of iGPS and laser tracker. International Journal of Advanced Manufacturing Technology. 2011;56:205-213. DOI: 10.1007/s00170-011-3166-0
2. Burge JH, Su P, Zhao C, Zobrist T. Use of a commercial laser tracker for optical alignment. Optical System Alignment and Tolerancing. Proc of SPIE. 2007;6676(66760E):1-12. DOI: 10.1117/12.736705
3. Huo D, Maropoulos PG, Cheng CH. The framework of the virtual laser tracker–a systematic approach to the assessment of error sources and uncertainty in laser tracker. Measurement. 2010:507-523. DOI: 10.01007/978-3-642-10430-5_39
4. Nubiola A, Bonev IA. Absolute calibration of an ABB IRB 1600 robot using a laser tracker. Robotics and Computer-Integrated Manufacturing. 2013;29(1):236-245. DOI: 10.1015/2012.06.004
5. Denavit J, Hartenberg RS. A kinematic notation for lowerpair mechanisms based on matrices. Trans. ASME Journal of Applied Mechanics. 1955;22:215-221
6. Baatz R, Bogena H, Franssen HH, Huisman J, Qu W, Montzka C, Vereecken H. Calibration of a catchment scale cosmic-ray probe network: A comparison of three parameterization methods. Journal of Hydrology. 2014;516:231-244. DOI: 10.1016/2014.02.026

7. Barazzetti L, Giussani A, Roncoroni F, Previtali M. Monitoring structure movement with laser tracking technology: Videometrics, Rare Imaging, and Applications XII; and Automated Visual Inspection. Proc. SPIE 8791. 2013:879106. DOI: 10.1117/12.2019997

8. Bargigli L, Gallegati M, Riccetti L, Russo A. Network analysis and calibration of the "leveraged network-based financial accelerator". Journal of Economic Behavior and Organization. 2014;99:109-125. DOI: 10.1016/2013.12.018

9. ASME B89.4.19-2006 Standard. Performance Evaluation of Laser-Based Spherical Coordinate Measurement Systems www.asme.org

10 Chapter

OPTIMIZING THE KINEMATICS OF SINGLE-SIDED LAPPING THROUGH STATISTICAL ANALYSIS OF TRAJECTORIES OF ABRASIVE PARTICLES

1. INTRODUCTION: AN OVERVIEW

In the present era, it is imperative that items are produced with a combination of superior quality and optimal efficiency. The importance of finishing technologies that can generate high-quality surfaces with low roughness and precise shape and dimensions is growing in today's society. These results can be achieved in the lapping process using relatively uncomplicated production methods. The lapping process, one of the oldest machining processes, is still widely used in precision manufacturing applications to achieve flatness tolerance and surface quality specifications. This method, known as machining, is utilized for shaping various materials including metals, alloys, glasses, natural substances like marble, granite, and basalt, materials employed in semiconductor technology, as well as carbon, graphite, diamond, and ceramics. It has found extensive application in numerous technical fields [1, 2]. The parts commonly treated using lapping include pneumatic and hydraulic components such as valve plates, slipper plates, seals, cylinder parts, and castings. It also includes pump parts like seal faces, rotating valves, and body castings. Additionally, transmission equipment like spacers, gears, shims, and clutch plates, as well as inspection equipment such as test blocks, micrometer anvils, optical flats, and surface plates, are processed using lapping. Cutting tools like tool tips and slitter blades, as well as automotive and aerospace parts like lock plates, gyro components, and seals, are also treated using this method [2, 3]. Abrasive machining processes are complex and unpredictable, with several contributing factors and consequences. The primary results of lapping are stock removal, roughness, and flatness, which are influenced by several factors that affect the quality of the operation. To achieve optimal quality and precision on worked surfaces, researchers should prioritize enhancing the lapping process through the examination of key process variables.

Typically, lapping operations can be categorized based on the active surface of the lapping tool. Peripheral lapping refers to the procedure where the tool axis and the workpiece surface are parallel to each other, while side lapping occurs when the tool axis and the workpiece surface are perpendicular to each other. To be more precise, the lapping procedures that are most commonly employed can be categorized based on the surfaces they produce, the kinematics of the operation, and the profiles of the tools utilized [4]. The two most renowned types of machinery are single-sided and double-sided lapping machines. Double-side surface lapping is regarded as the most precise technique for achieving parallelism and uniformity of size. This approach involves simultaneously lapping two parallel surfaces to provide optimal results. This particular type of machine tools is equipped with a planetary system [5]. Flat lapping often employs standard systems that utilize conditioning rings.

The active surface of the lapping plate exhibits form defects of either convexity or concavity due to wear [3, 6]. This greatly affects the precision of the workpieces' form. If the lapping plate is not perfectly flat, it should undergo re-conditioning. The kinematic approach can be used to repair tool form inaccuracies. The contact between workpieces and the tool is determined by the shape, size, and lapping kinematics of the workpieces. The distribution of abrasive particles on the lapping plate changes when the kinematic circumstances are altered, such as by positioning the workpieces at various radii, adjusting rotational velocities, or introducing extra motions of the conditioning ring. This chapter provides a description of atypical kinematic systems. The results of the simulations have demonstrated that the speed ratio k1 and the period ratio k2, which indicate the relationships between the three fundamental movements of unusual lapping systems, are significant elements that influence the distribution of trajectories. The parameters were optimized. The primary optimization criterion was the assumption of uniformity in tool wear.

2. MAIN FACTORS OF LAPPING PROCESS

The lapping system comprises multiple components, including the lapping plate, abrasives, lapped workpieces, kinematics, and the machine. The lapping process, which determines the quality of the product, the wear of the tool, and the efficiency of the process, is influenced by them [1, 4]. The lapping process can be influenced by two groups of input elements: controllable factors and non-controllable ones, sometimes known as noise factors. The first group encompasses factors such as machining parameters, working pressures and speeds, abrasive type, features of the work equipment, tool, machine, and duration of machining. The uncontrolled input variables encompass several factors such as external temperature, slurry distribution, system vibrations, internal tension, and more. Figure 1 provides a comprehensive description of the primary components involved in the lapping process.

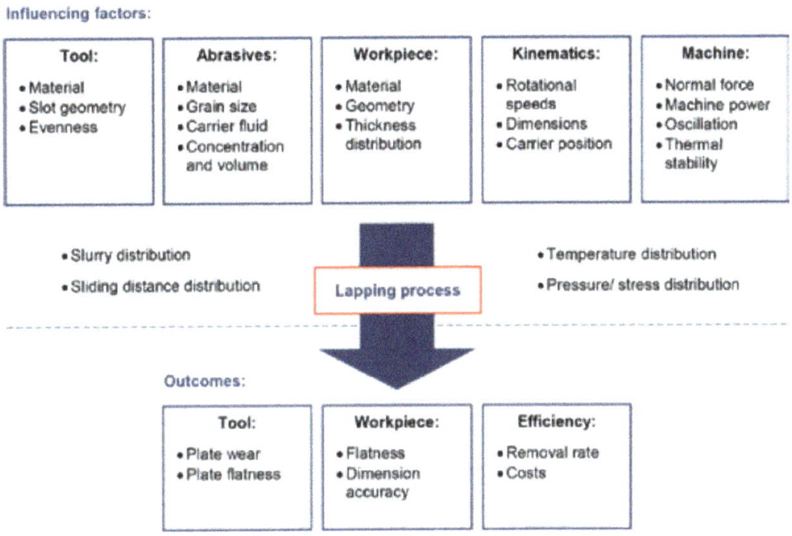

Figure 1. Main factors of lapping process.

First of the most important input parameters, which influence the surface formation and material removal in lapping process is the lapping plate, its geometry, hardness, etc. Generally, it could be established that workpieces are machined to a mirror image of the tool with respect to the flatness. During lapping process, abrasives play an important role on material removal based on size, shape, and hardness. It has been shown that material removal rate increases with increasing abrasive size up to a critical size at which it begins to stabilize and even decrease [7, 8]. Moreover, multiple studies have also shown that surface roughness improves with decreasing grain size. The properties of the abrasive vehicle, which is the medium which transports the abrasive across the workpiece, are important to how effective the abrasives are in material removal. In addition, the flatness, geometric and arrangement of workpieces have also effect on lapping results. The material hardness of workpieces has a crucial influence on a removal mechanism in abrasive processes, as well on efficient of lapping [9]. Surface formation of fragile workpieces, e.g., ceramics, occurs as a result of cracking propagation or by plastic deformation, when depth of cut is smaller than critical [10].

There has been a great amount of research conducted on the kinematics of lapping [2, 3, 6, 11, 12]. The lapping pressure and lapping speed can be seen as the main variables in the lapping process. They are closely interdependent with kinematic system and machine parameters.

3. ABRASIVE WEAR IN SINGLE-SIDED LAPPING PROCESS

One of the most important mechanisms of the lapping process is abrasive wear. The process describes the separation of material from surfaces in relative motions caused by protrusions and particles between two surfaces or fixed in one of them. In lapping process, abrasive grains are guided across the surface to be lapped and backed up by a lapping plate or conditioning rings. It is crucial to minimize friction and wear of the abrasive and to maximize abrasive wear of the workpiece [2, 3].

Abrasive wear is commonly classified into two types: two- and three-body abrasion. In two-body abrasion, particles are rigidly attached to the second body. When abrasive particles are loose and free to move, then we deal with three-body abrasion. Therefore, in a two-body abrasion, the abrasive grains can cut deeply into the workpiece due to the sliding motions and in the three-body abrasion, the particles spend part of time in cutting and part of time in rolling [3].

Also in the case of lapping process, several abrasives move into active positions and other grains leave the working gap. Only a specified part of all particles is able to enter the working gap with height H, which is appropriate to the lapping pressure F. Figure 2 presents abrasive wear in single-sided lapping process and different types of lapping grains being in the working gap between tool and workpieces [3]. As can be seen in Figure 2, abrasives play a crucial role on material removal based on size, shape, and hardness. For example, it can be noticed that grain A is too small and grain B is too big. It means that only some of the abrasives are active and they roll like grain C or slide like grain D. The example of passive grain is grain E, which does not stick to the affecting partner.

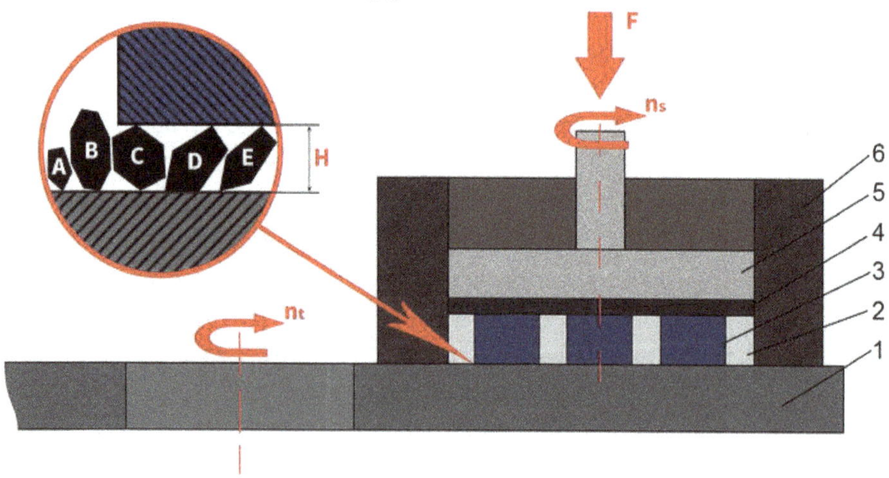

Figure 2. Single-sided lapping process: (1) lapping plate, (2) separator, (3) workpiece, (4) felt pad, (5) weight system, (6) conditioning ring.

Moreover, regular sliding grains in and out of the working gap cause the changes in the initial structure of the lapping plate. First, strongest loaded grains disintegrate into smaller and their particles are able to take part in the lapping process depending on the size and shape. Hence, the number of bigger grains is reduced, while the number of smaller grains is increased and the structure of the lapping tool is changed. The changing trajectories and velocity of abrasives cause some grains crash with each other or with grain fragments. It can be established that after crashes, grains are accumulating and can disintegrate at any time. The chain between a workpiece and lapping plate is fulfilled by active grains, which under their impact and edge transfer the normal forces into the surface of the affecting partner. Acting forces are proportional to the material volume. To reach a stationary working gap height, the amount of all normal forces transferred by active grains must be equal to the pressing force F [3].

4. KINEMATIC ANALYSIS OF STANDARD SINGLE-SIDE LAPPING SYSTEM

4.1. Single-sided lapping machine overview

Nowadays, there are many manufacturers that offer surface planarization technology with lapping kinematics. The most well-known are Peter Wolters, Lapmaster, Stähli, Engis, Hamai, Kemet, Mitsunaga and LamPlan. After a careful analysis of numerous researches and offers of many lapping machines producers, it has been emerged that most of lapping machines have standard kinematic system. Single-sided lapping machines differ in tool size, diameter of lapping plate, size and number of conditioning rings, type of pressure system (weighting or pneumatic system). More complex machines are equipped with forced drive option of conditioning rings. Such a system maintains a constant speed of workpieces.

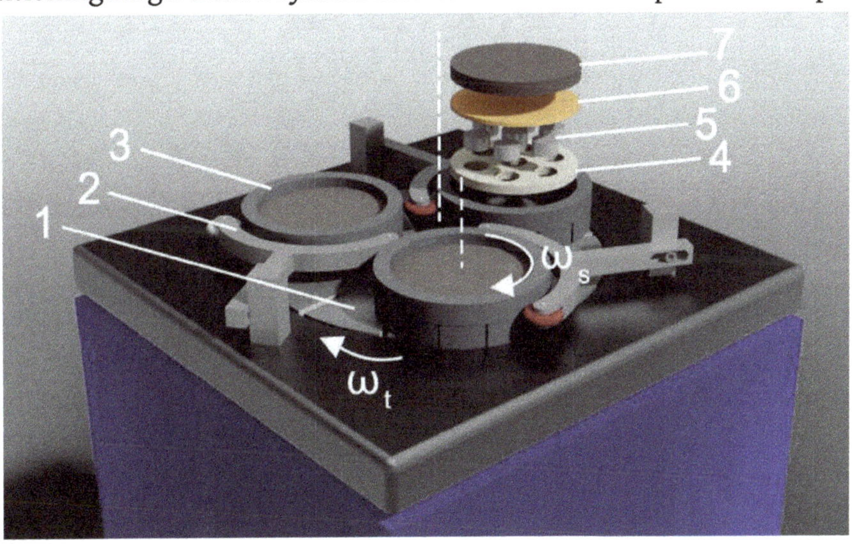

Figure 3. Single-sided lapping machine: (1) lapping plate, (2) roller forks, (3) conditioning ring, (4) separator, (5) workpieces, (6) pad, (7) weight.

Conventional single-sided lapping machine is shown in Figure 3. The key component of the machine is the tool, i.e., the annular-shaped lapping plate (1), on which the workpieces (3) are applied to. One machine usually has three to four conditioning rings (5). However, laboratory lapping machines that have one ring are also popular. The lapping plate (1) rotates with angular velocity ω_t and it drives conditioning rings, where separators (4) are placed allowing additional move of workpieces (3). Due to the frictional force, conditioning rings (5) rotate on the working surface of lapping plate (1) with angular velocity ω_s. This force depends on a radial position, velocity of conditioning rings, and friction conditions. The radial position (R) of conditioning rings can be controlled with roller forks (2). During the lapping process, a certain load is provided through felt pad (6) by weight disk (7) or pneumatic system. In this way, the parts are pressed against a film of abrasive slurry that is continuously supplied to the rotating lapping plate [2].

4.2. Kinematical fundamentals

Due to the fact that the kinematics of the lapping is affected by a number of factors related to the influence and properties of the workpiece—abrasive slurry—lapping plate system, in the following model considerations, it is assumed that the angular velocities of conditioning rings, separator and workpieces are identical. Moreover, conditioning rings role is to even the lapping plate (Figure 4).

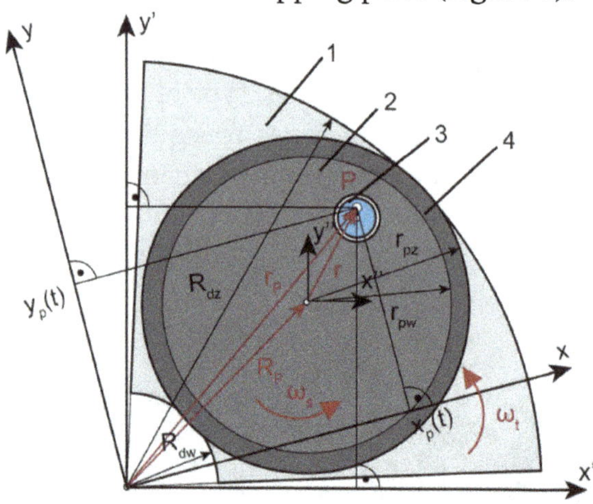

Figure 4. Kinematic diagram of single-sided lapping machine: (1) lapping plate, (2) separator, (3) workpiece, (4) conditioning ring.

The input parameters for analysis are: angular velocity of lapping plate ω_t and of conditioning rings ω_s, inner R_{dw} and outer R_{dz} diameter of lapping plate, radial position of conditioning ring $R_{p.}$

In order to model a lapping plate, the position of any point P belonging to a workpiece must be determined. It is possible to do this by a radius vector in two coordinate systems: absolute and relative, which is related with rotating tool.

The position of any point $P(r,\phi_p)$ belonging to a workpiece in single-sided lapping system are determined in x″y″ coordinate system, which is related to conditioning ring as:

$$x_p'' = -r \cdot \cos(\varphi_s) \tag{1}$$

$$y_p'' = -r \cdot \sin(\varphi_s) \tag{2}$$

The coordinates of point P in x′y′ coordinate system are:

$$x_p' = x_p'' \cdot \cos(\omega_s \cdot t) + y_p'' \sin(\omega_s \cdot t) + R_{Px} \tag{3}$$

$$y_p' = -x_p'' \cdot \sin(\omega_s \cdot t) + y_p'' \cdot \cos(\omega_s \cdot t) + R_{Py} \tag{4}$$

After transforming it to coordinate system related to rotating lapping plate:

$$x_p = x_p' \cdot \cos(\omega_t \cdot t) + y_p' \cdot \sin(\omega_t \cdot t) \tag{5}$$

$$y_p = -x_p' \cdot \sin(\omega_t \cdot t) + y_p' \cdot \cos(\omega_t \cdot t) \tag{6}$$

$$r_p(t) = \sqrt{x_p^2 + y_p^2} \tag{7}$$

The relative velocity v of point P is defined as the derivative of the position with respect to time:

$$v = \frac{dr_p}{dt} \tag{8}$$

4.3. Single abrasive trajectories analysis

Equations of any point belonging to a workpiece position in standard single-sided lapping system were implemented in MATLAB program. This program allows to analyze single-side lapping system and can be used to mark out cycloids paths, which can be treated as areas where the lapping plate wears by the grain placed in a specific location of a conditioning ring or workpiece.

In order to analyze single abrasive trajectories, an additional parameter must be defined, which is rotational speed ratio of the conditioning ring to the lapping plate:

$$k_1 = \frac{\omega_s}{\omega_t} \tag{9}$$

Analysis of multiple simulations leads to conclusions that the k_1 parameter determines the type of cycloid path. Results are shown in Figure 5. It can be observed that epicycloid trajectories are marked out when the value k_1 is less than 0. When the value k_1 is close to 0, stretched epicycloid then interlaced epicycloid are received. Pericycloids appear when the value k_1 is between 0 and 1. At k_1 bigger than 1, hypocycloids can be obtained. Initially, they are stretched hypocycloids; then, they transform to interlaced hypocycloids. Attention needs to be paid to the following trajectories: cardioid (k_1 = -1), concentric circle (k_1 = 0), eccentric circle (k_1 = 1) and concentric ellipse (k_1 = 2). Moreover, the cycloid curvature radius $R_p(t)$ is a periodic function of time t and its cycle equals:

$$T = \frac{2\pi}{\omega} \tag{10}$$

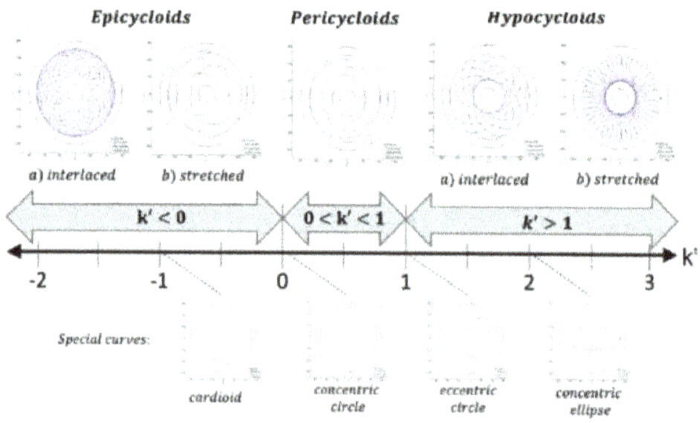

Figure 5. Path types depending on the rotational speed ratio k_1.

5. UNCONVENTIONAL LAPPING KINEMATICS SYSTEMS

Analysis of numerous researches and offers of many lapping machines producers leads to conclusion that most of lapping machines have the standard kinematic system. Unconventional lapping systems, where the conditioning ring performs an additional move such as radial, secant, swinging or off-center are presented in Figure 6. Simulations have shown that changing the kinematic system in single-sided lapping process causes a wide density variation of single abrasive trajectories [13].

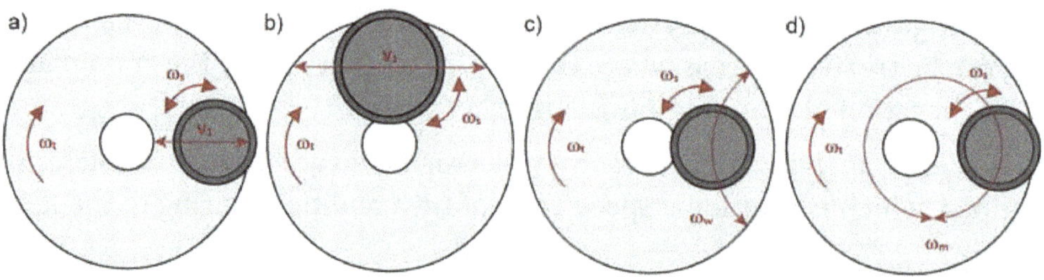

Figure 6. Examples of approach to kinematics system lapping plate—conditioning ring: (a) radial, (b) secant, (c) swinging, and (d) off-set.

The simulations allowed state that the most desirable system is that system, which allows to smoothly control the position of the conditioning ring on the lapping plate. It was pointed out that some systems are not very different from the standard kinematic system. Generated trajectories were almost identical. This is due to the fact that with such systems, the deflection of conditioning ring along the radius was much smaller, than in the case, for example, of the radial lapping system. In order to correct the flatness of lapping plate, the ring must be shifted toward or out of the center of the tool.

The detailed kinematic diagram of single-sided lapping system with the additional movement of conditioning ring along a chord is presented in Figure 7. It can be observed that in this idea, there is only one conditioning ring, which in addition to rotary motion performs a reciprocating motion between point A and B. The position of the conditioning ring is not constant as in conventional system, parameter $R_{Px}(t)$ changes the value in time. The distance from the center of the lapping plate to the chord equals R_{Py}.

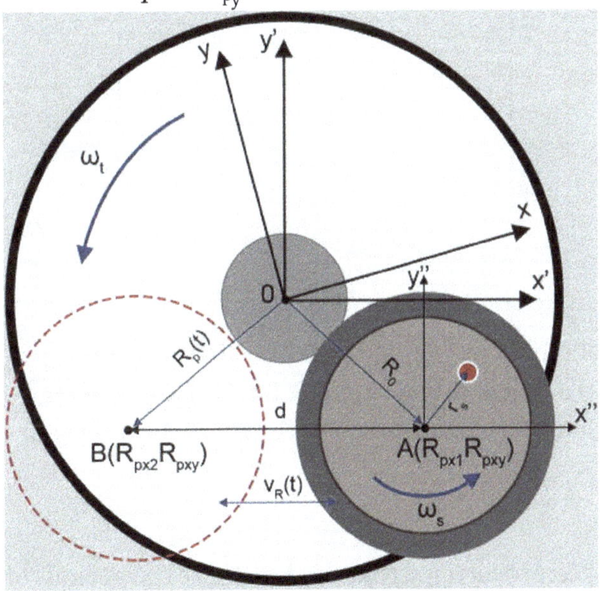

Figure 7. Diagram of single-sided lapping machine with reciprocating movement of conditioning ring.

The principle of kinematics calculation is identical as in case of standard single-sided lapping system. Due to additional motion of the conditioning ring, only Eq. (3) must be changed:

$$x_p' = x_p'' \cdot \cos(\omega_s \cdot t) + y_p'' \cdot \sin(\omega_s \cdot t) + R_{Px}(t) \tag{11}$$

Considering the reciprocating motion of the conditioning ring, R_{Px} is a function of time and it can be expressed as:

$$R_{Px} = R_o + \Delta R \qquad (12)$$

There are many ways in positioning control applications to move from one point to another. The most common are trapezoidal and S-curve motions. In trapezoidal move profile, velocity is changed linearly and acceleration makes abrupt changes [14]. This can be quantified in an S-curve motion profile by "jerk," which is the time derivative of acceleration. S-curve motion helps to smoothen the motion and reduce the possibility of exciting a vibration or oscillation [15]. A single stroke of reciprocating motion with S-curve profile is shown in Figure 8. The conditioning ring moves between two points, where T_R is the reciprocating period, T_A is the acceleration time of the uniformly accelerated motion and T_D is the deceleration time of the uniformly decelerated motion. The conditioning ring has a maximum velocity v_{Rmax} at the central range of the reciprocating stroke and uniformly accelerated and decelerated motions at the two ends of the stroke.

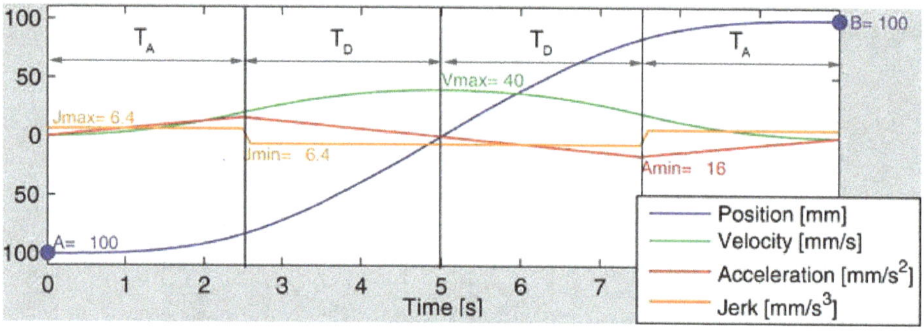

Figure 8. Single stroke of reciprocating motion with S-curve profile.

For the simplicity, the velocity expressions of another non-dimensional parameter was introduced and defined by Eq. (12). Parameter k_2 is the ratio of the reciprocating period to the lapping plate rotary motion period:

$$k_2 = \frac{T_R}{T_T} = \frac{d \cdot \omega_t}{v_R \cdot \pi} \qquad (13)$$

where d is the reciprocating stroke, T_R and T_T are the periods of the reciprocating motion of the conditioning ring and rotary motion of the lapping plate.

6. KINEMATIC OPTIMIZATION

Kinematic analysis showed that the basic kinematic parameters significantly affect the trajectory and velocity distributions. To improve the quality and flatness of the machined surfaces, the selection of parameter values, which are optimal was carried out. It was assumed that, the main optimization criterion is the uniformity of tool wear. In this section, the material removal rate and the lapping plate wear model are described in detail. Based on the lapping plate wear, its uniformity is calculated and the maximum value is sought.

6.1. Material removal rate

The volume of material removed by lapping process at a local position during every unit of time is an important part of study about lapping process. There are many models which describe material removal rate (MRR) during lapping, grinding or polishing. Several researchers worked on experiment and analytical models [2, 16, 17]. They show that the lapping process can be improved by optimizing both the machining efficiencies, and the consideration of the process parameter influences with surface lap height. Considerations about MRR are important because it is the only solution to provide the maintenance of reliability and lifetime of the produced workpieces.

One of the most common wear models in abrasives process is the tribological model developed by Preston, who related relative velocity and pressure to material removal rate (MRR) which is known as Preston's equation:

$$\frac{dH}{dt} = k \cdot p \cdot v \tag{14}$$

where H is a height of removed material in any point on the lapping plate, k is a constant relating to the process parameters, p is the applied pressure, and v is the relative velocity of the workpiece. The parameter k varies based on any modifications to the material removal process such as abrasive and slurry type, feed rate, and other process parameters [16].

However, some of the experiments have shown that in the case of relatively high or low velocities and pressure, linear relationship in Preston's equation does not hold true. Therefore, many modifications to the Preston's equation were proposed. Moreover, there are many formulations in literatures, which are based on the Preston's equation. Using the Preston's equation, Runnels and Eyman [18] proved that the MRR in the chemical-mechanical polishing process is related to the normal and shear stresses. Tseng and Wang [19] modified the Preston's equation to express the MRR as a function of the normal stress of the wafer, the relative velocity of the polishing and the elastic deformation of the abrasive grains. Nanz [20] provided new MRR equation considers the bending of pad and flow of slurry.

6.2. Lapping plate wear model

Intensity of lapping plate wear can be assumed as a contact intensity of the tool with the workpieces through the lapping abrasive grains. There are different methods for calculating contact intensity. This section assumes the method of calculating particles density of interpolated trajectories. Therefore, in order to simulate the trajectories and to count their distribution, Matlab program was designed.

An example of trajectories density determination steps is shown in Figure 9. Initially, location of random particles is generated within the conditioning ring.

Then, the trajectories of the particles are calculated with a use of kinematic equations. A set of points, which are equally spaced from each other, are generated with interpolation function. The lapping plate surface is divided into small squares with the same area. Finally, a statistics function is used to count the total number of points within each square of the lapping plate surface. The contact intensity can be developed for a profile of the tool. It allows to determine if the wear causes a concavity or the convexity. The area of the lapping plate must be divided into equal rings. A measure of points in an appropriate area is determined by equation:

$$D_i = \frac{n_i}{A_i - A_{i-1}} = \frac{n_i}{(2i-1)\cdot\pi\cdot r^2} \qquad (15)$$

where n is a points number in area A_i and r is rings width.

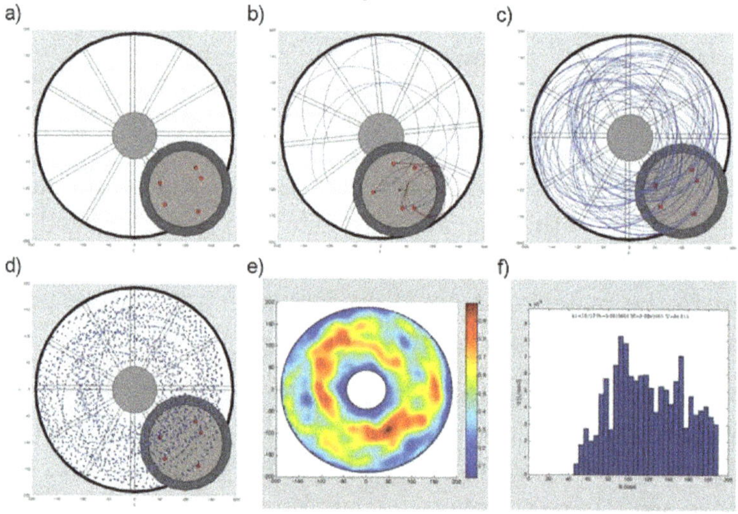

Figure 9. Trajectories density determination: (a) random generated particles, (b, c) trajectories generating, (d) trajectories interpolation, (e) density of trajectories, (f) profile density of trajectories.

To calculate the standard deviation S_D of all the values of D_i, which is given by:

$$S_D = \sqrt{\frac{\sum(D-\bar{D})^2}{N-1}} \quad (i = 1, 2, 3 \ldots N) \qquad (16)$$

where $\bar{D}$ represents the average of all the values and N is the total number of the divided rings or squares [18].

6.3. Parameters optimization

Trajectories distribution of particles in single-sided lapping process is significantly affected by rotational speed ratio of the conditioning ring to the lapping plate k_1 (Eq. (9)) in standard single-sided lapping system and in addition by period ratio of the reciprocating motion of the conditioning ring to rotary motion of the lapping plate k_2 (Eq. (13)) in system with additional movement of conditioning ring. To obtain a better uniformity base on trajectory simulations and consequently even lapping plate wear, it is important to optimize both parameters. The uniformity of tool wear was assumed as main optimization criterion. In order to describe the evenness of the lapping wear, the trajectories distribution uniformity is defined:

$$U = \left(1 - \frac{S_D}{\bar{D}}\right) \cdot 100\% \qquad (17)$$

where S_D is the standard deviation of trajectories density and $\bar{D}$ is the average value of the trajectories distribution.

It can be predicted that during the lapping process, there are more than 1 million active particles in the slurry and on the lapping plate. However, because of the calculation time, an appropriate particle number which can reflect the same regularity as the real number has to be determined. Figure 10 shows the influence of the amount of abrasive grains to the uniformity. Results are presented for standard single-sided lapping system, when k_1 = 0.45 and R_P = 125 mm. It can be observed that for 1000 randomly distributed particles, uniformity is stable and constant.

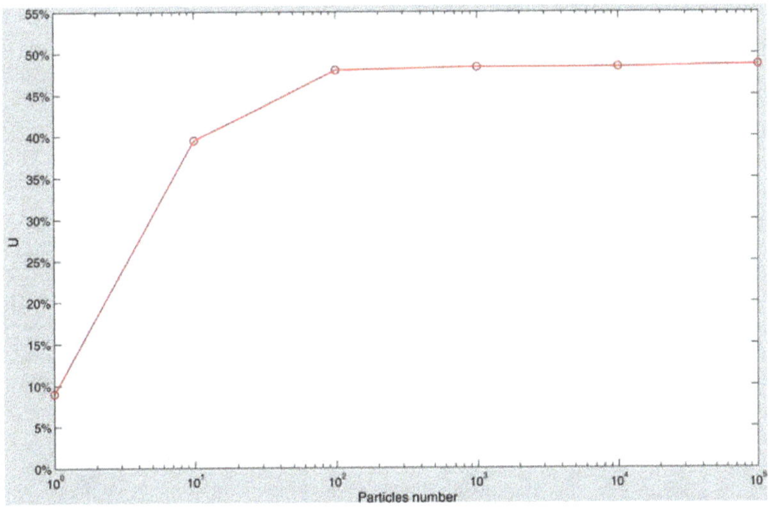

Figure 10. Uniformity of trajectories distribution produced by different number of random particles.

Trajectory density uniformity for standard single-sided lapping system and for single-sided lapping system with reciprocating motion of conditioning ring is presented in Figures 11 and 12. Trajectories were generated during 60 s on lapping plate with internal diameter of 88 mm and outer diameter of 350 mm. The maximum uniformity was obtained in conventional system, when k_1 = 0.6–0.9, and in non-standard lapping system, when k_1 = 0.65–0.75 and k_2 = 1–2. For single-sided lapping system with reciprocating motion of conditioning ring, the uniformity was about 10% higher.

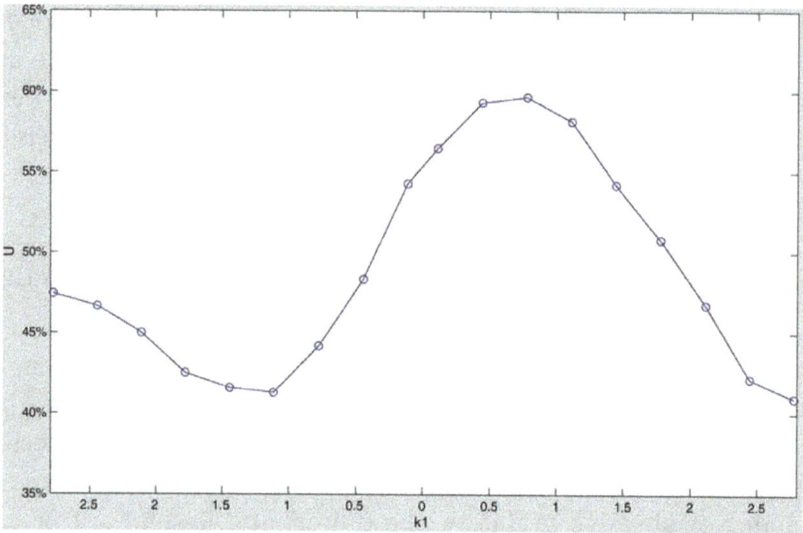

Figure 11. Trajectory density uniformity versus k_1 *for standard single-sided lapping system, 1000 random particles, radial position of conditioning R = 125 mm, simulation time t = 60 s.*

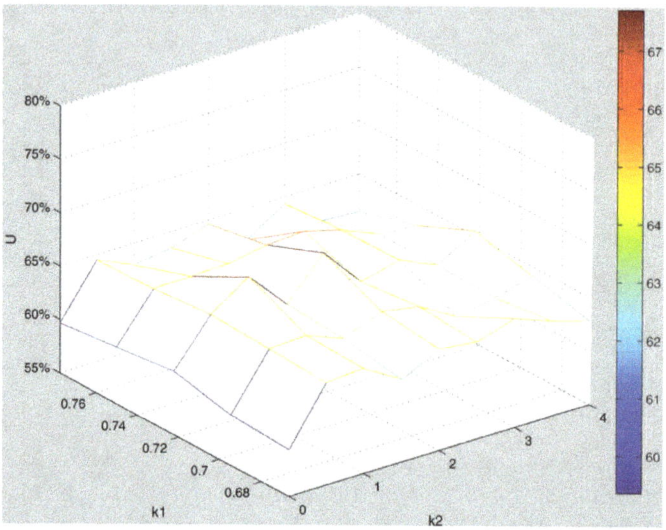

Figure 12. Trajectory density uniformity versus k_1 *and* k_2 *for single-sided lapping system with reciprocating motion of conditioning ring, 1000 random particles, simulation time t = 60 s.*

7. AUTOMATED LAPPING SYSTEM

The above considerations regarding complex kinematics of lapping process were the starting point to develop automated lapping system. Nowadays, machine tools are more efficient than in the past. Lapping machine manufacturers also improve their machines and supply their basic constructions with additional components. As a result of automation, some of the supporting operations can be eliminated.

Lapping machines are supplied with feeding tables, loading and unloading systems of conditioning rings, which form mini-production lines. However, it was emerged that the system where the ring is led by the manipulator during the machining is novel. In order to create universal mechanism that moves the conditioning ring at any path is complicated and in some cases impossible. Thanks to the robot that moves an effector from point to point, it is possible to change the ring trajectory at any moment and different conditions. Owing to this solution, it is possible to apply any lapping kinematics and use robot for supporting operations.

One of a number of challenges in designing automated manufacturing systems is the selection of a suitable robot. This problem has become more demanding due to increasing specifications and complexity of the robots. However, it was decided that the robot should perform material handling tasks and also lapping process. In initial selection, the articulated robots, which have four to five degrees of freedom and are powered by an electrical drive should be chosen for further evaluation. Furthermore, a continuous path or Point-to-Point control system is required.

The idea of how single-sided lapping machine and the robot working together presents Figure 13. The robot situated next to the lapping machine handle primarily sorted workpieces from the table to the separator, located in conditioning ring. Then robot grips the conditioning ring and shifts it onto the lapping plate, which starts to rotate. The machining is executed by the robot. It shifts the ring with workpieces in such a way to keep the flatness of the plate along the radius. After lapping process, robot shifts the conditioning ring onto collecting table and workpieces fall into the box with finished parts. Finally, the flatness of the lapping plate is controlled and is fixed in case when an error occurs.

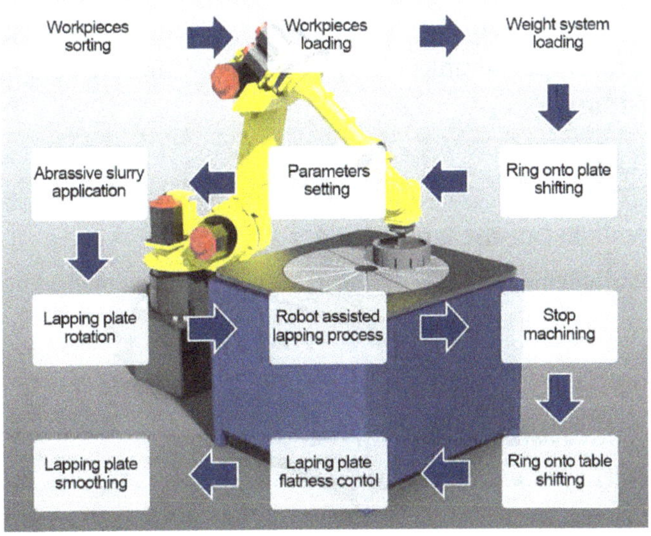

Figure 13. Idea of robotic single-sided lapping system.

8. CONCISE SUMMARY

The purpose of this chapter was to demonstrate the impact of several elements on the geometric outcomes of the single-sided lapping process. The mechanics of the executive system is a highly significant element. The flatness of the lapping plate's active surface significantly affects the accuracy of the lapped surfaces. Therefore, it is crucial to ensure the tool's flatness is maintained. The uneven wear of the lapping plate can be identified as the primary cause of its variation in flatness.

By selecting suitable kinematic parameters, it is possible to achieve a desirable distribution of contact density. This distribution plays a crucial role in determining wear and correcting the flatness mistake of the lapping plate. Consequently, the kinematic model of the single-sided system was thoroughly investigated and simulated using the Matlab program. Through simulations, it was noted that the distribution of abrasive particles on the lapping plate changes when the kinematic conditions are altered. This can be achieved by adjusting the placement of the workpieces at different radii, changing the rotational velocities, or introducing additional movements of the conditioning ring. Therefore, the verification of the impact of supplementary directing movements of the conditioning ring has been conducted. The trajectory distribution in unusual lapping systems is primarily influenced by two major factors: the speed ratio (k1) and the period ratio (k2). These ratios indicate the relationships between the three fundamental motions of the system. The objective of optimizing these parameters was to enhance the quality and evenness of the machined surfaces. The primary optimization criterion was the evenness of tool wear. The highest level of uniformity was achieved in the traditional system when the value of k1 ranged from 0.6 to 0.9. In the non-standard lapping system, the optimal values for k1 were between 0.65 and 0.75, while k2 ranged from 1 to 2. In the single-sided lapping system with reciprocating motion of the conditioning ring, the uniformity was approximately 10% more than in the conventional system.

NOMENCLATURE

ω_t	angular velocity of the lapping plate
ω_s	angular velocity of the conditioning rings
R_{dw}	inner diameter of lapping plate
R_{dz}	outer diameter of lapping plate
R_p	radial position of conditioning ring
r_P	radial position of examined points
t	machining time
H	wear removal rate

k	Preston's coefficient
p	force per unit area
v	lapping relative velocity
D_i	trajectories density
d	distance of reciprocating motion
v_R	linear velocity of reciprocating motion
T_R	reciprocating period
T_A	acceleration time
T_D	deceleration time
J_A	acceleration jerk
J_D	deceleration jerk
k_1	rotational speed ratio of the conditioning ring to the lapping plate
k_2	period ratio of the reciprocating motion of the conditioning ring to rotary motion of the lapping plate
S_D	standard deviation
U	uniformity of lapping plate wear

REFERENCES

1. Klocke F. Manufacturing Processes 2: Grinding, Honing, Lapping. Berlin, Heidelberg: Springer-Verlag; 2009

2. Barylski A. Obróbka powierzchni płaskich na docierarkach. Gdansk: Wydawnictwo Politechniki Gdańskiej; 2013

3. Uhlmann E, Ardelt T, Spur G. Influence of kinematics on the face grinding process on lapping machines. CIRP Annals - Manufacturing Technology. 1999;48(1):281-284. DOI: 10.1016/S0007-8506(07)63184-9

4. Marinescu ID, Uhlmann E, Doi T. Handbook of Lapping and Polishing. New York, Taylor & Francis Publishing House; 2006

5. Barylski A. Podstawy docierania jednotarczowego powierzchni płaskich. Gdańsk, Zeszyty Naukowe Politechniki Gdańskiej. Mechanika. 1992;67

6. Barylski A, Piotrowski N. Koncepcje niekonwencjonalnych układów kinematycznych docierania jednotarczowego z wykorzystaniem robota. Mechanik. 2013;78(8-9):36-33

7. Martin K. Neuere Erkenntnisse über den Hartmetallabtrag beim Läppen. Maschinenmarkt. 1973;79:103-104

8. Ardelt T. Verfahrensvergleich Planschleifen mit Planetenkinematik – Planparallelläppen. IDR. 2001;35(3):214-224

9. Neauport J, Destribats J, Maunier C, Ambard C, Cormont P, Pintault B, Rondeau O. Loose abrasive slurries for optical glass lapping. Applied Optics. 2010;49(30):5736-5745

10. Molenda J, Charchalis A. Wpływ twardości materiału obrabianego na jakość powierzchni po docieraniu. Logistyka. 2015;3:3363-3370

11. Ishikawa K, Ichikawa K, Suwabe H. A basic study on corrective techniques to recovering accuracy of deteriorated lapping plate. Japan Society of Precision Engineering. 1988;3:234-236

12. Satake U, Enomoto T, Fujii K, Hirose K. Optimization method for double-sided polishing process based on kinematical analysis. Procedia CIRP. 2106;41:870-874

13. Barylski A, Piotrowski N. New kinematic systems in single-sided lapping and their influence on lap wear. Applied Mechanics and Materials. 2016;831:14-24

14. Lewin C. Mathematics of motion control profiles. Performance Motion Devices. 2013; https://pdfs.semanticscholar.org/a229/fdba63d8d68abd09f70604d56cc07ee50f7d.pdf [Accessed: 10.25.2017]

15. Nguyen KD, Ng T-C, Chen I-M. On algorithms for planning S-curve motion profiles. International Journal of Advanced Robotic Systems. 2008;5(1):99-106

16. Evans J, Paul E, Dornfeld D, Lucca D, Byrne G, Tricard M, Klocke F, Dambon O, Mullany B. Material removal mechanisms in lapping and polishing. CIRP Annals - Manufacturing Technology. 2003;52(2):611-633

17. llez-Arriaga L, Téllez A. Correction of the Preston equation for low speeds. Applied Optics. 2007;46(9):1408-1410

18. Runnels SR, Eyman LM. Tribology analysis of chemical-mechanical polishing. Journal of Electrochemical Society. 1994;141(6):1698-1701. DOI: 10.1149/1.2054985

19. Tseng WT. Comparative study on the roles of velocity in the material removal rate during chemical mechanical polishing. Journal of the Electrochemical Society. 1999;146(5):1952-1959

20. Nanz G, Camilletti LE. Modeling of chemical-mechanical polishing: A review. IEEE Transactions on Semiconductor Manufacturing. 1995;8(4):382-389. DOI: 10.1109/66.475179

Chapter 11

AN OPTIMIZATION APPROACH IS PROPOSED FOR SOLVING THE INVERSE KINEMATIC PROBLEM

1. INTRODUCTION: AN OVERVIEW

A kinematic chain is a system of stiff or flexible links that are joined by joints or kinematic pairs, allowing for relative motion between the connected components. Manipulator kinematics can be classified into two categories: forward kinematics and inverse kinematics. The process of determining the position and orientation of a serial manipulator using forward kinematics is straightforward and mathematically uncomplicated. However, when it comes to inverse kinematics, there is no one answer. In general, inverse kinematics often yields several solutions. Therefore, finding a solution for inverse kinematics is quite challenging and requires significant processing resources. Implementing real-time control for any manipulator configuration is typically costly and time-consuming. The forward kinematics of a manipulator involves determining the position and orientation of the end effector by translating from joint space to Cartesian space. The reverse process, where the position and orientation of the end effector are determined from Cartesian space to joint space, is known as inverse kinematics. Calculating the optimal joint angles is crucial for ensuring that the end effector can accurately reach the required position and for the construction of the manipulator. Inverse kinematics solutions form the basis of several industrial applications. It is common in real-time environments to use joint variables to quickly modify the end effector. The forward kinematics of an industrial robot manipulator with n joints can be expressed as,

$$y(t) = f(\theta(t)) \tag{1}$$

where $\theta_i = \theta(t), i = 1, 2, 3, \ldots, n$ and position variables by $y_j = y(t), j = 1, 2, 3, \ldots, m$.

Inverse kinematics for n number of joints can be computed as,

$$\theta(t) = f'(y(t)) \tag{2}$$

The inverse kinematics solution of robot manipulators has been extensively studied and several solution schemes have been created in recent years due to the presence of multiple, nonlinear, and unpredictable solutions. Various strategies, such as iterative, algebraic, and geometric approaches, can be employed to solve inverse kinematics problems. [8] suggested an inverse kinematic solution using quaternion transformation. Researchers between the ages of 20 and 36 have suggested using quaternion algebra to solve the inverse kinematics problem for various configurations of robot manipulators. [35] introduced a quaternion approach to illustrate the kinematics and dynamics of rigid multi-body systems. [34] proposed an analytical solution for a 5-degree-of-freedom manipulator, taking into account singularity analysis. [11] provided a solution for the kinematics and dynamics of a flexible manipulator using quaternion-based methods. [14] presented a comprehensive derivation of inverse kinematics by utilizing exponential rotational matrices. However, traditional analytical and Jacobian-based inverse kinematics methods have been shown to be difficult and computationally intensive, making them unsuitable for real-time applications. Due to the aforementioned factors, multiple authors have utilized an optimization-based approach to solve inverse kinematics.

Optimization strategies are effective for solving the inverse kinematics problem for various configurations of manipulators and spatial mechanisms. Nonlinear kinematic issues can be solved using conventional methods like Newton-Raphson, while differential problems of manipulators can be computed using predictor-corrector type methods. However, a significant limitation of these systems is the occurrence of Singularity or sick condition, which leads to the production of local solutions. Furthermore, if the initial guess is not accurate, the approach becomes unstable and fails to converge to the optimal answer. Hence, newly discovered metaheuristic techniques can be employed to overcome the limitations of traditional optimization methods. The literature survey demonstrates that metaheuristic algorithms or bi-inspired optimization strategies are more advantageous in attaining global optimum solutions. The primary concern about these algorithms inspired by nature is the formulation of the objective function. These algorithms are direct search algorithms that do not rely on gradient or differentiation of the objective function. The metaheuristic algorithm is compared to heuristic algorithms based on their convergence rate, which has been demonstrated to be slower for heuristic-based techniques. Thus, employing metaheuristic techniques such as Genetic Algorithms (GA), Biogeography-Based Optimization (BBO), Teachers-Learner Based Optimization (TLBO), Artificial Bee Colony (ABC), Ant Colony Optimization (ACO), etc. would be appropriate for improving the convergence rate and obtaining a worldwide solution. The literature survey reveals that teaching learning based optimization

(TLBO) shares similarities with swarm based optimization, emphasizing the influence of learning methods from teacher to student and student to student. The population or swarm consists of a collective of pupils who acquire knowledge from either teachers or other students. If students acquire knowledge from a teacher, it is referred to as the teacher's phase. Similarly, when students learn from other students, it is known as the student phase. The output represents the outcome or academic performance of students. Thus, the number of subjects can be seen as the variables of the function, while grades or results represent the fitness value, which is a range between 5 and 6. Several other population-centered strategies have been successfully implemented and demonstrated efficacy [33].

Nevertheless, not all algorithms are appropriate for solving complicated problems, as demonstrated by Wolpert and Macready. Conversely, evolutionary strategy (ES) based methods such as GA, BBO, etc. yield superior outcomes for diverse situations and are likewise characterized as population-based metaheuristics. The numbers 16 and 28. Furthermore [22] suggested a modified evolutionary algorithm to solve the inverse kinematics problem of a duplicate manipulator. The system takes into account the minimizing of joint displacement error ($\Delta\theta$) and the positional error of the end effector. The inverse kinematic solution of the PUMA 560 robot was proposed by [32] utilizing the cyclic coordinate descent (CCD) and Broyden-Fletcher-Shanno (BFS) technique. [23] presented an inverse kinematics solution for a 4-degree-of-freedom PUMA manipulator by employing a genetic algorithm. This research employs two distinct objective functions that are predicated on end-effector displacement and joint variable rotations. [18] suggested use an evolutionary algorithm to predict the trajectory of a 3-degree-of-freedom revolute manipulator. [25] proposed a solution for inverse kinematics and trajectory planning for a D-joint robot manipulator using a deterministic global optimization method. [1] This paper presents a new global optimization approach for solving the inverse kinematics problem of a redundant manipulator. [4] This study presents a genetic programming-based approach to solve the inverse kinematics problem of the PUMA robot manipulator. This work use genetic programming to develop mathematical models based on provided direct kinematic equations. [17] suggested optimizing the design parameter, specifically the link length, for a 2-degree-of-freedom manipulator. [15] presented a method for solving the inverse kinematics problem of a 2-degree-of-freedom articulated robot manipulator using a real coding genetic algorithm. [19] introduced a solution technique for the inverse kinematics of a 3-degree-of-freedom redundant manipulator using the reach hierarchy method. [30] Suggested an inverse kinematic solution for a 3-degree-of-freedom PUMA manipulator to address significant displacement. This approach utilizes a genetic method that incorporates adaptive niching and clustering. [12] presented a solution for the inverse kinematics of a 6-degree-of-freedom MOTOMAN robot manipulator, which allows for precise positioning of the end-effector. This technique utilizes an adaptive genetic algorithm to achieve the optimal location of the end effector. [26] suggested a method for inverse kinematic and trajectory development of a humanoid arm manipulator. This method involves using forward recursion with backward cycle computation. [21] suggested a real-time optimization technique as a solution for the inverse kinematics of a 6R revolute manipulator. [24] introduced a kinematic solution that utilizes three distinct methods: the bee algorithm, a neural network that is then tuned by the bee algorithm, and an evolutionary approach. A kinematic solution for a 3-degree-of-freedom serial robot manipulator is provided, utilizing a real-time genetic algorithm. [13] provided an inverse kinematic solution for a 6-degree-of-freedom robot manipulator by employing an immune genetic algorithm. [9] suggested a traditional methodology, specifically a penalty function-based optimization method, for solving inverse kinematics (IK). While there are limited techniques available to handle difficult NP issues, they necessitate the use of a high-performance computing equipment and complex computer programming.

However, although optimization algorithms have been commonly used in the domain of multi-objective and NP-hard problems to achieve highly optimized solutions, the TLBO algorithm has not yet been applied to solve inverse kinematics problems and trajectory calculations for robot manipulators. Furthermore, the computational expense of obtaining the inverse kinematics solution using the implemented techniques has been compared without any specific adjustment of relevant parameters. The main objective of this study is to minimize the Euclidean distance of the end effector position by solving the inverse kinematics problem for a 5R robot manipulator. We will compare the solutions achieved using Genetic Algorithm (GA) and Teaching-Learning-Based Optimization (TLBO). The outcomes of all algorithms are calculated using inverse kinematics equations and yield the resulting error for data statistics. Put simply, the coordinates of the end-effector are used as an input to calculate the joint angles. The 4th order spline formula is used to generate the trajectory of the end effector and the corresponding joint angles of a robotic arm. This is done utilizing the TLBO, GA, and quaternion methods. The subsequent arrangement of the paper is as follows: Section 2 focuses on the mathematical modeling of the 5R robot manipulator. It provides a detailed derivation of the forward and inverse kinematics of the 5R manipulator using quaternion algebra. Section 3 focuses on the construction of the inverse kinematic goal function for a 5R manipulator. The experimental results gained from simulations are thoroughly addressed in Section 5.

2. MATHEMATICAL MODELING USING THE QUATERNION VECTOR TECHNIQUE

This section focuses on the mathematical modeling of quaternion vector algebra and its use in deriving inverse kinematics equations. Quaternion vector methods are effective for rotating and translating a point, line, or other objects, regardless of the homogeneous transformation matrix, while still referencing the origin coordinate system. The process of interpolating a series of rotations and translations using Euler's angle method is quite intricate. Put simply, the variables exist within an isotropic space, which can be described as the surface topology of a sphere and possesses complex characteristics. This section provides a concise explanation of quaternion mathematics to evaluate references and establish a foundation for the mathematical derivation of inverse kinematics.

2.1. Rotation and translation from quaternion

The above discussions gives the importance of quaternions and the necessity of it. The quaternion rotation and translation are lies in four dimensional space therefore it is quite difficult to represent here or to imagine. Figure 1 represents the rotation through quaternion and Eq. (3) describes rotation of a point in a space mathematically.

$$h = \cos\left(\frac{\theta}{2}\right) + i^{*} \sin\left(\frac{\theta}{2}\right) + j^{*} \sin\left(\frac{\theta}{2}\right) + k^{*} \sin\left(\frac{\theta}{2}\right) \quad (3)$$

The 4-dimensional space, imagination of fourth axis is quite complex. Therefore, in Figure 1 a unit distance point around axis (X, Y, and Z) is given and which traces a circle. When this rotation circle is projected on a plane then the point P_1 can be seen rotated through angle θ to point P_3 which crosses the mid-point P_2. Therefore P_1 point is transforming to P_3 following by straight line makes cos(θ/2) and sin(θ/2).

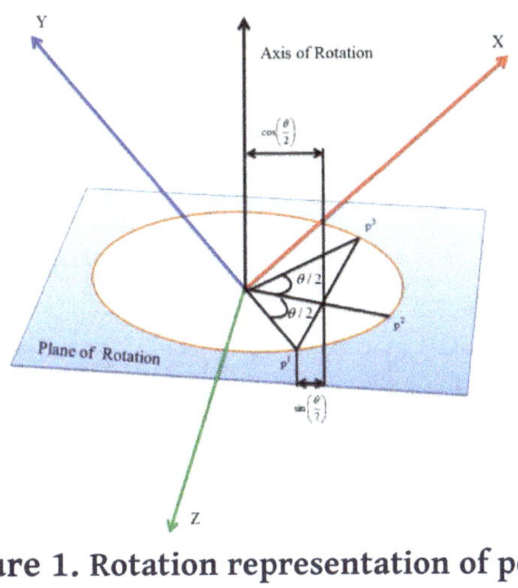

Figure 1. Rotation representation of point.

Now two quaternions can be represented on the basis of above discussed concept. If there is subsequent rotation of two quaternion h_1 and h_2 then the composite rotations $h_1{}^{*}h_2$ can be given by From Figure 1, p^1 is the point vector representing initial position and p^3 is the point vector final condition to be transformed. Therefore,

$$p^3 = h_2{}^{*}\left(h_1{}^{*}p^{1*}h_1^{-1}\right)^{*}h_2^{-1}$$

$$= (h_2{}^{*}h_1)^{*}p^{1*}\left(h_1^{-1*}h_2^{-1}\right) \quad (4)$$

$$= (h_2{}^{*}h_1)^{*}p^{1*}\left(h_2^{-1} * h_1^{-1}\right)$$

Now pure translations t_r can be done by quaternion operator that is given below,

$$t_r = h + p^1 \quad (5)$$

Quaternion transform can be given by,

$$p^2 = h^{*}p^{1*}h^{-1} \quad (6)$$

Therefore, an expression for the inverse of a quaternion can be given as,

$$H^{-1} = \left[< h^{-1} >, < -h^{-1} \otimes P \otimes h >\right] \quad (7)$$

where $-h^{-1} \otimes P \otimes h = -P + [-2k(j \times (-P)) + 2j \times (j \times (-P))]$ quaternion product, which is defined in the most general form for two quaternions $h_1 = (r_1, j_1)$, and $h_2 = (r_2, j_2)$ as

$$h_1 \otimes h_2 = \left[r_1 r_2 - j_1 \bullet j_2 \;\; r_1 j_2 + r_2 j_1 + j_2 \times j_1\right] \quad (8)$$

where $j_1 \bullet j_2$ and $j_2 \times j_1$ denote dot and cross products, between the 3-dimensional vectors j_1 and j_2. Clearly, quaternion multiplication is not commutative. The set of elements {±1, ±i, ±j, ±k} is known as the quaternion group of order 8 in multiplication.

Similarly vector transformation multiplication can be given as

$$H_1 \otimes H_2 = (h_1, P^1) \otimes (h_2, P^2) = h_1{}^{*}h_2, \;\; h_1{}^{*}P^2{}^{*}h_1^{-1} + P^1 \quad (9)$$

where, $h_1{}^{*}P^2{}^{*}h_1^{-1} = P^2 + 2r_1(j_1 \times P^2) + 2j_1 \times (j_1 \times P^2)$

2.2. Quaternion derivation for 5R manipulator kinematics

The configuration and base coordinate frame attachment of 5R manipulator is given in Figure 2(a) and MATLAB plot of 5R manipulator is presented in (b). Where $\theta_1, \theta_2, \theta_3, \theta_4$ and θ_5 joint angles for are articulated arm and d_1, d_2 and d_3 are the link offset. a_1, and a_2 represents link lengths.

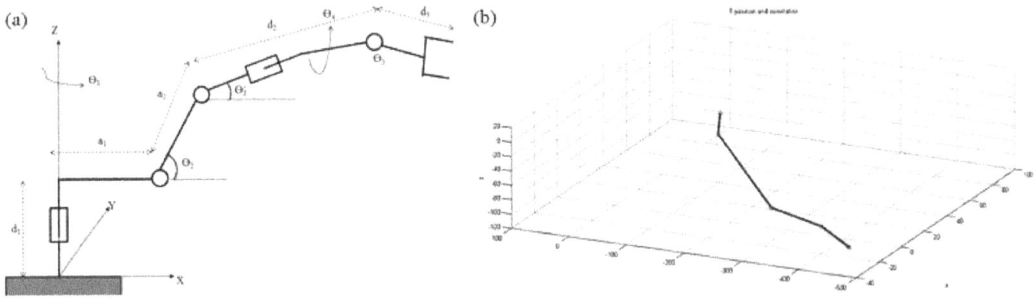

Figure 2. (a) Base frame and model of 5R manipulator; (b) configuration of 5R manipulator.

Now quaternion for successive transformation of each joint can be calculated from the Eq. (3) as follows,

$$H_1 = \left[< \overline{C_1} + \overline{S_1}\,\widehat{k} >, < a_1 C_1 \widehat{i} + a_1 S_1 \widehat{j} + d_1 \widehat{k} > \right] \tag{10}$$

$$H_2 = \left[< \overline{C_2} + \overline{S_2}\,\widehat{j} >, < -a_2 S_2 \widehat{i} - a_2 C_2 \widehat{k} > \right] \tag{11}$$

$$H_3 = \left[< \overline{C_3} + \overline{S_3}\,\widehat{j} >, < -d_4 S_3 \widehat{i} - d_4 C_3 \widehat{k} > \right] \tag{12}$$

$$H_4 = \left[< \overline{C_4} + \overline{S_4}\,\widehat{i} >, < d_4 \widehat{i} > \right] \tag{13}$$

$$H_5 = \left[< \overline{C_5} + \overline{S_5}\,\widehat{j} >, < -d_6 S_5 \widehat{i} - d_6 C_5 \widehat{k} > \right] \tag{14}$$

Inverse of a dual quaternion can be calculated by Eq. (8),

$$H_1^{-1} = \left[< \overline{C_1} - \overline{S_1}\,\widehat{k} >, < -a_1 \widehat{i} > \right] \tag{15}$$

$$H_2^{-1} = \left[< \overline{C_2} - \overline{S_2}\,\widehat{j} >, < a_2 \widehat{k} > \right] \tag{16}$$

$$H_3^{-1} = \left[< \overline{C_3} - \overline{S_3}\,\widehat{j} >, < d_4 \widehat{k} > \right] \tag{17}$$

$$H_4^{-1} = \left[< \overline{C_4} - \overline{S_4 \widehat{i}} >, < -d_4 \widehat{i} > \right] \tag{18}$$

$$H_5^{-1} = \left[< \overline{C_5} - \overline{S_5 \widehat{j}} >, < d_6 \widehat{k} > \right] \tag{19}$$

$$Q_i = H_i \otimes H_{i+1} \ldots\ldots H_n \tag{20}$$

Where in case of 5R manipulator arm n = 5. Now calculating quaternion vector products using Eq. (20)

$$Q_i = H_i \otimes H_{i+1} \ldots\ldots H_n$$

$$Q_5 = H_5 = \left[< \overline{C_5} - \overline{S_5 \hat{j}} >, < -d_6 S_5 \hat{i} - d_6 C_5 \hat{k} >\right]$$

$$Q_4 = H_4 \otimes Q_5 = \left[< \overline{C_4} + \overline{S_4 \hat{i}} >, < d_4 \hat{i} >\right] \otimes \left[< \overline{C_5} + \overline{S_5 \hat{j}} >, < -d_6 S_5 \hat{i} - d_6 C_5 \hat{k} > \right] \quad (21)$$

$$Q_4 = \left[< \overline{C_4 C_5} + \overline{S_4 C_5 \hat{i}} + \overline{C_4 S_5 \hat{j}} + \overline{S_4 S_5 \hat{k}} >, < (d_4 - d_6 S_5)\hat{i} + d_6 C_5 S_4 \hat{j} - d_6 C_4 C_5 \hat{k} > \right] \quad (22)$$

$$Q_3 = H_3 \otimes Q_4 = \left[< \overline{C_3} + \overline{S_3 \hat{j}} >, < -d_4 S_3 \hat{i} - d_4 C_3 \hat{k} >\right] \otimes \left[< \overline{C_4 C_5} + \overline{S_4 C_5 \hat{i}} + \overline{C_4 S_5 \hat{j}} + \overline{S_4 S_5 \hat{k}} >, < (d_4 - d_6 S_5)\hat{i} + d_6 C_5 S_4 \hat{j} - d_6 C_4 C_5 \hat{k} > \right] \quad (23)$$

$$Q_3 = \left[< \overline{C_4 C_{3+5}} + \overline{S_4 C_{3-5} \hat{i}} + \overline{C_4 S_{3+5} \hat{j}} - \overline{S_4 S_{3+5} \hat{k}} >, < (d_4 + d_4 - d_6 C_4 C_5 S_3 - d_4 C_3 - d_6 S_5 C_3 - d_4 S_3)\hat{i} + (d_6 C_5 S_4 \hat{j} + (-d_4 S_3 + d_6 S_5 S_3 - d_6 C_4 C_5 C_3 - d_4 \hat{C}_3)k > \right] \quad (24)$$

$$Q_2 = H_2 \otimes Q_3 = \left[< \overline{C_2} + \overline{S_2 \hat{j}} >, < -a_2 S_2 \hat{i} - a_2 C_2 \hat{k} >\right] \otimes \left[< \overline{C_4 C_{3+5}} + \overline{S_4 C_{3-5} \hat{i}} + \overline{C_4 S_{3+5} \hat{j}} - \overline{S_4 S_{3+5} \hat{k}} >, < (d_4 + d_4 - d_6 C_4 C_5 S_3 - d_4 C_3 - d_6 S_5 C_3 - d_4 S_3)\hat{i} + (d_6 C_5 S_4 \hat{j} + (-d_4 S_3 + d_6 S_5 S_3 - d_6 C_4 C_5 C_3 - d_4 \hat{C}_3)k > \right] \quad (25)$$

Therefore,

$$Q_2 = \left[< \left(\overline{C_2}\, \overline{C_4 C_{3+5}} - \overline{S_2 C_4 S_{3+5}}\right) + \left(\overline{C_2 S_4 C_{3-5}} - \overline{S_2 S_4 S_{3+5}}\right)\hat{i} + \left(\overline{C_2 C_4 S_{3+5}} + \overline{S_2 C_4 C_{3+5}}\right)\hat{j} + \left(-\overline{C_2 S_4 S_{3+5}} - \overline{S_2 S_4 S_{3-5}}\right)\hat{k} >, < (-a_2 S_2 + d_4 C_2 + d_4 C_2 - d_4 C_{2-3} + d_6 S_5 C_{2+3} - d_4 S_{2+3} - d_6 C_4 C_5 S_{2+3} - d_4 S_3)\hat{i} + (d_6 C_5 S_4 \hat{j} + \left(-a_2 C_2 - d_4 S_2 - d_4 S_2 + d_6 C_4 C_5 C_{2+3} + d_4 S_{2-3} + d_6 S_5 S_{2+3} + d_4 \hat{C}_{2-3}\right)k >\right] \quad (26)$$

$$Q_1 = H_1 \otimes Q_2 = \left[< \overline{C_1} + \overline{S_1}\hat{k} >, < a_1 C_1 \hat{i} + a_1 S_1 \hat{j} + d_1 \hat{k} >\right] \otimes$$
$$\left[< \overline{(C_2 C_4 C_{3+5}} - \overline{S_2 C_4 S_{3+5}}) + (\overline{C_2 S_4 C_{3-5}} - \overline{S_2 S_4 S_{3+5}})\hat{i}\right.$$
$$+(\overline{C_2 C_4 S_{3+5}} + \overline{S_2 C_4 C_{3+5}})\hat{j} + (-\overline{C_2 S_4 S_{3+5}} - \overline{S_2 S_4 S_{3-5}})\hat{k} >,$$
$$< (-a_2 S_2 + d_4 C_2 + d_4 C_2 - d_4 C_{2-3} + d_6 S_5 C_{2+3} - d_4 S_{2+3}$$
$$-d_6 C_4 C_5 S_{2+3} - d_4 S_3)\hat{i} + (d_6 C_5 S_4 \hat{j} +$$
$$\left.(-a_2 C_2 - d_4 S_2 - d_4 S_2 + d_6 C_4 C_5 C_{2+3} + d_4 S_{2-3} + d_6 S_5 S_{2+3} + d_4 \hat{C}_{2-3})k >\right] \tag{27}$$

Therefore,

$$Q_1 = \left[< \overline{(C_1 C_4 C_{2+3+5}} + \overline{S_1 S_4 S_{2+3-5}}) + (\overline{C_1 S_4 C_{2+3-5}} - \overline{S_1 C_4 S_{2+3+5}})\hat{i}\right.$$
$$+(\overline{C_1 C_4 S_{2+3+5}} + \overline{S_1 S_4 C_{2+3-5}})\hat{j} + (\overline{S_1 C_4 C_{2+3+5}} - \overline{C_1 S_4 S_{2+3-5}})\hat{k} >,$$
$$< (a_1 C_1 - a_2 S_2 + d_4 C_2 C_1 + d_4 C_2 C_1 - d_4 C_{2-3} C_1 + d_6 S_5 C_{2+3} C_1 - d_4 S_{2+3} C_1$$
$$+ d_6 C_4 C_5 S_{2+3} C_1 - d_6 C_5 S_4 S_1)\hat{i} + (d_6 C_5 S_4 + d_6 C_5 S_4 C_1 - a_2 S_2$$
$$+ a_2 S_2 C_1 - a_2 S_2 S_1 + d_4 C_2 S_1 + d_4 C_2 S_1 - d_4 C_{2-3} S_1 + d_6 S_5 C_{2+3} S_1$$
$$- d_4 S_{2+3} S_1 - d_6 C_4 C_5 S_{2+3} S_1)\hat{j} + (d_1 - a_2 C_2 - d_4 S_2 - d_4 S_2$$
$$\left.+ d_6 C_4 C_5 C_{2+3} + d_4 S_{2-3} + d_6 S_5 S_{2+3} + d_4 \hat{C}_{2-3})k >\right] \tag{28}$$

$$O_{j+1} = H_j^{-1} \otimes O_j \tag{29}$$

Now calculating vector pair of quaternion using Eq. (29), to solve the inverse kinematics problem, the transformation quaternion of end effector of robot manipulator can be defined as

$$[R_{be}, T_{be}] = O_1 = \left[< w + a\hat{i} + b\hat{j} + c\hat{k} >, < X\hat{i} + Y\hat{j} + Z\hat{k} >\right] \tag{30}$$

Now using Eq. (29), O_2 will be given by,

$$O_2 = H_1^{-1} \otimes O_1$$
$$O_2 = \left[< \overline{C_1} - \overline{S_1}\hat{k} >, < -a_1 \hat{i} >\right] \otimes$$
$$\left[< w + a\hat{i} + b\hat{j} + c\hat{k} >, < X\hat{i} + Y\hat{j} + Z\hat{k} >\right] \tag{31}$$
$$O_2 = [< (o_{21}) + (o_{22})\hat{i} + (o_{23})\hat{j} + (o_{24})\hat{k} >,$$
$$< (o_{25})\hat{i} + (o_{26})\hat{j} + o_{27}\hat{k} >]$$

where, $o_{21} = w\overline{C_1} + c\overline{S_1}$, $o_{22} = a\overline{C_1} + b\overline{S_1}$, $o_{23} = b\overline{C_1} - a\overline{S_1}$, $o_{24} = c\overline{C_1} - w\overline{S_1}$, $o_{25} = X\overline{C_1} - a_1$- $Y\overline{S_1}$, $o_{26} = Y\overline{C_1} - X\overline{S_1}$, $o_{27} = Z$.

Now,

$$O_3 = H_2^{-1} \otimes O_2$$
$$O_3 = \left[< o_{31} + o_{32}\hat{i} + o_{33}\hat{j} + o_{34}\hat{k} >, < o_{35}\hat{i} + o_{36}\hat{j} + p_z\hat{k}\right] \tag{32}$$

where, $o_{31} = \overline{C_2}o_{21} + c(\overline{C_2S_2}) - w(\overline{S_2S_1})$, $o_{32} = \overline{C_2}o_{22} + \overline{S_2}o_{23}$, $o_{33} = \overline{C_2}o_{23} - \overline{S_2}o_{22}$, $o_{34} = \overline{C_2}o_{24} - \overline{S_2}o_{21}$, $o_{35} = -ZS_2 + XC_1C_2 + YS_1C_2 - a_1C_2$, $o_{36} = YC_1 - XS_1$, $o_{37} = a_2 - ZC_2 + XC_1S_2 + YS_1S_2 - a_1S_2$.

$$O_4 = \left[< o_{41} + o_{42}\hat{i} + o_{43}\hat{j} + o_{44}\hat{k} >, < o_{45}\hat{i} + o_{27}\hat{j} + o_{47}\hat{k}\right] \tag{33}$$

where,

$$o_{41} = \overline{C_{2+3}}o_{21} + \overline{S_{2+3}}o_{23}$$
$$o_{42} = \overline{C_{2+3}}o_{22} - \overline{S_{2+3}}o_{24}$$
$$o_{43} = \overline{C_{2+3}}o_{23} - \overline{S_{2+3}}o_{21}$$
$$o_{44} = \overline{C_{2+3}}o_{24} - \overline{S_{2+3}}o_{22}$$
$$o_{45} = -ZS_2 - a_2S_3 - XC_1S_2S_3 - YS_1S_2S_3 + a_1S_2S_3 -$$
$$ZC_2S_3 + Z - ZC_3 + XC_1C_2C_3 + XS_1C_2C_3 - a_1C_2C_3$$
$$o_{46} = YC_1 - XS_1$$
$$o_{47} = -ZS_2S_3 + a_2C_3 + XC_1C_2S_3 + YS_1C_2S_3 -$$
$$a_1C_2S_3 + ZC_2C_3 + XC_1S_2C_3 + YS_1S_2C_3 - a_2S_2C_3$$

Therefore, all the joint variables can be calculated by equating quaternion vector products and quaternion vector pairs i.e. Q_1, Q_2 and Q_3 to O_1, O_2 and O_3 respectively.

$$\Rightarrow \theta_1 = a\tan\left[\frac{Y - u_y}{X - u_x} \times \frac{(a_1 + d_4C_2 + d_4C_2 - d_4C_{2-3} + d_6S_5C_{2+3} - d_4S_{2+3} + d_6C_4C_5S_{2+3})}{(-a_2S_2 + d_4C_2 + d_4C_2 - d_4C_{2-3} + d_6S_5C_{2+3} - d_4S_{2+3} - d_6C_4C_5S_{2+3})}\right] \tag{34}$$

$$\theta_2 = a\tan 2\left[\left(\frac{Z + v_x}{d_4}\right), \mp\sqrt{1 - \left\{\left(\frac{Z + v_x}{d_4}\right)\right\}^2}\right] \tag{35}$$

where, $d_4S_2 + d_4S_2 - v_x = Z$, $d_4S_2 + d_4S_2 = Z + v_x$, $S_2 = \left(\frac{Z+v_x}{d_4}\right)$

$$\theta_3 = a\tan 2\left[\begin{array}{l} m\sqrt{1 - \left\{\left(\frac{-XS_2 + XC_1C_2 + YS_1C_2 - a_1C_2 - v_y}{(-d_4 - d_6S_5)}\right)\right\}^2}, \\ \left(\frac{-ZS_2 + XC_1C_2 + YS_1C_2 - a_1C_2 - v_y}{(-d_4 - d_6S_5)}\right), \end{array}\right] \tag{36}$$

$$\theta_4 = a\tan 2\left[m\sqrt{1-\left\{\left(\frac{-ZS_2S_3 + a_2C_3 + XC_1C_2S_3 + YS_1C_2S_3 - a_1C_2S_3 + ZC_2C_3 + XC_1S_2C_3 + YS_1S_2C_3 - a_2S_2C_3}{-d_6C_5}\right)\right\}^2}, \left(\frac{-ZS_2S_3 + a_2C_3 + XC_1C_2S_3 + YS_1C_2S_3 - a_1C_2S_3 + ZC_2C_3 + XC_1S_2C_3 + YS_1S_2C_3 - a_2S_2C_3}{-d_6C_5}\right), \right] \tag{37}$$

$$\theta_5 = a\tan 2\left[\left(\frac{-XS_2 - a_2S_3 - XC_1S_2S_3 - YS_1S_2S_3 + a_1S_2S_3 - ZC_2S_3 + Z - ZC_3 + XC_1C_2C_3 + YS_1C_2C_3 - a_1C_2C_3 - d_4}{-d_6}\right), m\sqrt{1-\left\{\left(\frac{-ZS_2 - a_2S_3 - XC_1S_2S_3 - YS_1S_2S_3 + a_1S_2S_3 - ZC_2S_3 + Z - ZC_3 + XC_1C_2C_3 + YS_1C_2C_3 - a_1C_2C_3 - d_4}{-d_6}\right)\right\}^2} \right] \tag{38}$$

3. INVERSE KINEMATIC SOLUTION SCHEME

In this section optimization algorithms are selected for computation of inverse kinematics solution of 5R manipulator. However, there are various types of optimization algorithms existed and can produce the desired IK solution, the major necessity is to achieve global optimum solution with fast convergence rate. Therefore, selection of appropriate optimization algorithm is important for fitness evaluations and GA is so far best known tool, but on the other hand TLBO has also proven its efficiency and performance. Finally selection of optimization algorithms has been made on global searching point, computational cost and quality of the result.

3.1. Optimization approach to solve inverse kinematics

Any Optimization algorithms which are capable of solving various multimodal functions can be implemented to find out the inverse kinematic solutions. The fitness function is given by the Eq. (46) fitness function F(x). Each individual represents a joint variable solution of the inverse kinematic problem for adopted population based metaheuristic algorithm. All individuals moving in D-dimensional search space and sharing the information to find out best fitness value of the function. Each individual contains set of joint angles ($\theta_1, \theta_2, \theta_3, \theta_4$ and θ_5) of 5R manipulator. The optimum set of joint angle can be find by using appropriate optimization algorithm from given desired position of end effector (X, Y, Z). In case of inverse kinematics of 5R manipulator multiple solutions exist for the single position of the end effector so it is required to find out the best set of joint angle in order to minimize whole movement of manipulator.

For the optimization of joint angle rotation of robot manipulator, one can define objective function or fitness function from joint angle rotation difference and other can be defined from end effector position displacement. These are known as joint angle error and positional function method [3, 7, 10, 29].

3.1.1. Position based function

The current position of the manipulator is described by (39):

$$P_c = [X_c, \ \ Y_c, \ \ Z_c] \tag{39}$$

Desired position of end effector can be denoted by (40):

$$P_d = [X_d, \ \ Y_d, \ \ Z_d] \tag{40}$$

Current position of end effector will be compared with the desired position P_d. General equation for the fitness function is given in Eq. (41) that is based on the distance norm of homogeneous Euclidian distance between the current positions to the desired position of end effector P_d evaluated by number of iterations.

$$P_{\min} = \|P_d - P_c(i)\|^2 \tag{41}$$

Current position P_c can be evaluated from Eqs. (34) through (38). Now putting the value of P_c on Eq. (42)

$$P_{\min} = \sqrt{(X_d - X_c(i))^2 + (Y_d - Y_c(i))^2 + (Z_d - Z_c(i))^2} \tag{42}$$

3.1.2. Joint angle error

Corresponding joint error can be given by the difference between current set of joint variables to the final required angles.

$$\theta_c = (\theta_{c1}, \theta_{c2}, \theta_{c3}, \theta_{c4}\ \theta_{c5}) \tag{43}$$

$$\theta_d = (\theta_{d1}, \theta_{d2}, \theta_{d3}, \theta_{d4}\ \theta_{d5}) \tag{44}$$

Therefore using square norm the objective function can be given as

$$\theta_{\min} = \|\theta_d - \theta_c\|^2 \tag{45}$$

Subjected to joint limits

$$\theta_1 \in [\theta_{1\min}, \ \ \theta_{1\max}]$$

$$\theta_2 \in [\theta_{2\min}, \ \ \theta_{2\max}]$$

$$\theta_3 \in [\theta_{3\min}, \ \ \theta_{3\max}]$$

$$\theta_4 \in [\theta_{4\min}, \ \ \theta_{4\max}]$$

$$\theta_5 \in [\theta_{5\min}, \ \ \theta_{5\max}]$$

Now overall error minimization can be given by using Eqs. (42) and (45),

$$F(x)_{\min} = \lambda\left[\sqrt{(X_d - X_c(i))^2 + (Y_d - Y_c(i))^2 + (Z_d - Z_c(i))^2}\right] + \left[\|\theta_d - \theta_c\|^2\right] \tag{46}$$

where λ is proportional weight factor for the minimization of the problem and calculation of the entire joint angles base on constraint can be achieved using fitness function (46). The performance of considered algorithm is checked with the parameters: a_1 = 60 mm, a_2 = 145 mm, d_1 = 150 mm, d_2 = 125 mm, d_3 = 130 mm. Upper and lower limit of five joint angles are: $\theta_1=[0,180]$; $\theta_2=[0,150]$; $\theta_3=[0,150]$, $\theta_4=[0,85]$ and $\theta_5=[15,45]$.

4. RESULTS AND DISCUSSIONS

The inverse kinematics of a 5-R manipulator has been computed using TLBO and GA algorithms. A comparison of the results has been made based on the quality and performance. Table 1 provides the coordinates of the end effector in five random positions, along with their corresponding inverse kinematics solutions. The current work is being carried out using MATLAB R2013a. The data sets are acquired from equations (34) to (38). The data sets are produced utilizing quaternion vector-based inverse kinematics equations, as outlined in Table 2. The obtained data sets are utilized to compare the IK solution using the adopted Genetic Algorithm (GA) and Teaching-Learning-Based Optimization (TLBO). Table 3 presents a comparison of the fitness function and the joint variables acquired using the TLBO and GA algorithms.

Table 1. Five different positions and joint variables.

Positions	Joint angles				
	θ_1	θ_2	θ_3	θ_4	θ_5
P1(-76.09, 54.36, -61.94)	84.559	77.518	101.74	30.616	38.697
P2(89.69, 192.55, 90.87)	84.791	97.25	130.44	50.771	36.428
P3(-4.24, 94.08, 97.55)	18.384	78.688	35.234	77.708	34.889
P4(29.10,154.02, -31.52)	104.43	115.47	124.11	7.3372	33.774
P5(-184.33, -43.21, 8.27)	39.177	107.13	97.052	65.672	15.374

Table 2. Desired joint variables determined through quaternion algebra.

SN	Position of joints determined through quaternion algebra							
	θ_1	θ_2	θ_3	θ_4	θ_5	X	Y	Z
1	112.5641	47.3165	8.2447	65.8373	39.8977	-186.6903	183.0670	-14.7039
2	153.1316	21.9812	126.9031	57.2629	30.4168	-92.6981	32.1423	157.3316
3	66.1779	143.2985	14.6124	73.6231	41.5228	-131.5420	-22.3866	-32.2155
4	57.6085	119.6396	104.5818	71.1946	33.8225	-10.7684	111.7435	77.4862
5	31.4308	2.9242	71.5757	63.0358	39.8749	64.7966	172.0372	151.5714
6	124.3702	116.7337	102.4999	53.1482	22.4807	-111.8590	-59.6708	60.8590
7	89.1765	13.1827	101.6747	80.3340	29.4704	-76.9533	96.2813	121.3505
8	5.6698	30.9685	57.2308	29.3079	29.6421	174.3873	107.6283	143.1839

SN	Position of joints determined through quaternion algebra							
	θ_1	θ_2	θ_3	θ_4	θ_5	X	Y	Z
9	131.5857	108.7086	92.8278	5.6664	36.4826	-104.6410	109.7511	40.5523
10	32.8579	102.3539	138.8770	26.4141	33.7466	146.4984	48.7416	54.4041
11	134.1878	70.4224	26.3511	82.2471	44.0566	-188.7864	15.4108	-53.9823

Table 3. TLBO results for joint variable and function value.

Positions	TLBO joint angles					Function value
	θ_1	θ_2	θ_3	θ_4	θ_5	
P1(-76.09, 54.36, -61.94)	86.598	72.165	72.165	40.894	30.459	0
P2(89.69, 192.55, 90.87)	83.874	69.895	69.895	39.607	30.76	0
P3(-4.24, 94.08, 97.55)	84.512	70.427	70.427	39.909	30.686	0
P4(29.10,154.02, -31.52)	85.566	71.305	71.305	40.406	30.53	0
P5(-184.33, -43.21, 8.27)	87.818	73.181	73.181	41.469	30.364	0
Positions	**GA Joint angles**					**Function value**
	θ_1	θ_2	θ_3	θ_4	θ_5	
P1(-76.09, 54.36, -61.94)	60.619	49.504	58.384	62.281	27.903	0
P2(89.69, 192.55, 90.87)	88.293	34.091	14.439	15.241	51.738	0
P3(-4.24, 94.08, 97.55)	55.004	49.274	63.942	47.842	33.633	0
P4(29.10,154.02, -31.52)	72.594	22.689	68.297	85.886	27.044	0.0137
P5(-184.33, -43.21, 8.27)	25.669	70.588	31.341	66.807	52.884	0

This work does not employ specific optimization of various parameters in the Genetic Algorithm (GA) and Teaching-Learning-Based Optimization (TLBO) algorithm. For future research, it is recommended to conduct a sensitivity analysis in order to obtain more accurate and improved outcomes. According to Table 3, TLBO produced superior solutions for position 4 compared to GA, based on the evaluation of the fitness function. Various distance-based norms exist, one of which is the Euclidean distance norm. It is employed in this case to determine the minimal distance between the positions of the end effector. If the distance between two points becomes 0 or less than 0.001, the fitness function will converge to the optimal or global minimum value. It is evident that the achieved fitness value is lower than the specified distance norm, indicating that the utilization of these algorithms is advantageous and of high quality.

Figures 3–7 represent the optimal fitness function value and corresponding joint variables for position 1. The following data illustrates the effectiveness of the implemented algorithms for solving the inverse kinematics problem of a 5-R manipulator. The objective function evaluation converges to zero error for both the GA and TLBO methods. However, at position 4, the GA algorithm produces an error of 0.013. It indicates that GA has lower performance in comparison to TLBO. The results acquired using GA are presented in Figures 8–12, showcasing the convergence and histogram graph. The joint angles obtained are in radians, which are subsequently translated into degrees and listed in Table 3. The GA results are produced using the MATLAB toolbox, which indicates a lack of convergence in a single run. From Figures 8–12, it is evident that there are various solutions obtained for joint angles for a single position. Similarly, there are multiple evaluations of the fitness function. The optimal fitness function obtained here utilizing the termination criteria and the accompanying joint variables has been selected for comparison.

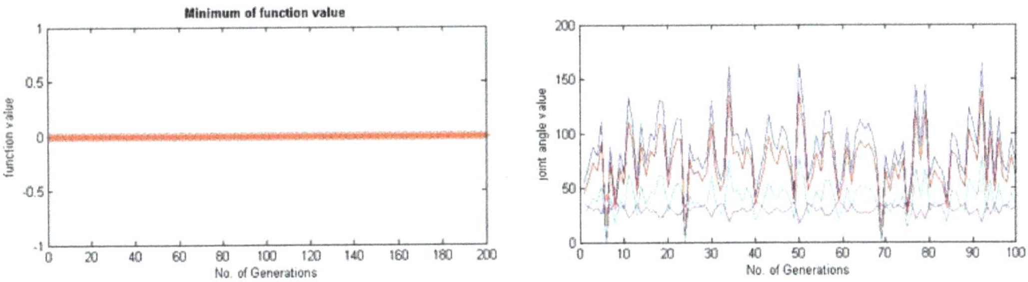

Figure 3. Joint variables and fitness function value for position P1.

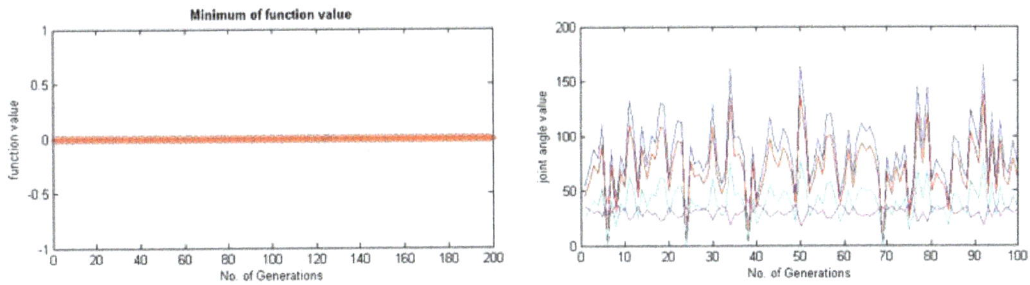

Figure 4. Joint variables and fitness function value for position P2.

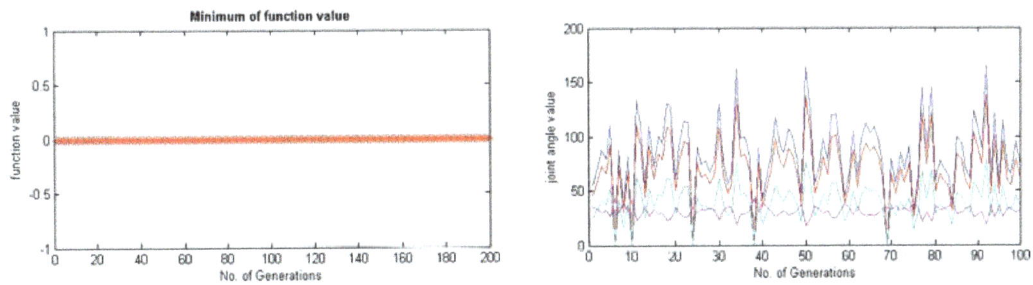

Figure 5. Joint variables and fitness function value for position P3.

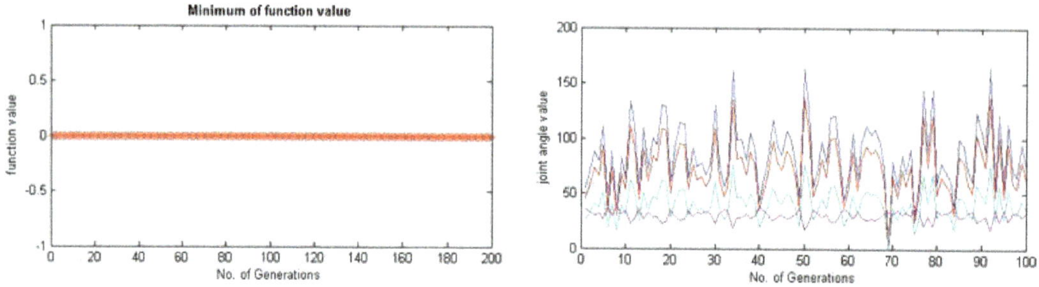

Figure 6. Joint variables and fitness function value for position P4.

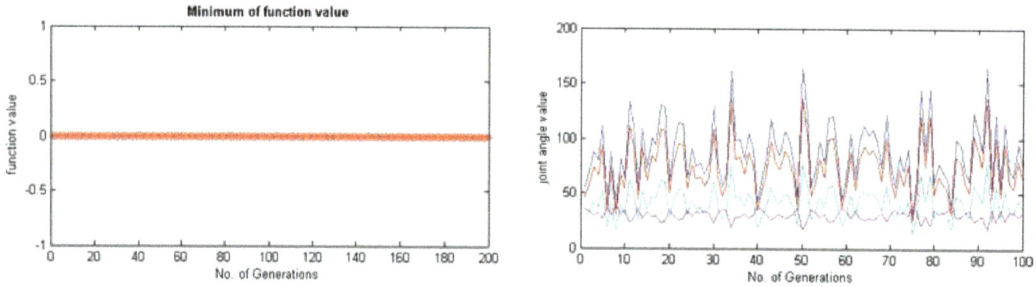

Figure 7. Joint variables and fitness function value for position P5.

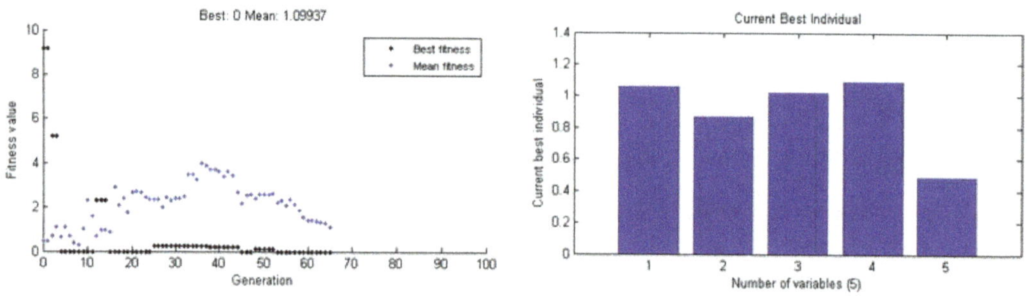

Figure 8. Joint variables and fitness function value for position P1.

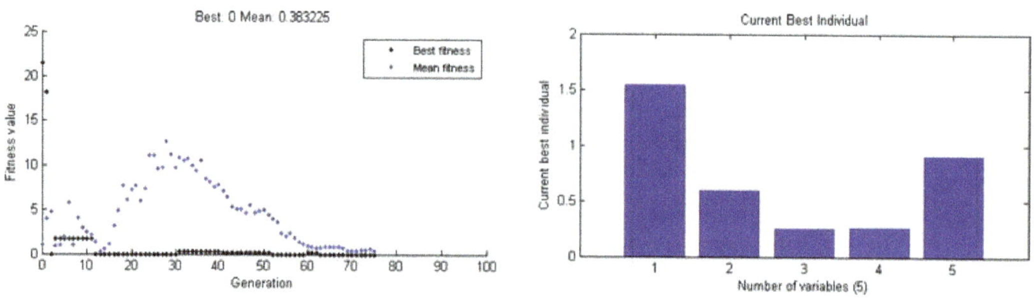

Figure 9. Joint variables and fitness function value for position P2.

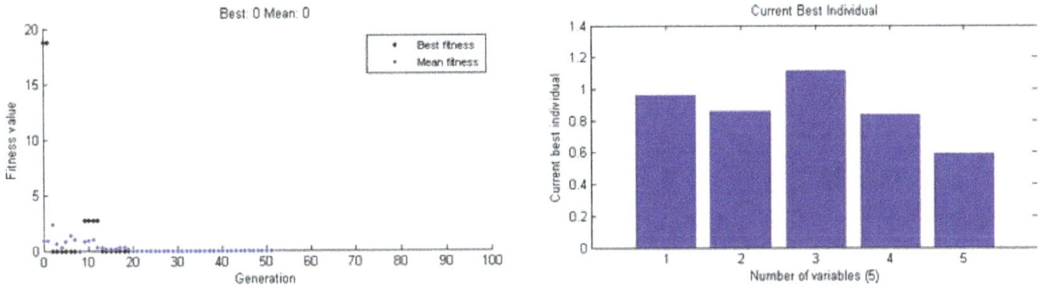

Figure 10. Joint variables and fitness function value for position P3.

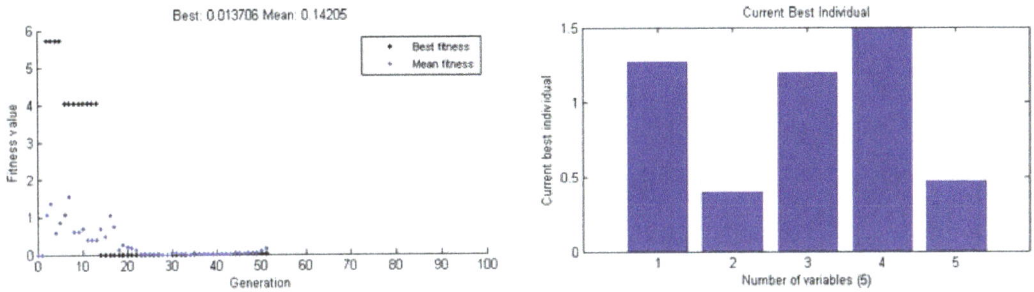

Figure 11. Joint variables and fitness function value for position P4.

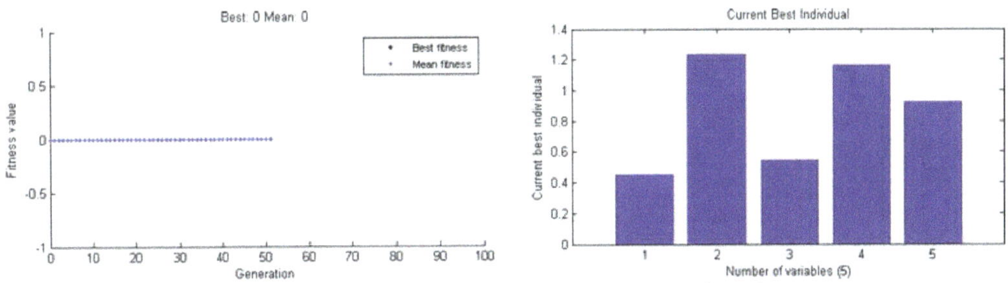

Figure 12. Joint variables and fitness function value for position P5.

The proposed work is conducted on a computer with a dual-core processor and 4 GB of RAM. It has been observed that the convergence of the solution for Genetic Algorithm (GA) is occurring more quickly than for Teaching-Learning-Based Optimization (TLBO) and quaternion algebra, while requiring less computation time. The joint angles trajectory obtained using a 4th order cubic spline is shown in Figure 13. The joint variables are utilized to compute the joint space trajectory for TLBO and GA, as depicted in Figures 14 and 15, using the inverse kinematic solution. The trajectory generating process took a total of 6 seconds, while the overall processing time required to complete this trajectory was 5.674 seconds. The execution time for TLBO is 15.671 seconds, which exceeds that of GA, which is 7.932 seconds. Based on the computational cost, the Genetic Algorithm (GA) outperforms the Teaching-Learning-Based Optimization (TLBO). Additionally, the use of quaternion algebra resulted in the least amount of time taken, as shown in Table 4.

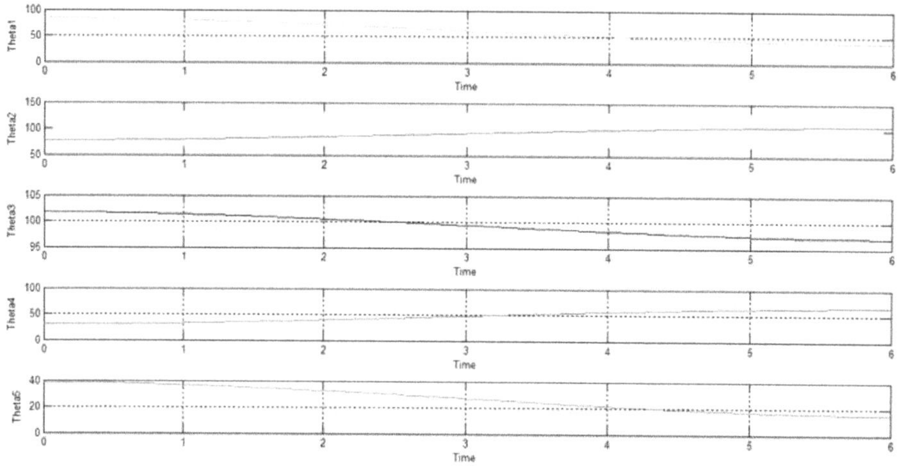

Figure 13. Trajectory of joint angle for quaternion.

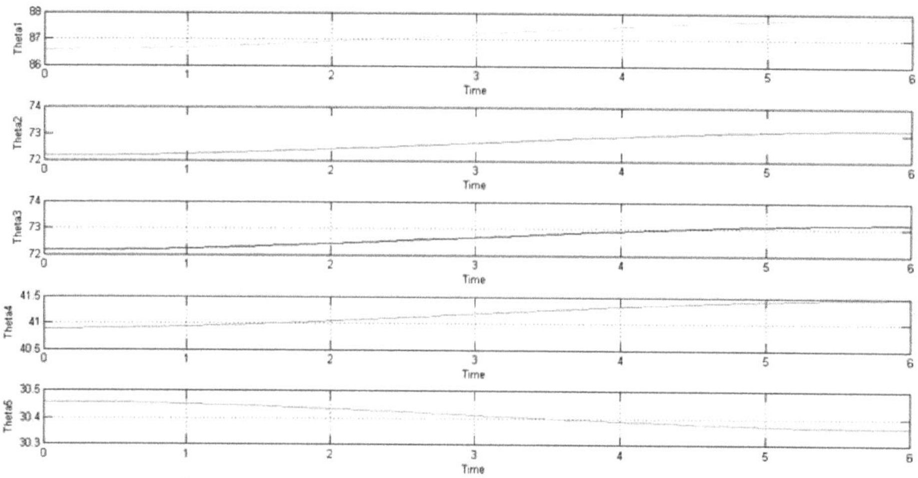

Figure 14. Trajectory of joint angle for TLBO.

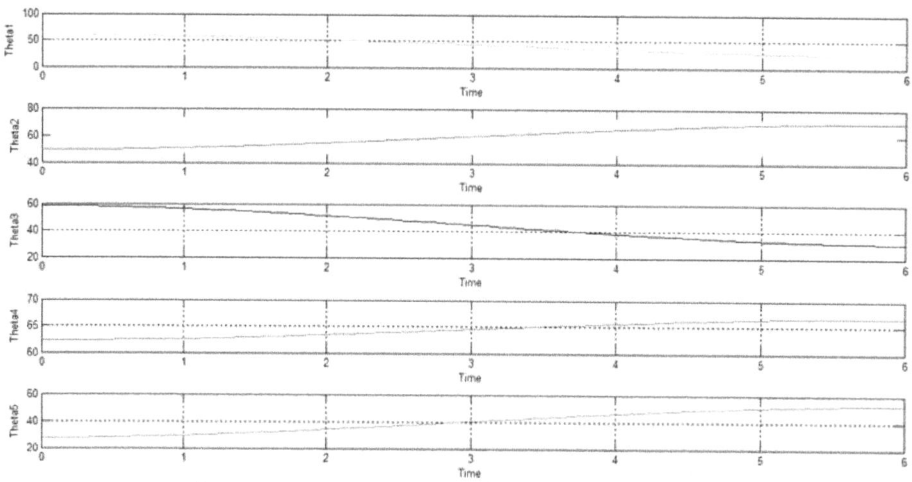

Figure 15. Trajectory of joint angle for GA.

Table 4. Computational time for inverse kinematic evaluations.

SN	Method	Computational time
1	TLBO	15.671 s
2	GA	7.932 s
3	Quaternion	0.993 s

5. CONCISE SUMMARY

This research addresses the issue of optimizing the positional and angular error of the end effector in a 5R robot manipulator. The optimization is achieved utilizing two algorithms, namely TLBO (Teaching-Learning-Based Optimization) and GA (Genetic Algorithm). Metaheuristic algorithms such as Particle Swarm Optimization (PSO), Genetic Algorithms (GA), Artificial Bee Colony (ABC), and others have been employed in diverse areas of industrial robotics. However, the primary challenge lies in effectively addressing the inverse kinematic problem for any given configuration of robot manipulators. The majority of optimization methods are typically employed for numerical solutions. However, it has been noted that these numerical solutions fail to provide a solution when the manipulator is in an ill-conditioned state. In addition, it has been noted that conventional optimization approaches tend to converge towards local minima. Hence, this study utilizes global optimization techniques such as TLBO and GA. Upon examining the outcomes, it is evident that the convergence rate of the used optimization algorithms is higher, and the complexity remains unaffected by the manipulator configuration. Despite the efforts of numerous academics to achieve a worldwide solution, the computational costs associated with the problem remain high. Therefore, one potential option to address the issue of computational cost is to utilize a quaternion objective function.

The selected algorithms are very suitable for both limited and unconstrained issues. In order to assess the efficacy of the algorithms under consideration, a comparison has been conducted using quaternion algebra. Table 3 presents a comparison of the results obtained from the algorithm that was adopted and the quaternion solutions proposed for a 5-R manipulator. This study examined the forward and inverse kinematic equations in order to formulate the objective function for the Teaching-Learning-Based Optimization (TLBO) and Genetic Algorithm (GA). These implemented algorithms have demonstrated the potential to achieve faster convergence and produce the global optimal solution for the given problem. In the future, it is advisable to consider adjusting the various parameters of Genetic Algorithms (GA) and Teaching-Learning-Based Optimization (TLBO) in order to prevent being stuck at local minimum points. Hybridization of these methods can be suggested and utilized for solving IK problems.

REFERENCES

1. Ahuactzin J, Gupta K. Completeness Results for a Point-to-Point Inverse Kinematics Algorithm. Detroit, MI: IEEE. Vol. 2. p. 1526-1531
2. Albert F, Koh S, Tiong S, Chen C. Inverse Kinematic Solution in Handling 3R Manipulator via Real-Time Genetic Algorithm. IEEE. p. 1-6
3. Ayyıldız M, Çetinkaya K. Comparison of four different heuristic optimization algorithms for the inverse kinematics solution of a real 4-DOF serial robot manipulator. Neural Computing and Applications. 2015. DOI: 10.1007/s00521-015-1898-8
4. Chapelle F, Bidaud P. A Closed Form for Inverse Kinematics Approximation of General 6R Manipulators Using Genetic Programming. 2001;4:3364-3369
5. Chen C, Her M, Hung Y, Karkoub M. Approximating a robot inverse kinematics solution using fuzzy logic tuned by genetic algorithms. The International Journal of Advanced Manufacturing Technology. 2002;20:375-380. DOI: 10.1007/s001700200166
6. Dulęba I, Opałka M. A comparison of Jacobian-based methods of inverse kinematics for serial robot manipulators. International Journal of Applied Mathematics and Computer Science. 2013. DOI: 10.2478/amcs-2013-0028
7. Eiben A, Smith J. Introduction to Evolutionary Computing. New York: Springer; 2003
8. Funda J, Taylor R, Paul R. On homogeneous transforms, quaternions, and computational efficiency. IEEE Transactions on Robotics and Automation. 1990;6:382-388. DOI: 10.1109/70.56658
9. Galicki M. Control-based solution to inverse kinematics for mobile manipulators using penalty functions. Journal of Intelligent and Robotic Systems. 2005;42:213-238. DOI: 10.1007/s10846-004-7196-9
10. Gan D, Liao Q, Wei S. Dual quaternion-based inverse kinematics of the general spatial 7R mechanism. Proceedings of the Institution of Mechanical Engineers, Part C: Journal of Mechanical Engineering Science. 2008;222:1593-1598. DOI: 10.1243/09544062jmes1082
11. Geradin M, Cardona A. Kinematics and dynamics of rigid and flexible mechanisms using finite elements and quaternion algebra. Computational Mechanics. 1988;4:115-135. DOI: 10.1007/bf00282414
12. He G, Hongming G, Zhang G, Wu L. Using Adaptive Genetic Algorithm to the Placement of Serial Robot Manipulator. Islamabad: IEEE; 2006. p. 1-6

13. Huang H, Xu S, Hsu H. Hybrid Taguchi DNA swarm intelligence for optimal inverse kinematics redundancy resolution of six-DOF humanoid robot arms. Mathematical Problems in Engineering. 2014;2014:1-9. DOI: 10.1155/2014/358269

14. Husty M, Pfurner M, Schröcker H. A new and efficient algorithm for the inverse kinematics of a general serial 6R manipulator. Mechanism and Machine Theory. 2007;42:66-81. DOI: 10.1016/j.mechmachtheory.2006.02.001

15. Kalra P, Mahapatra P, Aggarwal D. On the Solution of Multimodal Robot Inverse Kinematic Functions using Real-coded Genetic Algorithms. IEEE. 2003;2:1840-1845

16. Kennedy J, Eberhart R, Shi Y. Swarm Intelligence. San Francisco: Morgan Kaufmann Publishers; 2001

17. Khatami S, Sassani F. Isotropic Robotic Design Optimization of Manipulators Using a Genetic Algorithm Method. Vancouver: IEEE; 2002. p. 562-567

18. Kim S, Kim J. Optimal Trajectory Planning of a Redundant Manipulator Using Evolutionary Programming. Nagoya: IEEE; 1996. p. 738-743

19. Korein B. Techniques for generating the goal-directed motion of articulated structures. IEEE Computer Graphics and Applications. 1982;2:71-81. DOI: 10.1109/mcg.1982.1674498

20. Kucuk S, Bingul Z. Inverse kinematics solutions for industrial robot manipulators with offset wrists. Applied Mathematical Modelling. 2014;38:1983-1999. DOI: 10.1016/j.apm.2013.10.014

21. Liu S, Zhu S. An optimized real time algorithm for the inverse kinematics of general 6R robots. In: Control and Automation, 2007. ICCA 2007. IEEE International Conference on. IEEE, Guangzhou. 2007. pp. 2080–2084

22. Nearchou A. Solving the inverse kinematics problem of redundant robots operating in complex environments via a modified genetic algorithm. Mechanism and Machine Theory. 1998;33:273-292. DOI: 10.1016/s0094-114x(97)00034-7

23. Parker J, Khoogar A, Goldberg D 1989, Inverse kinematics of redundant robots using genetic algorithms. Proceedings of IEEE International Conference on Robotics and Automation

24. Pham D, Castellani M, Fahmy A. Learning the Inverse Kinematics of a Robot Manipulator using the Bees Algorithm. Daejeon: IEEE; 2008. p. 493-498

25. Piazzi A, Visioli A. A Global Optimization Approach to Trajectory Planning for Industrial Robots. Grenoble: IEEE. 1997;3:1553-1559

26. Rajpar A, Zhang W, Jia D. Object Manipulation of Humanoid Robot Based on Combined Optimization Approach. Harbin: IEEE; 2007. p. 1148-1153

27. Rao R, Savsani V, Vakharia D. Teaching–learning-based optimization: A novel method for constrained mechanical design optimization problems. Computer-Aided Design. 2011;43:303-315. DOI: 10.1016/j.cad.2010.12.015

28. Rocke D, Michalewicz Z. Genetic algorithms + data structures = evolution programs. Journal of the American Statistical Association. 2000;95:347. DOI: 10.2307/2669583

29. Sun L, Lee R, Lu W, Luk L. Modelling and simulation of the intervertebral movements of the lumbar spine using an inverse kinematic algorithm. Medical & Biological Engineering & Computing. 2004;42:740-746. DOI: 10.1007/bf02345206

30. Tabandeh S, Clark C, Melek W 2006, A Genetic Algorithm Approach to solve for Multiple Solutions of Inverse Kinematics using Adaptive Niching and Clustering. In: IEEE, pp. 1815–1822

31. Vrongistinos K, Wang Y, Hwang Y. Quaternion smoothing on three-dimensional kinematics data. Medicine & Science in Sports & Exercise. 2001;33:S84. DOI: 10.1097/00005768-200105001-00480

32. Wang L, Chen C. A combined optimization method for solving the inverse kinematics problems of mechanical manipulators. IEEE Transactions on Robotics and Automation. 1991;7:489-499. DOI: 10.1109/70.86079

33. Wolpert D, Macready W. No free lunch theorems for optimization. IEEE Transactions on Evolutionary Computation. 1997;1:67-82. DOI: 10.1109/4235.585893

34. Xu D, Acosta Calderon C, Gan J. An analysis of the inverse kinematics for a 5-DOF manipulator. International Journal of Automation and Computing. 2005;2:114-124. DOI: 10.1007/s11633-005-0114-1

35. Zoric N, Lazarevic M, Simonovic A. Multi-body kinematics and dynamics in terms of quaternions: Langrange formulation in covariant form – Rodriguez approach. FME Transactions. 2010;38:19-28

36. Zu D. Efficient inverse kinematic solution for redundant manipulators. JME. 2005;41:71. DOI: 10.3901/jme.2005.06.071

12

Chapter

AN ANALYSIS AND APPLICATION OF A RANDOM MULTI-TRAJECTORY GENERATION METHOD FOR ONLINE EMERGENCY THREAT MANAGEMENT IN THE PATH PLANNING ALGORITHM

1. INTRODUCTION: AN OVERVIEW

The various applications of robotics have resulted in a significant advancement in robot path planning over the past few years [1, 2]. In order to achieve autonomy, there has been extensive research conducted on issues such as path planning, motion planning, and online obstacle management. Unmanned Aerial Vehicles (UAVs), a robot platform that is simple to access, have been utilized in a growing number of commercial and research applications in recent years. UAV autonomy is the capacity to online plan a trajectory from a starting position to a destination while satisfying the kinematic constraints, including obstacle avoidance and trajectory planning [3]. When a UAV is following the intended path, emergency threat management (ETM) is one of the most difficult challenges to overcome in robot path planning. This is due to the possibility of a sudden threat bursting into view or dynamic obstacles being detected on the line. The following attributes should be taken into account by UAVs in such circumstances:

1. Time efficiency: The most important requirement for ETM algorithm is time efficiency. For general ETM, the configuration is periodically updated, such as heuristic algorithm A* [4], which it is computationally intensive if the map is represented with high resolution. In order to guarantee safety, ETM requires real-time performance.

2. Kinematic feasibility: Kinematic feasibility denotes that the output of the planner meets the kinematic constraints of the robot as well as the environment. The constraints include: (a) Path smoothness: The planner is required to output kinematic smooth path, sometimes even kinodynamically feasible as well. Thus, the path should meet the state of art tracking constraints, and enables low tracking error for UAV; (b) Minimum cost of switching: The strategy of handling the threat, especially ET, is to find the cost minimum path by generating a new path or multiple paths besides the initial one. The cost for choosing the best path should take the dynamic constraints, energy consumption and time performance into consideration.
3. Specific requirements: UAVs have already been applied to many areas, such as inspection, photography, and monitoring. They have to meet some specific requirements according to environments and system constraints. For example, best pose based illumination of tunnel inspection for crack and spalling [5], and stable tracking with obstacle avoidance as UAV photography [6] which should be able to keep stable capturing even during the flying.

Development with open robot platform [7] and field implementation [8] has witnessed the promising performance of Sampling Based (SB) methods. SB algorithms (SBA) have the advantages for planning in high dimensional space, and it is with the ability to deal with multiple classes of path or motion planning problem in both static and dynamic environment [9]. Rapidly-exploring random trees (RRTs) are single query methods which obtain Voronoi biased property and only generate homotopy paths simultaneously [12]. Although it proposes to solve the multiple degrees of freedom (DOF) operating problems in known static environments [10, 11], SBA shows great performance of dealing with any kind of path or motion planning problem in complex environments for unmanned ground robots or aerial robots.

In this paper, we introduce two biased-sampling methods: obstacle biased and Homologous Classes (HC) biased, which are used to conduct path planning. We have previously addressed the obstacle biased path method in [13] through a demonstration of UAVs. The ET problem is addressed by the HC classed biased approach, which generates alternative paths for dynamic online switching. In order to guarantee singularity between paths, HC implements an online dynamic reconfiguration approach that endeavors to generate an increased number of paths with distinct obstacle references. Therefore, when confronted with ET, it is capable of conducting alternative transitioning online. Guiding Attraction based Random Tree (GART) is the name of the obstacle-biased planning method, while Multi-GART (MGART) is the name of the HC-biased planning method. The UAVs are capable of comprehending the clearance region, and we regard the environment as known a priori. The proposed methods are effectively evaluated through comparative simulations and experiments.

2. PRELIMINARY MATERIALS

2.1. Homology and homotopy

For path planner, the purpose is to find a feasible path p_f (cost minimum or complete) from the initial position to the goal position in the workspace $W \in R^{\{n\}}$, n denotes the dimension of space the robots locate. A general cost function can be represented as:

$$J = \int c_{energy} + c_{time} + c_{threat} \, dt \quad (1)$$

The $c_{energy,ctime,cthreat}$ denote energy cost, time consumption, and threat respectively. These costs are not fixed, since the energy cost can be path length, and time consumption can change according to the velocity limitation. For cost constrained path planner, the goal is to find the asymptotic optimal rather than the completeness solution. Then, more than one path can be found during the process, and the paths can be homotopic or belong to different homotopic classes (or called homology).

It is illustrated in Figure 1. Given a continuous map $H: I \times I \rightarrow \Gamma$ or $H: h(s,t) = h_t(s)$, Γ denotes the topological space and $I = [0,1]$ is the unit interval. The obstacle regions are labeled with R_{o1}, $x_{initial=h(0,t)}$ denotes start point, $x_{inter=h(1,t)}$ denotes the goal position, x_{inter} denotes an inter node for obstacle avoidance. For the continuous deformation, given $h(s,0) = \pi_0$, $h(s,1) = \pi_1$, the path can be continuously mapping through π_0 to π_1 with $t \in [0,1]$. For any path deformed between, they are homotopic with π_0, π_1 if and only if they stay in the closed loop $\pi_0 \coprod -\pi_1$, where the closed loop cannot collide with any obstacle region.

Definition 1—Homotopic Paths: It denotes the equivalence class of all paths under continuous mapping $H: h(s,t) = h_t(s)$, which locates in the closed loop formed $h(s,0) \coprod -h(s,1)$. Any path in the set can be continuously deformed into any other without colliding with any obstacle in the space. For all paths in the set, they are with the same fixed end points.

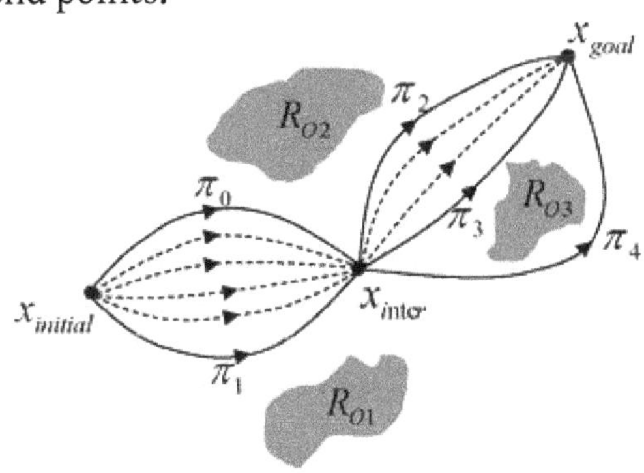

Figure 1. Homotopic and homologous classes and paths.

We can conclude that π_2 and π_3 belong to the same homotopic class. However, we can find path π_4, which shares the same start and ending node, cannot be continuously deformed to π_3 due to the isolation of the obstacle. It means $(\pi_3 \cup -\pi_4) \cap R_{o3} \neq \Theta$. In such case, we call π_3 and π_4 are homologous, and they belong to different homotopic classes.

Definition 2—Homologous Paths: Paths, which follows the same continuous mapping $H: h(s,t) = h_t(s)$, cannot form a closed loop by $h(s,0) \coprod -h(s,1)$. The homologous paths belong to different homotopic classes.

2.2. Problem statement

Path planning follows a common procedure to perform trial and error process under empirical constraint to achieve completeness. The problem of path planning does not only solve a problem for exploration optimization, but also try to model the environment with a best descriptor as discussed in [13]. Let us take a look again with the problem of path planning which can be represented as:

$$\mathrm{H} = \{\mathrm{h(s)} | \mathrm{x}_{\mathrm{initial}} = h(0), \mathrm{x}_{\mathrm{initial}} = h(1), s \in [0,1]\} \tag{2}$$

The path h(s) (homologous) should stay in obstacle free region R_{free}, that is, $h(s) \in R_{free}$. Usually, the path is piecewise continuously, and it can also be smoothed to obtain first order continuous thus to ensure kinematics continuous [14]. Besides the exploration to achieve completeness (in Eq. (1)), the obstacle modeling method is also important and affect the planning results.

To solve this problem, this paper proposed a multi-path online switching approach, that is, the path planner can find alternative homologous-paths. Then, this paper designs an online fast switching strategy. For multiple path planner, it aims at finding as many paths as possible,

$$\mathrm{H}_{\mathrm{alter}} = \cup_{i=1} h^i(x(t), u(t)) \tag{3}$$

H_{alter} denotes the set of all the alternative paths $h^i(x(t), u(t))$, x(t) denotes the state, and u(t) denotes the control. However, the mission planner cannot use all the planed paths for online switching, it should find the reasonable paths without redundancy. We propose the follow rule,

$$\mathrm{H}_{\mathrm{reason}} = \{h^i | h_i \underset{\neq}{H} h_j, \nabla \mathrm{i} \neq \mathrm{j}\} \tag{4}$$

H_{reason} denotes the paths set where any two paths are not homotopic to each other, $H\neq$ denotes non-homotopy. Now, we have the paths which keep distinguishable from each other with different obstacles sequence surrounding.

3. RAPIDLY EXPLORING RANDOM TREE PATH PLANNER

This section aims to explain the fundamental science behind the rapidly-exploring random tree (RRT [12]), which we then use as the basis for our innovative technique to enhance active exploration in complex contexts. SBAs are search algorithms that use random sampling and collision checking to incrementally extend the search space. They were initially introduced to address planning problems in high-dimensional spaces. These methods possess the advantages of simultaneously taking into account the control and kinematics limitations. They are capable of gradually building a tree from the initial state to the goal state by continually sampling and expanding.

It is shown Figure 2, the whole tree graph by exploration is represented as G_T, the black solid dot denotes the valid state within step accessibility under kinematics constraints, and the black solid lines connect each parent state with child state for extension. Every step, a new sample g_{sample} will be generated randomly. It should be cleared to all that the initial random sampling does not mean a fixed connection, that is, the random sampling can be a direction for extending. Then, the random sample g_{sample} tries to find the nearest state in the tree for connection under minimum Euclidean metric, Where g_i is an element of all valid states set G_T.

$$g_{parent} = \sup \min_{g_i \in G_T} E\left(g_i, g_{sample}\right) \tag{5}$$

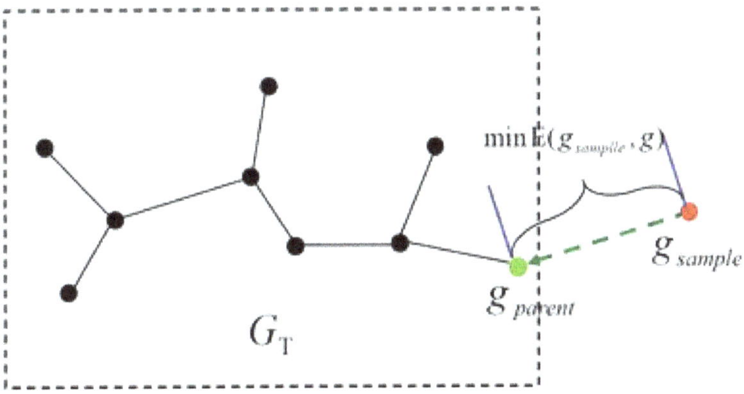

Figure 2. RRTs propagate by applying the minimal cost criterion to connect the newly sampled guard to the previous tree.

3.1. RRT connection with kinematic constraints

For RRT planner, given a system with state (xẋ ,xẏ ,θ)xẋxẏθ, and a general form of system model:

$$\begin{cases} \dot{X_x} = f^x(x_x, u_x) \\ \dot{x_y} = f^y(x_x, u_y) \\ \dot{\theta} = f^\theta(\theta, u_\theta) \end{cases} \tag{6}$$

It can extend with simply random sampling with control inputs $[u_x, u_y, u_\theta]$. The random sample has to follow the kinematics constraints. Given the robot system, the differential constraints can be represented as a set of implicit equations as g(x,ẋ)=0gxẋ=0, and it can be further represented as:

$$\dot{x} = f(x, u) \quad (7)$$

Here, x denotes the state, and $u \in U$ denotes the valid control in allowable controls set. Given the parent state $g_{parent}(t)$, the time step follows a Δt limits. Then, the control inputs vary with $u = \{u(t') | t \le t' \le t + \Delta t\}$. To compute $x(t+\Delta t)$, we can follow a typical procedure as [12]. It should be noted that the planner should extend toward the newly sampled g_{sample}. The planner first computes the possible region of reachability from current state x(t):

$$x(t + \Delta t) \in [x(t) + f(x(t), u(t) - \Delta t \cdot \epsilon), x(t) + f(x(t), u(t) + \Delta t \cdot \epsilon)] \quad (8)$$

where ϵ is the maximum first order factor of control input. RRT now picks a new state along the direction from parent to new sample, that is, $g_{new} \in [x(t) + f(x(t), u(t) - \Delta t \cdot \epsilon), x(t) + f(x(t), u(t) + \Delta t \cdot \epsilon)]$ and $g_{new} = g_{parent} + \delta(g_{sample} - g_{parent})$ with $\delta \in [0,1]$.

3.2. Voronoi biased incremental search

Before discussing the Voronoi biased property of the SBAs, let first introduce some basic notation. Given a set of points $S = \{s_i | i = 1, 2, ..., n\}$ in a n-dimension space X. For any two distinct points s_p and s_q in set S, the dominant region (Voronoi region) of s_p over s_q is defined as the region where any inside point should be closer to s_p over s_q, that is,

$$R_{-s_p} = \left\{ \chi \in \left| |s_p - \chi|^L < |s_q - \chi|^L \right. \right\} \quad (9)$$

Where χ is the dominant region corresponding to s_p, $||^L$ denotes the Lebesgue measurement. In a normal case, any point s_i has its own dominant region with,

$$R_{s_i} = \left\{ \chi \in \left| |s_i - \chi|^L < |s_j - \chi|^L, \text{ for all } i \neq j \right. \right\} \quad (10)$$

Normally, random sampling of RRT follows a Monte-Carlo Method [15] to perform an uniformly sampling in a n-dimensional space under Lebesgue measurement. We can look back at the beginning of Section 3, the new sampled node tries to connect to the nearest node under Euclidean metric. We can now analyze the problem in another perspective that given g_{parent} and g_s, they connect to the same origin g_o. Then, a new sample g_{sample} is generated randomly following a Monte-Carlo process. In order to explore and reach the goal, g_{sample} tries to connect to the tree following the metric defined in Eq. (5). It means that g_{parent} and g_s can be connected for expansion under minimum distance principle, then g_{sample} has to be assigned to the dominant region which subjects to a closer point (the Voronoi region). Under this principle, g_{parent} and g_s can acquire new resource for extension with the ability to keep distinct region and extending their branches.

A typical Voronoi-biased exploration using sampling can be seen in Figure 3, where each branch keeps distinct with each other to form a star network like structure and it behaves the same for heuristic informed RRT [16]. Here, unlike the dominant region of a point, RRT branch can be also treated as a distinct individual with its own Voronoi region for acquiring the extending resource.

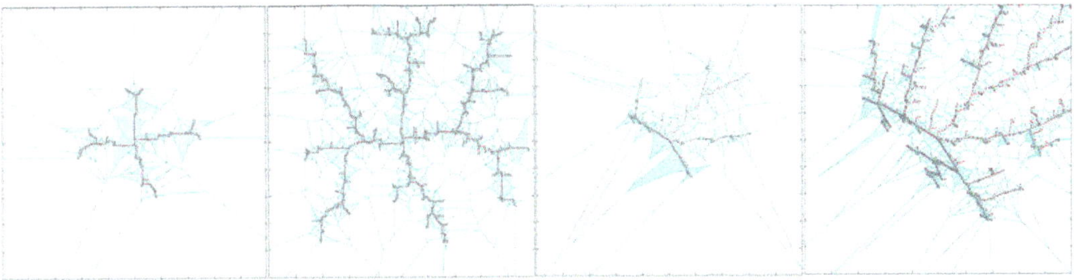

RRT: 200 Iterations RRT: 1000 Iterations hRRT: 200 Iterations hRRT: 1000 Iterations

Figure 3. Results of incremental exploration of RRT and hRRT [16] after 200 and 1000 iterations, respectively.

4. OBSTACLE AND HOMOLOGY BIASED PATH PLANNER

In this section, we provide two methods to address the primary challenges of managing a cluttered environment and processing online eye tracking (ET) data. These methods include an obstacle-biased approach and a homology-biased approach. The process of detecting collisions during the incremental exploration is time-intensive, and it adheres to a standardized protocol to ensure safety. The step validation of each new sampling state offers directional information about the distribution of obstacles.

4.1. Obstacle biased path planner under kinematic constraints

SBAs mostly deploy the general idea of generating random samples for incremental exploration, and the sample locating in obstacle region will be discarded since it is time consuming and no benefits for increasing the performance of exploring. We firstly deployed a simple idea which was proved to have much higher time performance then RRT and RRT* in [17].

This paper introduces an obstacle biased algorithm, using obstacle information to help generating more samples for connection. It is shown in Figure 4, the newly sampled states $x^{s}_{1,x}{}^{s}_{1}$ tries to connect to the nearest state in the tree. However, x^s_1 leads toward the obstacle region, x^s_2 locates in obstacle region. To use the obstacle information, this paper proposes an active exploring method, that is, inner propulsion and outer attraction.

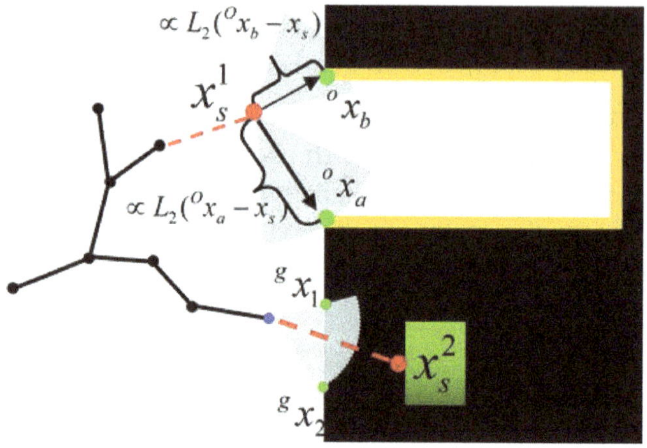

Figure 4. Obstacle biased SBA uses the obstacle location as input, with inner propulsion and outer attraction, to generate more samples for exploration. $x^s_{1,x}{}^s_2$ ***are new samples, the black region denotes obstacle region.***

For outer attraction, new sample xs1x1s performs a collision checking, and find the nearest nodes $^{ox}{}_a, ^{ox}{}_b$. We define that the further the obstacle to the sample, the more attraction it can support, that is, the attraction is proportional to the distance between obstacle and the sample using L2-norm L_2. The sample then re-allocation by add a obstacle biased attraction as:

$$x_1^{s*} = x_1^s + k\left[\left({}^o x_a - x_1^s\right) + \left({}^o x_b - x_1^s\right)\right] \tag{11}$$

Where k is a constant to adjust the shifting percentage of the attraction vector.

The inner sample in collision with the obstacle is regarded to provide guiding information for the algorithm. This paper tries to find two more states $^{gx}{}_1, ^{gx}{}_2$ within kinematic reachable region (discussed with Eq. (8)), it tries to find out the first two safe state with two directions which are out of obstacle region in the kinematic reachable region (the light blue fan-shaped region). Then, the two newly generated samples $^{gx}{}_1, ^{gx}{}_2$ follows principle Eq. (11) to redistribute to the final position, and connect to the tree.

By using the two proposed approaches, we can generate more useful samples for extending, especially, the samples generate around the edge of the obstacles with the ability to perform more active exploration in cluttered environments. Besides, the outer attraction redistributes the samples toward the narrow corridor between the obstacle, which thus increases the probability of finding safe path through such obstacle crowed region.

4.2. Homology biased

We assume any path $h_t(s)$ generated using SBA is consisted by a set of nodes $h_{t(s)=\{ht|ht(s'}$),s′∈[0,1]}, as it is illustrated in Figure 5(a) that exploring tree is consist of the red nodes. Each red node is regarded as distinct with other nodes in the tree, with a distinct dominant region, i.e. Voronoi region. Thus, a path $h_t(s)$, which is consisted a set of states from the initial state to the goal, can be isolated with each other with a distinct region $V(b_T)$ combined by all Voronoi regions of the states.

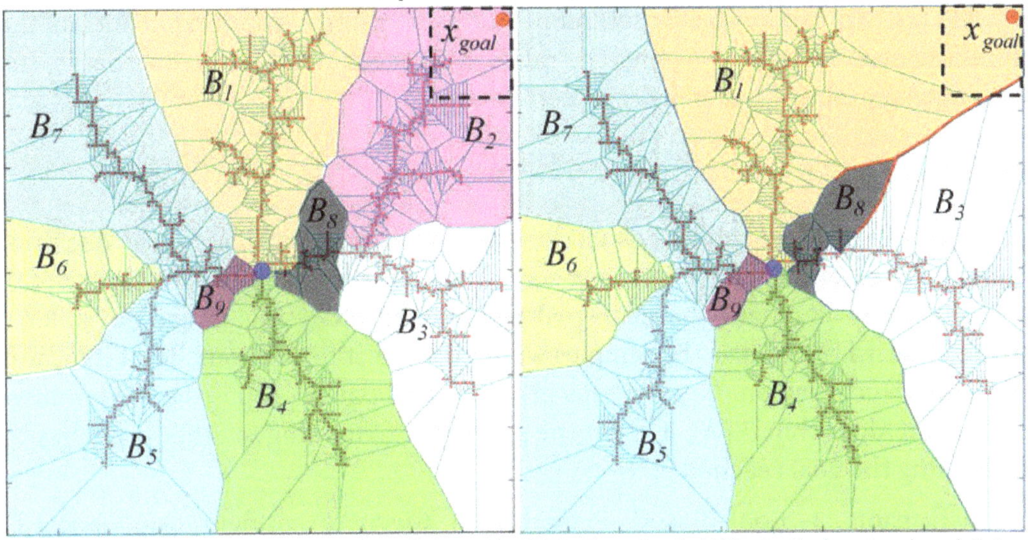

(a) RRT Voronoi biased regions, each region keep distinct with each other

(b) The Extending-Forbidden method removes branch B_2 to enable exploration of B_1, B_8, B_3 toward the goal

Figure 5. The extending-forbidden algorithm (EFA) tries to find all the states along the goal reached branch at each goal reaching checking step, such as branch B_2. Then, EFA sets the flag of the states to be inactive, switching the extending probability to the nearby branches.

The region dominant property differs the path with each other, where a SB tree with multiple paths (the path here may not connect to the goal, but they keep distinct with each other from the initial position) can be described by a set of branches $B_T = \{b_T(1), b_T(2), ..., b_T(n)\}$. For each branch, it consists of a list of states which connect with each other to form tree structure. In the tree, the end state relies on the parent state for extending as well as trajectory tracking.

4.2.1. Extending-forbidden algorithm

The path planner performs exploration following the Monte-Carlo approach. Given a configuration space C in a topological space (the rectangle region as illustrated in Figure 5), we denote the Lebesgue Measurement of the configuration space as $L^*|C|$. Then we can get the Lebesgue measurement of each branch $b_T(i)$ of the tree using the same metric. Authors in [18] proved that the dispersion of n random nodes sampled uniformly and independently in $V(b_T(i))$ is,

$$D = \sup_{b_T(i)} \left| \frac{\psi(b_t(i) : n)}{n} - \frac{L^*|V(b_T(i))|}{L*|C|} \right| = o\left(\left(\frac{\log(n)}{n} \right)^{\frac{1}{d}} \right) \tag{12}$$

Where $\psi(b_t(i))$ denotes the number of samples m, $1 \le m \le n$, that lies in the sub-branch $b_T(i)$, n is the number of all the sampling, d is the dimension of the configuration space. D denotes the difference between the ration of sampling probability and ration of space Lebesgue measurement, which follows the knowledge that Monte-Carlo method performs a uniform sampling. It means the sampling probability approaches the ratio of Lebesgue measurements, that is, the exploration probability can be represented as:

$$P_{b_T(i)} = \frac{L^*|V(b_T(i))|}{L^*|C|} \tag{13}$$

However, the probability of exploring in the configuration space does not benefit the extending bias toward the goal. Let us still take a look at Figure 5(a), the branch B_2 dominant the near-goal region, and other region are not able to extend toward the goal as the samples will not connect to the branches if it locates in the near-goal region. To solve this problem, this paper proposes an Extending-Forbidden Algorithm (EFA), it shifts the source for extending to other branches by forbidding the goal reached path.

Definition 3—Goal-biased Probability: Given a configuration space C, the exploring tree T and all its branches which are main branches B_T and its corresponding sub-branches. The goal-biased event denotes a branch can exploring toward the goal. If a goal region can be represented as G_r, and Voronoi($G_r 2 B_T$(i)) is the region that belongs to goal region and the Voronoi region of branch $B_T(i)$. Then, the nominal goal biased probability of branch $B_T(i)$ toward G_r is:

$$P_G^*(B_T(i)) = \frac{L^*|voronoi(G_r 2 B_T(i))|}{L^*|C|} \tag{14}$$

And the real goal biased probability is normalized value of all branches, that is,

$$P_G(B_T(i)) = \frac{P_G^*(B_T(i))}{\sum_{j=1}^{n} P_G^*(B_T(j))} \tag{15}$$

Definition 4—Long Sub-branch (LSB) and Short Sub-branch (SSB): Given a tree T and all its Voronoi distinct main branches B_T. Then, we can define a length threshold δ_B. For all end vertices in each main branch, we calculate the length l_{sb} from end to the goal reach reached path (any state which firstly reached). If the length $l_{sb} > \delta_B$, then we call it Long Sub-branch (LSB). If $l_{sb} \le \delta_B$, we call the sub-branch as Short Sub-branch (SSB).

It should be noted that the threshold is very empirical in Definition 4, and it is decided based on the configuration space and the kinematic constraints. In Figure 6, we set it as 15 meters, then we have SSB_1, SSB_2, SSB_3 as SSB, and LSB_1 as LSB. The reason why we have this definition is that we cannot shift all the extending resource to the neighbor branches, and we have the hypothesis that the SSB must has lower probability of finding a new path even given the resource for extending. For example, the main Voronoi dominant branches B_T keep distinct with each other, and each main branch has the probability $P_{bT}(i)$ for exploration in the configuration space. After the tree stops extending at a certain iteration, we can have the results as illustrated in Figure 5(a). Since goal region is in Voronoi region of branch B_2, then we know that we have the goal biased probability as $P_G(B(2)) = 1$. One branch B_2 reached the goal region which is represented with dotted back rectangle, EFA searches the SSBs and LSBs based on Definition 4, and it labels states and executing forbidden. Then, we have a resource shifted Voronoi graph as illustrated in Figure 5(b), where we can see that branch B_1 and B_3 obtains the Voronoi region which belongs to branch B_2. The two branches also obtain the rectangle goal region, that is, their goal biased probabilities are bigger than zero, $P_G(B(1)) > 0, P_G(B(3)) > 0$.

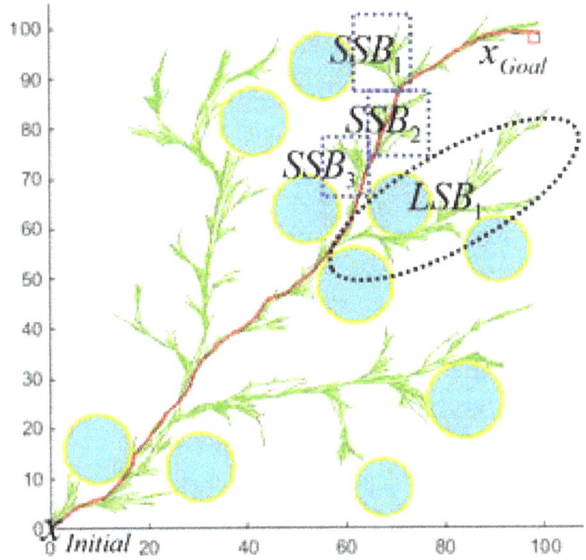

Figure 6. An intuitive example of SSB and LSB in an exploring tree, which are generated using threshold principle as defined in Definition 4.

Since EFA can shift the goal biased by extending resource to other branches, while not all paths can obtain such resource. The following truths are hold:

1. The increasing of goal biased probability ensures the generation of a feasible path toward the goal, but not all branches with goal biased probability can reach the goal at the same time. Only one can reach the goal because of Voronoi dominant probability, thus the general SBA cannot find multiple paths.
2. The efficiency of generating multiple paths mainly depends on the environment adaptability of random exploring algorithms. For random exploring algorithm, their merits of generating multiple branches enable the generation of multiple paths.

4.2.2. Reasonable alternative reference chosen

The proposed MGART is able to perform extending-forbidden toward multiple paths, as the random exploring property guarantees completeness and diversity. However, the quality of explored paths cannot be guaranteed, particularly a large number of homotopic paths are generated. This paper proposes an approach to generate the reasonable path, and we analysis under the hypothesis that the environment is highly cluttered and it is not practical to set threshold for path planner to choose the best homological paths.

Definition 5—Reasonable Alternative Paths: Consider two homotopic paths $h(\pi_1)$ and $h(\pi_2)$ in a configuration space C. The surrounding obstacle information along each path are $\beth(h(\pi 1))$ and $\beth(h(\pi 2))$. The reasonable alternative path exists if and only the surrounding information of the two paths are not the same, such that, $\beth(h(\pi 1)) \neq \beth(h(\pi 2))$.

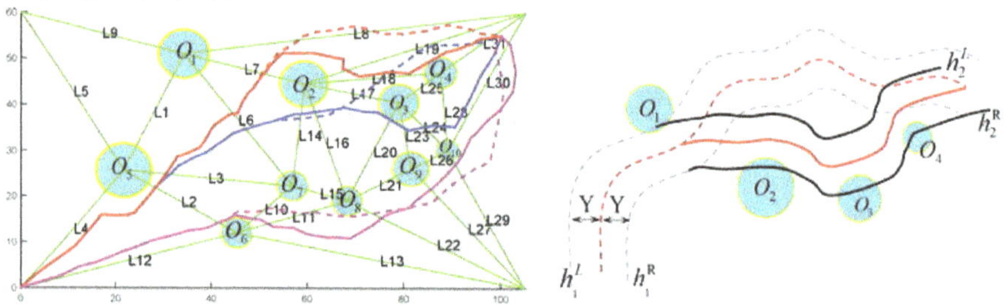

(a) An example of multiple paths generated by path planner, the green edges are generated using DT. The surounding information uses the edges for labeling.

(b) The information parameter also considers the surounding obstacles within the sensing range, thus to decide whether homotopic paths should keep.

Figure 7. An illustration of surrounding information used to find reasonable alternative path for online emergency threat management. The information parameter consists of edge information and obstacle surrounding information within sensing envelop.

Given the sensing range of a robot as Y, and the obstacles set $O=\{o_1, o_2, ..., o_n\}$. For path $h(\pi_1)$, it consists of a set of discrete states $X(\pi_1)=\{x(\pi_1)_1, x(\pi_1)_2, ..., x(\pi_1)_n\}$. In this paper, we assume that any obstacle o_i can be described with a circle or ellipse centered at o^c_i, then we can build a Delaunay Triangulation [19] connection (DTc) using the obstacle centers, the initial state, and the goal state. For DTc, it can generate a network like structure, and each two states have at most one connection. It is illustrated in Figure 8(a) that the green edges are the valid connections, with labels to distinct with each other. For any path, if the path intersects with an edge, the edge information should be added to the information factor, such as the solid red path intersects with edge L7, then $L7 \in \beth(h(\pi 1)) L7 \in \beth h \pi 1$. The edge labeling method can guarantee the uniqueness toward homology, while we note that homotopic paths can also be used to perform emergency threat management. It is represented in Figure 7(b), the solid red path h_s and the dotted red path h_D are homotopic to each other. Given the sensing range Υ, we have the sensing envelop which are dotted purple lines $h^L_{1,h}{}^R_1$ for h_D and solid black lines $h^L_{2,h}{}^R_2$ for

h_s, indicating the maximum detection range for emergency threat. Then, we have the {o1}⊂ℶ(hD) and {o1,o2,o3,o4}⊂ℶ(hS), thus we have h_s and h_D both regarded as reasonable alternative paths for online threat management.

The informative approach discussed thus can help to label each path in a configuration space, such as the results listed in Table 1 of paths in Figure 7(a). Then according to Definition 5, we can find the label of each path. For any several paths which have the same label, we choose the shortest path and use as the candidate for online fast switching.

Table 1. The path information parameter is consist of two parts, which are edge information using DTc and obstacle information using sensing envelop, respectively.

Path	Path surrounding information	
	Edge information	Obstacle information
Dotted red	L2.L3.L6.L7.L8.L19.L31	O_5, O_2
Solid red	L2.L3.L6.L7.L8.L19.L31	$O_5, O_{2.} O_3, O_4$
Dotted blue	L2.L3.L6.L14.L16.L17.L18.L31	$O_5, O_{2.} O_3, O_4$
Solid blue	L2.L3.L6.L14.L16.L20.L23.L24	$O_5, O_{2.} O_3$
Dotted pink	L2.L10.L11.L22.L27.L29.L30	$O_6, O_{7.} O_8$
Solid pink	L2.L10.L11.L22.L27.L29.L30	$O_6, O_{9.} O_{10.} O_3$

4.3. Emergency threat management

The reasonable alternative path set H_{RAP} provides a network with cluttered environment adaptivity. The concept of visibility was discussed in [20], where the cycle information is used to enable fast deformation for motion planning. Visibility is defined as:

$$V : \begin{cases} [0,1] \times [0,1] \rightarrow \{0,1\} \\ V(x_t, x_{t'}) = 1, \textit{if } \mathcal{L}_{link}(x_t, x_{t'}) \in C_{free} \subset C \\ V(x_t, x_{t'}) = 0, \textit{if } \mathcal{L}_{link}(x_t, x_{t'}) \in C_{obs} \subset C \end{cases} \tag{16}$$

Where x_0 denotes a state of a path, Llink-link is the connection of two states, C_{free} is the free space and C_{obs} is the obstacle region. A visual illustration is provided in Figure 8(a), where visibility can only in obstacle free region.

It is noted that the visibility in this paper means a possible connection to switch from one path to another for emergency threat management. For switching with visibility, given all the reasonable alternative paths H_{RAP}, the algorithm performs exploration for visibility state at each UAV state x_{UAV} among H_{RAP}. The algorithm then outputs the visible guards (states) x_{RAP} as illustrated in Figure 8(b).

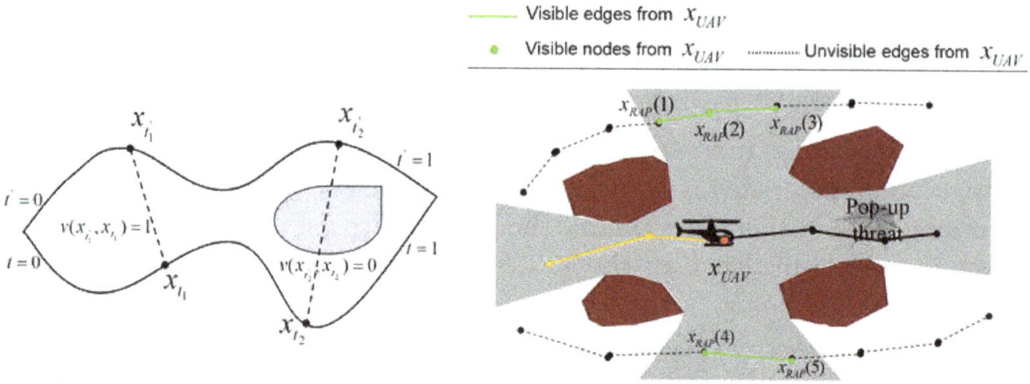

(a) Visibility concept illustration. $x()$ denotes the state of a path. Purple region denotes the obstacle.

(b) UAV is able to find visiable nodes of the multiple paths, and then pick the visible nodes out for switching. The UAV is also able to remember the taveled path as represented in yellow.

Figure 8. For online switching, UAV should follow the visible-node selection algorithm (a) to explore the possible switching route, then it switches to the cost minimum path for ETM (b).

To avoid the pop-up threat (or dynamic threat), UAV must select one entry guard from the visible guard set to reconnect to another pre-planned path to the goal. To validate the best connection, that is, the entry point and the entry connection, this paper applies the heuristic:

$$x^* = \underset{i,j}{\operatorname{argmin}}\ C_{FE}\left(x_{UAV}, H_{RAP(i)}\left(x_{RAP(j)}\right)\right) + C_{TC}\left(H_{RAP(i)}\left(x_{RAP(j)}\right)\right) \tag{17}$$

Using a simple cost based metric, where C_{FE} is the forward energy cost which is the distance and the turning cost from UAV position x_{UAV} to entry state $x_{RAP(j)}$ and the path from entry state to the goal $H_{RAP(i)}(x_{RAP(j)})$. The turning cost is the integration of heading angle difference at each state, which denotes the smoothness of the planned path. C_{TC} is the threat cost, the integration of inverse distance between state and obstacles. Using such approach, the algorithm can find five visible states $x_{RAP} = \{x_{RAP}(1), \ldots, x_{RAP}(5)\}$ as illustrated in Figure 8(b). Then, it tries to find the best entry state using the minimum cost principle (Eq. (17)).

We further consider a situation that there may have no visible states at current location. The paper proposes to use a long-term memory approach to handle this problem, that is, the travel path should be stored in memory, such as the orange edges and states illustrated in Figure 8(b). In the meanwhile, the method stores the visible state along the traveled path. Then, the UAV has to fly back to find a cost minimum path toward the goal if it confronts with pop-threat and has no visible states.

5. EXPERIMENT AND DISCUSSION

In this section, we highlight the performance of the obstacle biased and homology biased path planner with the ability of emergency threat management (avoiding pop-up and dynamic threat online). In the section, we will discuss the following points: (1) How the threshold of EFA affects the performance of MGART. (2) The time performance and reliability of reasonable alternative chosen algorithm. (3) The online emergency threat management performance. The algorithm is implemented using MatLab 2016b on a laptop computer with a 2.6 GHz Intel Core I5 processor.

5.1. Comparative simulation of multiple path exploration

We design three different scenarios, which are non-obstacle scenario, rounded obstacle crowed scenario, and irregular polygons crowed scenario, to perform comparative simulations. All the scenarios are 2D with 100*65 m^2 space, and obstacles randomly generated.

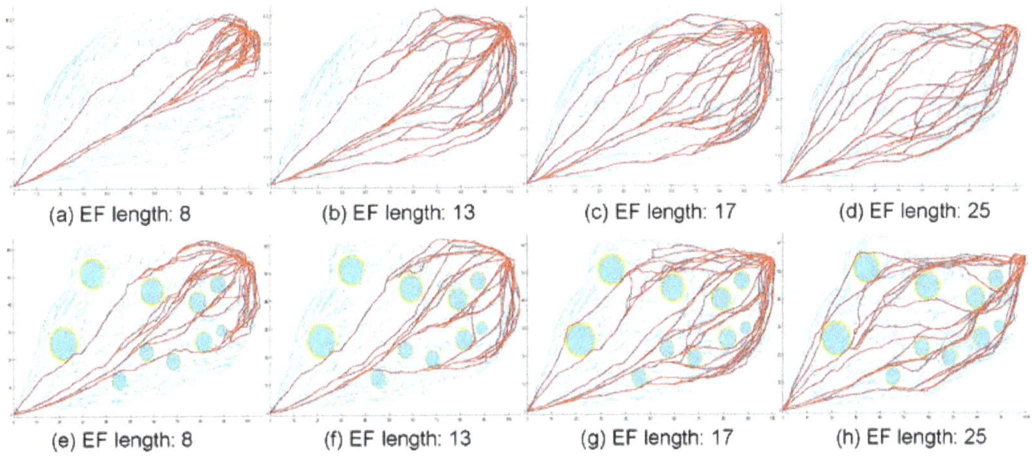

Figure 9. Illustration of alternative paths generated by MGART vary with representative backward EF length. (a)–(d) denotes the results in non-obstacles scenario, (e)–(h) denotes the results in obstacle crowed.

For scenario 1, it is a non-obstacle environment, and we set the variable threshold as a set with value {3,5,7,8,11,13,15,18,20,25,28} for representation. As we know the length of EFA threshold affects the goal biased probability, which directly decide the area of the newly obtained Voronoi region of the neighbor branches, we design a set of comparative experiments to study the effects between EF length and RAPs. An intuitive result of the relationship between planned paths and EF length after 10,000 iterations are provided in Figure 9(a)–(d). MGART can find 37 paths after 10,000 iterations if EF length is set as 3 step-length, and the number decreases to 22 if the EF length is set as 28. The reason is that the longer the EF length, the further the neighbor branches can obtain the goal biased resource. Thus, the neighbor branches need more steps to exploring toward the goal, that is, less paths will be achieved with better homology performance. As we can see that the paths in Figure 9(d) have a better homology performance than Figure 9(a). The same EF length variation experiment is also deployed in scenario 2, and results are shown in Figure 9(e)–(h). For RAPs, it is the same with the results in scene 1 that RAPs decrease with the increasing of the EF length. However, as the increasing of EF length enables more branches to explore toward the goal as well as increasing the homologous paths (see in Figure 9(e)–(h)), the number of the RAPs increasing with the increasing of the EF length. The statistic relation between the RAPs and the EF length is illustrated in Figure 10, which further proves the conclusion.

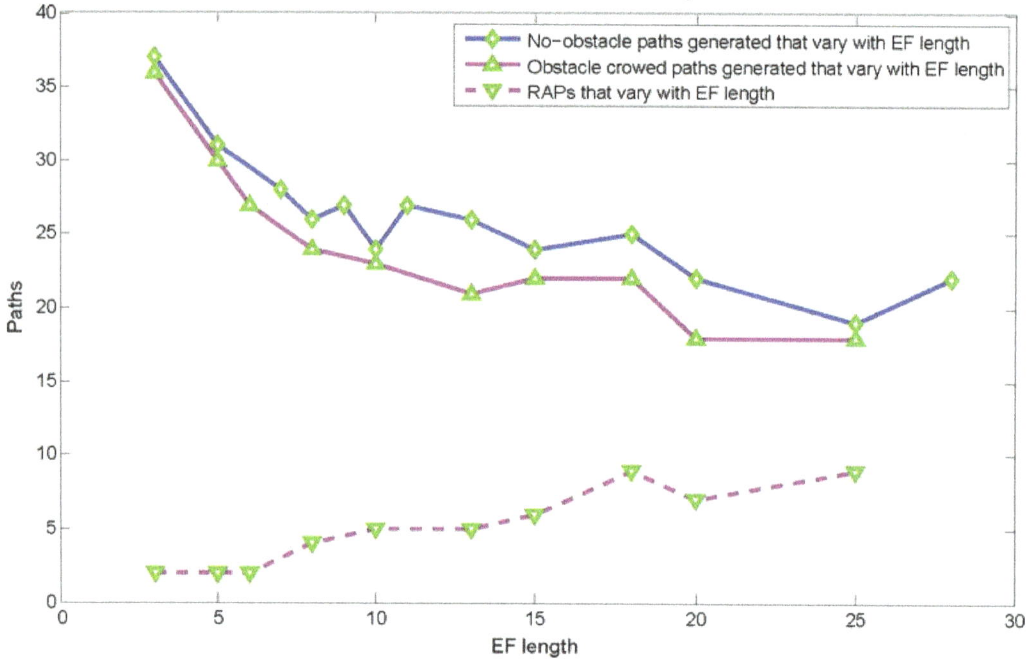

Figure 10. Relation between EF backward length and APs and RAPs with two scenarios. Here the solid diamond line denotes the relation of scene 1, and the triangle lines denotes the results of scene 2.

The EFA can be used to any SBAs by shifting the goal biased resource to achieve multiple RAPs for online switching. This paper compares the performance between MGART and MRRT* in three scenarios with 10,000 iterations. We compare the efficiency of generating a path, RAPs, average time for finding a path, and average time for any RAP are compared in Table 2. GART has a better performance in both path exploration and RAP generation, such that MGART can find at least 3 times of the number of paths toward the goal than MRRT*. Because GART introduces the environmental information to speed up the exploring process, the results prove that MGART is more efficient in finding RAPs, which is almost 100% faster than MRRT*. For time performance, we can see that MGART also outperforms MRRT* with at least 3 times advantage.

Table 2. Detailed comparison of planning efficiency between MH-GART and MH-RRT* in all three scenarios.

Scenario	Algorithm	Paths after 10,000 iterations	RAPs after 10,000 iterations	Average time for a feasible path	Average time for a RAP
Scenario 1	MGART	48	/	2.081	/
	MMRRT*	12	/	7.825	/
Scenario 2	MGART	43	9	2.093	8.213
	MRRT*	11	5	7.169	15.772
Scenario 3	MGART	34	10	2.257	7.674
	MRRT*	11	5	6.789	14.935

Besides comparison of the time performance of finding online switching paths (that is RAPs), we also pay attention to the quality of the path generated. The average lengths and standard deviation of the length of all paths in each scene are illustrated in Figure 11. The average length of the paths that generated by MGART and MRRT* are illustrated in Figure 11(a), we can see that MGART has a strong convergence performance than MRRT*. The standard deviation of the lengths is shown in Figure 11(b), results demonstrate that MGART is more likely to find paths with smaller fluctuation as well as smaller cost.

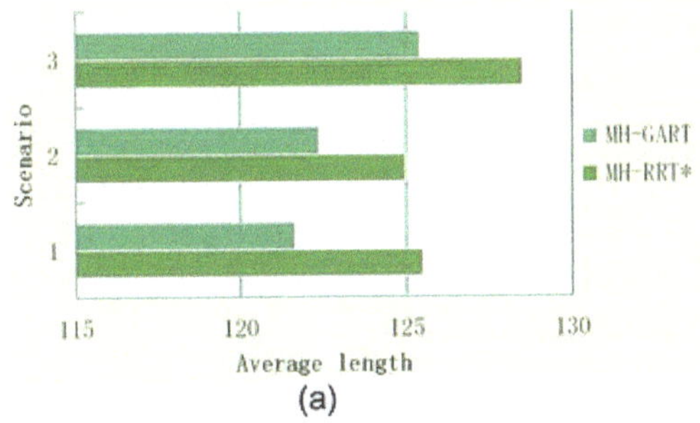

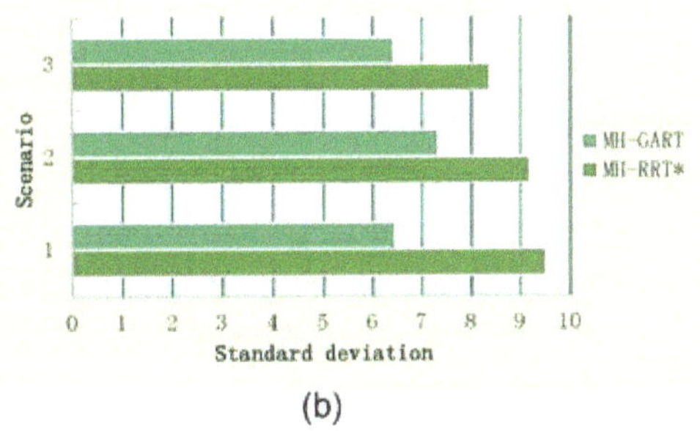

Figure 11. Comparison of (a) average length and (b) standard deviation of the APs generated by MGART and M-RRT* in three scenarios.

5.2. Performance of reasonable alternative path chosen

We also test the path labeling algorithm, that is, the surrounding information pursuing using DTc and sensing envelop, which is used to obtain the reasonable alternative paths under Definition 5. It is should be noted that the under the definition, any two paths do not have the same information parameter, which enables fast switching when facing pop-up threat. As the path label method guarantees the unique labeling of all the paths, only the paths which stretch in a parallel way and within the same sensing envelop have the same labels.

The results of simulation after 10,000 iterations in scenario 2 and 3 are provided in Table 3. For each single path, the time needed for labeling the path mainly depends on the area, dimension, the complexity of the configuration space. For our tested with area 100*60, the average time for acquiring the information for labeling 0.078 s (see in Table 3). The average time needs for RAP pursing of our cases is 0.139 s.

Table 3. Time performance of proposed method in two scenarios.

Scenario 2		Scenario 3	
Time for labeling (s)	Time for RAPs (s)	Time for labeling (s)	Time for RAPs (s)
0.0721	0.146	0.0842	0.132

5.3. Experiments of emergency threat management

MGART can be used for 3D and 2D pop-up threat management, and the 3D environments can be easily segmented by DT. We evaluate the performance of our method in both 2D and 3D environments, and we also compared the time performance.

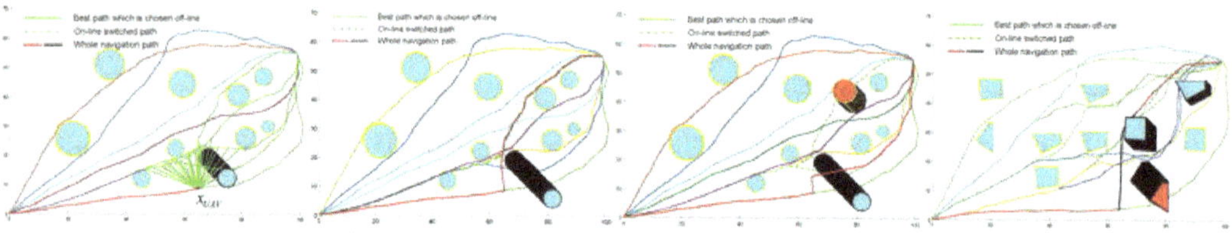

(a) Visible states (b) ETM with one obstacle (c) ETM with two obstacles (d) ETM with three obstacles

Figure 12. Tests of on-line switching to avoid dynamic threats using MGART in 2D scenarios. (a) Robot detects moving threat at position x_{UAV}, then it evaluates all its visible neighbor RAPs (the green lines) to choose the switching path. (b) Complete navigation of avoiding one moving threat, the red path is the navigation path. (c) Test of avoiding two moving threats, the black and red circles are threats. (d) Tests of avoiding three moving threats.

For 2D environments, we implement three tests with different number of dynamic threats. The RAP chosen algorithm works when robot realizes that the path will collide with the pop-up threat, that is, robot at position x_{UAV} detects the moving threat (see in Figure 12(a)). The simulation setting is illustrated in Table 4, where the robot speed is 10 m/s and the moving threat can be detected within 10 m detection range. Thus, the robot has less than 1 s to re-plan a path and executing to avoid the obstacle. RAP chosen algorithm first evaluates all its neighbor RAPs (the green lines) around the robot, and chooses the cost minimal and collision free path based on principle Eq. (17) (the dotted green path in Figure 12(b)). It is noted that Figure 12(b)–(d) are results of using MGART to avoid one, two, and three moving threats, respectively. The black parts along the navigation path denote the position where threat is detected by robot. We also execute test in 3D environment (see in Figure 13) with pup-up and moving threats. The on-line switching is supposed to be used for aerial robots in 3D, thus Dublin's Curves is used when switching from current position to safe path.

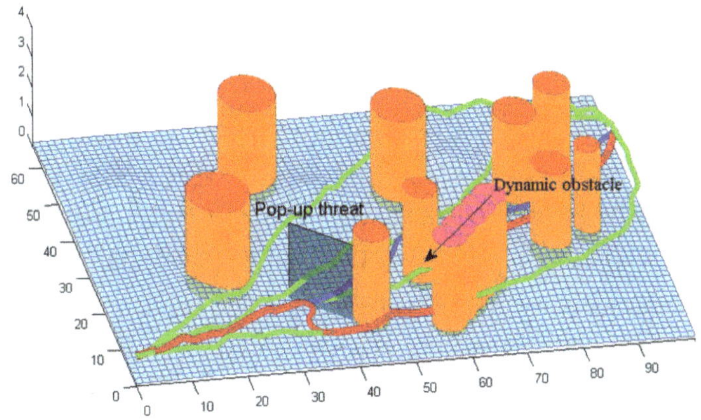

Figure 13. The experiment of using MGAT for 3D emergency threat management case, where pop-up threat and moving threat are exist in the environment.

Table 4. Setting of simulation for ETM.

Robot speed	10 m/s
Detection range	10 m
Speed of cyan obstacle (in Figure 12(b)–(c))	1.4 m/s
Speed of red obstacles (in Figure 12(c)–(d))	0.8 m/s
Speed of cyan obstacle (in Figure 12(d))	0.8 m/s
Online switching range	20 m

Throughout all of our experiments, we analyze the temporal efficiency of each transition in order to evade the current perilous circumstances. Regarding the avoidance of a moving threat (as shown in Figure 12(b)), the required time to switch to another RAP is 0.0507 seconds, and the total duration of navigation is 13.14 seconds at a speed of 10 m/s. In the instance of two dynamic threats, as shown in Figure 12(c), the total navigation time is 13.32 seconds, with 0.0912 seconds spent specifically on avoiding the second hazard. In scenario 3, we implemented a prolonged time for the threat, as depicted in Figure 12(d). The presence of two cyan hazards hinders the progress of the robot along the blue path, necessitating multiple switches as it navigates the black path. The average duration is a maximum of 0.15 seconds, which can be reduced when implemented on a robot's platform using C++ programming. The total navigation time is 13.5 seconds.

6. CONCISE SUMMARY

The primary innovation of this study is the introduction of an online EMT planner, which addresses the occurrence of pop-up threats and shifting obstacles while tracking a pre-planned course. We provide a novel method for multiple path planning called MGART. This approach builds upon the existing GART algorithm by incorporating a 'Extending Forbidden' algorithm. This algorithm effectively redistributes the likelihood of reaching the objective towards neighboring branches around the branch that has already reached the target. The algorithm demonstrates the advantageous qualities of GART and the capability to navigate across complex situations. Additionally, it ensures asymptotically optimal performance and completeness. Furthermore, the method demonstrates the ability to provide numerous paths without employing variant cost principles. Instead, it relies solely on the EFA threshold, allowing for the option of online dynamical switching.

In the future, we aim to conduct research on the topic of online visual location and environment perception, which has not been adequately addressed in this study. We want to implement cognitive sensing and autonomous capabilities for robots.

REFERENCES

1. Goerzen C, Kong Z, Mettler B. A survey of motion planning algorithms from the perspective of autonomous UAV guidance. Journal of Intelligent and Robotic Systems. 2010;57(1–4):65
2. Yang L, Qi J, Song D, Xiao J, Han J, Xia Y. Survey of robot 3D path planning algorithms. Journal of Control Science and Engineering. 2016;5(2016):1-22
3. Chen H, Chang K, Agate CS. UAV path planning with tangent-plus-Lyapunov vector field guidance and obstacle avoidance. IEEE Transactions on Aerospace and Electronic Systems. 2013;49(2):840-856
4. Davoodi M, Panahi F, Mohades A, Hashemi SN. Multi-objective path planning in discrete space. Applied Soft Computing. 2013;13(1):709-720
5. Rathbun D, Kragelund S, Pongpunwattana A, Capozzi B. An evolution based path planning algorithm for autonomous motion of a UAV through uncertain environments. In: Proceeding of IEEE Digital Avionics Systems Conference. Vol. 2; 2002. pp. 8D2-1-8D2-12
6. Valenti RG, Jian Y-D, Ni K, Xiao J. An autonomous flyer photographer. In: Proc. of 2016 IEEE International Conference on Cyber Technology in Automation, Control, and Intelligent Systems (CYBER); 2016. pp. 273-278
7. Coleman D, Sucan I, Chitta S, Correll N. Reducing the barrier to entry of complex robotic software: A moveit! case study. arXiv preprint. 2014;1404(3785):1-14
8. Yang K, Gan SK, Sukkarieh S. A Gaussian process-based RRT planner for the exploration of an unknown and cluttered environment with a UAV. Advanced Robotics. 2013;27(6):431-443

9. Yoshida E, Kanehiro F. Reactive robot motion using path replanning and deformation. In: IEEE International Conference on Robotics and Automation (ICRA); 2011. pp. 5456-5462
10. Lindemann SR, LaValle SM. Incrementally reducing dispersion by increasing Voronoi bias in RRTs. In: IEEE International Conference on Robotics and Automation. Vol. 4; 2004. pp. 3251-3257
11. Kavraki L, Latombe J-C. Randomized preprocessing of configuration for fast path planning. In: Proc. of IEEE International Conference on Robotics and Automation; 1994 May 8. pp. 2138-2145
12. LaValle SM, Kuffner JJ Jr. Randomized kinodynamic planning. The International Journal of Robotics Research. 2001;20(5):378-400
13. Yang L, Xiao J, Qi J, Yang L, Wang L, Han J. GART: An environment-guided path planner for robots in crowded environments under kinodynamic constraints. International Journal of Advanced Robotic Systems. 2016;13(6):1-18
14. Yang L, Song D, Xiao J, Han J, Yang L, Cao Y. Generation of dynamically feasible and collision free trajectory by applying six-order Bezier curve and local optimal reshaping. In: Proc. of IEEE/RSJ International Conference on Intelligent Robots and Systems (IROS); 2015 Sep 28. pp. 643-648
15. Karaman S, Frazzoli E. Sampling-based algorithms for optimal motion planning. The International Journal of Robotics Research. 2011;30(7):846-894
16. Urmson C, Simmons R. Approaches for heuristically biasing RRT growth. In: IEEE/RSJ International Conference on Intelligent Robots and Systems. Vol. 2; 2003, October. pp. 1178-1183
17. Pan J, Manocha D. Fast probabilistic collision checking for sampling-based motion planning using locality-sensitive hashing. The International Journal of Robotics Research. 2016;35(12):1477-1496
18. Yang L, Qi J, Jiang Z, Song D, Han J, Xiao J. Guiding attraction based random tree path planning under uncertainty: Dedicate for UAV. In: IEEE International Conference on Mechatronics and Automation (ICMA); 2014 Aug 3. pp. 1182-1187
19. Levcopoulos C, Krznaric D. Quasi-greedy triangulations approximating the minimum weight triangulation. Journal of Algorithms. 1998;27(2):303-338
20. Jaillet L, Siméon T. Path deformation roadmaps: Compact graphs with useful cycles for motion planning. The International Journal of Robotics Research. 2008;27(11–12):1175-1188

INDEX

www.ingramcontent.com/pod-product-compliance
Lightning Source LLC
LaVergne TN
LVHW062228200926
848484LV00006B/1502
9781787153820